PRAISE FOR *AMERICAN HEMP*

"Hemp has an amazing and positive story. Throughout history it has been relied upon to clothe, feed, and shelter people all over the world, and recently it has also been used medically to reduce suffering from a wide variety of maladies. But beginning in 1937 with the passage of the Marijuana Tax Act, our own United States Government wrongly began the process of vilifying it by linking it to marijuana. Yes, the plants are similar, but hemp has never caused any harm whatsoever to anyone! But finally, hemp is now making a comeback. And one of the guiding lights who is showing the way is Jen Hobbs in her book *American Hemp.* The title says it all. A fascinating manifesto on the economic, environmental, and health benefits of the hemp industry. Read this book, and you will agree!"

—**Judge James P. Gray**, Superior Court of Orange County (Ret.)

"This ancient plant may be a bright light leading the way to bring over 30,000 products to market and help farmers prosper well into the future. Jen Hobbs's book is a must read for all true hemp believers!"

—**Pam Ellison**, founder of Minnesota Frozen Farm Forum

"This book is a must-read for the licensed and non-licensed hemp cultivators and organic farmers looking for a crop with ultimate sustainability. The important history of hemp and the many uses of the sativa plant are examined. The research is already here, the facts are in."

—**Paul Frank**, CEO of Hemp Solutions of Minnesota

"Hemp has so much potential to help farmers, create jobs, spark innovation and new technologies, and make existing products better, stronger, and more sustainable. Thank you to Jen Hobbs f[illegible] ate and emphasize the importance of hemp for o[illegible]

—**Erica McBride Sta**[illegible] al Hemp Association

"I ask questions for a livin[illegible] and have often wondered why hemp is illegal to grow. Why aren't we growing this plant and using it for so many things? Why do we grow way too much of one crop and make another illegal? What can hemp do for you? Well, in this awesome book Jen Hobbs offers all the answers we need to make this happen."

—**Pete Dominick**, comedian and Sirius/XM radio host

AMERICAN HEMP

HOW GROWING OUR NEWEST CASH CROP CAN IMPROVE OUR HEALTH, CLEAN OUR ENVIRONMENT, AND SLOW CLIMATE CHANGE

JEN HOBBS

Skyhorse Publishing

Skyhorse Publishing books may be purchased in bulk at special discounts for sales promotion, corporate gifts, fund-raising, or educational purposes. Special editions can also be created to specifications. For details, contact the Special Sales Department, Skyhorse Publishing, 307 West 36th Street, 11th Floor, New York, NY 10018 or info@skyhorsepublishing.com.

Skyhorse® and Skyhorse Publishing® are registered trademarks of Skyhorse Publishing, Inc.®, a Delaware corporation.

Visit our website at www.skyhorsepublishing.com.

10 9 8 7 6 5 4 3 2 1

Library of Congress Cataloging-in-Publication Data is available on file.

Cover design by Brian Peterson
Cover illustrations by gettyimages

Print ISBN: 978-1-5107-4329-8
Ebook ISBN: 978-1-5107-4330-4

Printed in the United States of America

For my daughter, who's too young to understand why the world is the way it is,
but who isn't too young to deserve a better future (we all do).

TABLE OF CONTENTS

FOREWORD BY JESSE VENTURA

In 2016, I wrote *Jesse Ventura's Marijuana Manifesto* with Jen Hobbs. The book outlined my philosophies on marijuana, detailed the impact of the war on drugs, and laid out the scientific benefits of cannabis. Although we focused on many of hemp's beneficial properties and how it is different from marijuana, we didn't go into all of its aspects. Hemp needed its own manifesto, which is why I'm pleased to write the foreword to Jen's *American Hemp*.

After Jen and I published *Jesse Ventura's Marijuana Manifesto*, more and more states have moved toward marijuana legalization—over half the US population now lives in a state that has legalized marijuana in some way—*plus* Congress finally legalized hemp in the 2018 Farm Bill by defining it as an agricultural crop, which we all know it has always been.

Colorado was the first state to legalize and grow hemp—that was back in 2012. The fact that it took Congress six years to grow a pair and remove hemp from the Controlled Substances Act is unfortunately typical to say the least. But what really gets me is the contradictory attitude states have had when they chose to legalize marijuana, but not hemp.

It's common knowledge that you can't get high by smoking hemp. Smoke as much of it as you want. It doesn't have enough THC in it to affect you. Although this sets hemp apart from marijuana, it boggles my mind that states didn't pass ballot measures to legalize hemp *and* marijuana at the same time! Why did states go through all the trouble to legalize marijuana, but kept hemp illegal until years later?

Take Oregon for example. Oregon was the first state to decriminalize marijuana. In 1973 the penalty for up to one ounce of weed was reduced from a criminal charge (including jail time) to a $100 fine; it became fully legal in 2014. The state legalized hemp in 2009, but the Department of Agriculture didn't actually give out any licenses to grow it until 2015—after the 2014 Farm Bill, the legislation that allowed states to start their own hemp research programs. Granted, many states were in legal battles with the DEA at this time because the agency was confiscating imported hempseed, but if a state has already allowed marijuana seeds to grow, why on earth is there an issue with allowing hemp seeds? Furthermore, as of January 1, 2018, I can legally buy marijuana, a *recreational* drug, in dispensaries throughout the state of California—and medical marijuana has been legal in California since 1996—but as of December 2018, there weren't any legal hemp farms operating in the state. Where does common sense come into this? And why did hemp lose the popularity contest? Is hemp not sexy enough? Do the thousands of known uses for hemp just bore America to tears?

Stranger still was always the federal government's classification of hemp and marijuana. Again, it doesn't take a scientist to tell us that hemp and marijuana are two plants that look similar but are not the same, yet the federal government classified them as if they were exactly the same. Until the 2018 Farm Bill, they were both classified as a Schedule I narcotic, the most dangerous drug classification (right up there with heroin) and getting caught with either plant by a federal agent could have come with the strictest legal consequences.

Even though I find it hard to believe anyone in the DEA could make these claims about hemp and marijuana with a straight face, the federal government did us a huge favor when it comes to this one-size-fits-all classification. Yes, we all know hemp and marijuana are not the same, but *legally speaking* they were being treated exactly the same under the Controlled Substances Act, so why not legalize both of them at the same time? Why did states put the marijuana issue ahead of the hemp issue if the federal government is treating them as the same substance? Congress made a huge error by not legalizing both in the 2018 Farm Bill, and now they're turning the entire legalization process into a political issue.

When it comes to the 115th Congress's voting records, the GOP—which held a majority in both houses until the 2018 midterm elections—blocked a number of commonsense, marijuana-friendly bills. Most Republicans voted no to increasing military veterans' access to medical cannabis, no to ending the 280E tax code (the IRS doesn't allow marijuana companies to take any deductions or business expenses on their tax returns), and no to allowing marijuana business access to banking services.[1] The only reason hemp was legalized in the 2018 Farm Bill was because Republican Senate House Majority Leader Mitch McConnell had been advocating for it for years.

McConnell is pro-hemp because Kentucky needs another cash crop now that tobacco doesn't pay the bills. He recognized the biggest bang for Kentucky's buck was to legalize all aspects of hemp—including CBD—because CBD is the biggest moneymaker right now for hemp farmers. Thanks to Mitch McConnell, the Republicans will now go down in history as the party that brought back hemp and added a whole new industry to the economy. So where does that leave the Democrats?

Democrats now realize they have no choice but to embrace marijuana legalization if they want to win major elections in 2020. Now that Democrats have regained control of the House in 2019, it'll be interesting to see if they actually take note of Democrat Earl Blumenauer's "Blueprint to Legalize Marijuana," a step-by-step process for the 116th Congress to federally legalize weed.

"There's no question: cannabis prohibition will end," Bluemenauer stated in his memo. "Democrats should lead the way."[2]

I don't really care who leads the way, as long as somebody is successful in legalizing all aspects of cannabis, releasing all the nonviolent drug offenders from prison, and expunging their records. However, the timing here is all too coincidental. Politicians never do anything out of the goodness of their hearts. For the first time in America's history, outgoing Congress members are leaving Capitol Hill and heading to the advisory panels of major marijuana corporations.

Canadian-based Tilray, a global leader in the medical cannabis industry, welcomed two prominent politicians to their international advisory board in December 2018: Howard Dean (former Democratic National Committee chair and former Vermont governor) and Michael Steele (former Republican National Committee chair and Maryland's lieutenant governor). Incidentally, Howard Dean has had a long history of opposing marijuana legislation. Even when he was seeking the Democrats' presidential nomination in 2004, he wouldn't commit to ending federal raids against medical cannabis facilities. While Steele has been vocal about ending federal cannabis prohibition, the Republicans have obviously never embraced this reform, and he didn't make marijuana legalization a priority while he was chair of the RNC. Also jumping on the marijuana bandwagon is John Boehner, the former Republican speaker of the House of Representatives. After opposing marijuana legalization his entire political career, in April 2018, he joined the advisory board of the cannabis investment company Acreage Holdings.

"When you look at the number of people in our state and federal penitentiaries, who are there for possession of small amounts of cannabis, you begin to really scratch your head," a 2018 *Bloomberg* article quoted John Boehner as saying.[3] "We have literally filled up our jails with people who are nonviolent and frankly do not belong there."

Boehner, this is not a head-scratcher. While anyone can have a change of heart, we all know politicians—including *you*—are all about dollar signs, not common sense. The cannabis industry is expanding, and politicians are looking to cash in. They were lining their pockets to keep marijuana illegal, now they're putting themselves in a position to line their pockets again, all while appearing to be on the right side of history. Canada has legalized both hemp and marijuana, but before the United States can legalize cannabis completely, politicians want to know how much money they can get out of it. It's that simple. So while Representative Earl Blumenauer may very well appear to be progressive by putting a blueprint together for House Democrats to legalize marijuana, he's really only taking note of which way the wind is blowing on the issue (and lining up future employment opportunities of his own).

Don't get me wrong. I'm glad the federal government legalized hemp within my lifetime. When I was running for governor of Minnesota in 1998, I spoke about the Hemp for Victory campaign from World War II and spoke about how Minnesota farmers could increase their paychecks by growing hemp. I also openly admitted to smoking marijuana on national television when it wasn't popular to do so. Meanwhile, John Boehner was in Congress at the time, and Howard Dean was the governor of Vermont. To think that they opposed cannabis back then and they're now welcomed into the industry with open arms? That's the real head-scratcher. Then again, every major corporation needs lobbyists—or *advisers*—and who could be better than former politicians? These Washington insiders know how to talk to (and persuade) current elected officials who still oppose cannabis legalization.

Yes, the 2018 Farm Bill has the potential to open the floodgates to more and more marijuana legalization. But I can't help but think about all those lives ruined—all those nonviolent drug offenders winding up in prison. The same politicians who passed the legislation that locked everyone up are now turning around and deciding *after* they leave office that it's time to move the needle on cannabis prohibition. Not because enough states have passed laws. Not because enough Gallup polls have shown that over half the US population isn't scared to admit it is time to legalize cannabis. Not because the drug war—which cost over a trillion dollars—hasn't actually been effective at doing anything to decrease drug use. Not because we have an opioid crisis in America right now due to *legal* prescription pain pills. But because these hypocrites know how much money they can make once cannabis is fully legal.

And here's another case in point: In the 2018 midterm elections, Missouri legalized medical marijuana by passing Constitutional Amendment 2. The ballot measure imposed a 4 percent tax on marijuana when it is sold to the consumer, and the funds from that tax go to health and care services for military veterans. Well, by December 2018, a bill was already drafted by a Democrat to legalize recreational marijuana. Why? Not because anyone pressured lawmakers. Not because ending cannabis prohibition is the right thing to do. But because the medical marijuana amendment only gives money to veterans. The recreational marijuana industry in Colorado has proven itself to be far too lucrative, so obviously the time has come to legalize it *and tax the hell out of it* and fill the coffers!

Money aside, marijuana is only used for one purpose—whether people are getting high for fun or for medicinal reasons, it doesn't matter. Everyone experiences the same euphoric effect that comes from using pot. Yet, hemp has endless purposes and possibilities, including the solution to man-made problems: it can reduce our carbon footprint, it can draw out toxins such as lead from our soil, it can feed us, it can clothe us, it can create biofuel and remove our dependency on foreign oil, it can heal the sick, it can be used to build sustainable housing, it can decrease our dependence on paper, plastics, and nonbiodegradable products. Frankly, even if Congress only legalized hemp in the 2018 Farm Bill for economic reasons and didn't consider the other positive outcomes, it doesn't matter—we finally have the ability to put our health and the health of our planet first, and that means a hell of a lot more than making money.

As everyone who's read *Jesse Ventura's Marijuana Manifesto* knows, the thirteen colonies grew hemp. When Benjamin Franklin tied a key to a kite string during a lightning storm, the string he used to control the kite's direction was made from hemp. The first American flags were made from homespun hemp, and many of our nation's founding documents were written on hemp parchment paper. So, what could be more American than this crop? Now that it's safe to be associated with hemp because it's legal, you'll see more and more politicians throwing their support behind it, and they just might get hired for an advisory board role as a reward for doing so.

I'll leave you with this last bit of American hemp history to consider: On February 24, 1794, George Washington wrote a letter to his farm manager at Mount Vernon and instructed him to plant more hemp:[4]

> *I am very glad to hear that the Gardener has saved so much of the St foin seed, & that of the India Hemp. Make the most you can of both, by sowing them again in drills. Where to sow the first I am a little at a loss (as Hares are very destructive to it) but think, as the Lucern which was sown broad in the Inclosure by the Spring, has come to nothing; as the ground is good; and probably as free from Hares as any other place, it might as well be put there; as I am very desirous of getting into a full stock of seed as soon as possible. Let the ground be well prepared, and the Seed (St foin) be sown in April. The Hemp may be sown any where.*

Over the years, this paragraph written by our nation's first president has been paraphrased and shortened and put on bumper stickers and memes all over the internet to read: "Make the most of the Indian hemp seed, and sow it everywhere!"

If there was ever a time to make the most of American hemp, our newest cash crop, the time is now. The blueprint is here; you're reading it. If we take Washington's advice, we can improve our health, clean our environment, slow climate change, and so much more.

NOTES

1. Tom Angell, "Analysis: GOP Congress Has Blocked Dozens Of Marijuana Amendments," Politics, *Marijuana Moment*, July 10, 2018, https://www.marijuanamoment.net/analysis-gop-congress-has-blocked-dozens-of-marijuana-amendments/.
2. Tom Angell, "Congressman Issues 'Blueprint To Legalize Marijuana' For Democratic House In 2019," Policy, *Forbes*, Oct. 17, 2018, https://www.forbes.com/sites/tomangell/2018/10/17/congressman-issues-blueprint-to-legalize-marijuana-for-democratic-house-in-2019/#5f455cd93aaf.
3. Jennifer Kaplan,"Ex-Speaker John Boehner Joins Marijuana Firm's Advisory Board," Business, *Bloomberg,* April 11, 2018, https://www.bloomberg.com/news/articles/2018-04-11/ex-speaker-john-boehner-joins-marijuana-firm-s-advisory-board.
4. George Washington, "From George Washington to William Pearce, 24 February 1794," *National Archives,* accessed Jan 23, 2019, https://founders.archives.gov/documents/Washington/05-15-02-0210.

INTRODUCTION: THE 2018 FARM BILL LEGALIZES HEMP

I think it's an important new development in American agriculture. There's plenty of hemp around; it's just coming from other countries. Why in the world would we want a lot of it to not come from here?

—Mitch McConnell, Senate Majority Leader[1]

I started writing *American Hemp* in December 2017. Every year since 2015, Congress has been unsuccessful in passing the Industrial Hemp Farming Act, designed to legalize hemp. When the legislation was included in the 2018 Farm Bill, I assumed it would be used as a bargaining chip and eventually cut.

Much to my surprise, Congress passed that behemoth bill on December 10, 2018 and simultaneously removed federal restrictions on hemp. And then, in the midst of a government shutdown, President Donald Trump signed the bill and officially made hemp legal.

Here was Congress removing a substance from the Controlled Substances Act, and there were no objections. Here was Congress declaring a Schedule I narcotic as an agricultural commodity, something we were told time and time again wasn't even possible because the DEA and FDA had the power to classify drugs and determine what is and isn't legal and why, *and there were no objections.*

The previous Farm Bill (passsed in 2014) granted states the authority to begin their own hemp research programs, and as of 2018, the number of states participating had grown significantly. Once politicians did the math, they realized if hemp-CBD was classified as a commodity rather than a Schedule 1 narcotic, farmers could make some serious money. So perhaps that was the leverage necessary to finally treat hemp as an agricultural crop in the 2018 Farm Bill? Regardless, here are the sections of the 2018 Farm Bill that fully legalized hemp: [2]

- **Sec. 297A: Definitions**

In the new Farm Bill, hemp is defined as *Cannabis sativa* L, a plant with a THC concentration of no more than .3 percent on a dry weight basis. This includes "all derivatives, extracts, cannabinoids, isomers, acids, salts, and salts of isomers, whether growing or not."

- **Sec. 12619: Conforming Changes to Controlled Substances Act (CSA)**

Section 102(16) of the CSA, which previously listed hemp as a Schedule I narcotic, has now been completely amended. After the Farm Bill passed, *marihuana*, as it is defined in the CSA, no longer includes hemp. They are now separate under federal law, with hemp being defined as an agricultural commodity. All aspects of hemp—including CBD derived from hemp—is now legal under federal law. This means hemp businesses can deduct expenses when they file their taxes and can even be publicly traded on the stock exchange, just like any other industry.

- **Sec. 11106: Insurance Period**

Now that farmers can legally grow hemp under federal law, they can receive crop insurance under the Federal Crop Insurance Act, which means hemp farmers can also utilize federal banks to open accounts and apply for loans—something they were unable to do previously due to hemp's CSA classification.

- **Sec. 11121: Reimbursement of Research, Development, and Maintenance Costs**

Hemp farmers involved in research projects can now be reimbursed by corporations funding the study. This includes those in the hemp pilot program.

- **Sec. 10114: Interstate Commerce**

The Farm Bill will not prohibit the interstate commerce of hemp or hemp products. Therefore, hemp can be grown in any state and then be shipped to another state to be processed or sold in a retail capacity. Prior to this, federal law stated hemp that was being grown legally and domestically had to stay within state lines. This is a huge win for the industry, considering even wine manufacturers can't distribute to all fifty states due to interstate commerce laws.

HEMP-CBD SAVES KENTUCKY FARMING

Senate Majority Leader Mitch McConnell has been a strong, public supporter of hemp. Kentucky farmers needed another crop to grow, now that the demand for tobacco has gone down significantly. According to the U.S. Department of Agriculture (USDA), over 600,000 acres of tobacco were harvested in Kentucky in 1919, but by 2018, the state produced less than 100,000 acres.[3] Soybeans have replaced tobacco as the leading crop in the state, but according to a Kentucky farmer interviewed in *Quartz*, the return on investment isn't as substantial as hemp. An acre of soybeans yields about $500, while an acre of hemp being grown for CBD has the potential to bring as much as $30,000 per acre. While this appears to be an overly optimistic figure, it turns out that earning $30,000 per acre is not only accurate, but a conservative estimate.

According to the Kentucky growers surveyed by *Hemp Business Daily,* a pound of dried CBD flower went for about $20–$50/pound in 2018 (depending on the quality of CBD content).[4] Most CBD farms yield about one pound per plant, and they can fit up to 2,500 plants per acre,[5] so on the low end of the spectrum, one acre could very well produce $30,000–$50,000.

Bringing that math full circle, Kentucky has 75,800 farms, and the average size is 169 acres (the national average for a farm is 444 acres).[6] Now, I'm not a hemp farmer, nor do I live in Kentucky, but I'd be pretty annoyed if I was growing 169 acres of soybeans, making approximately $84,500 every harvest season (prior to all my expenses), knowing full well I could be making at least $5,070,000 on the same exact piece of land if only the dumbasses in Congress would just pass a law. Sure, that amount doesn't reflect net pay, but even after all expenses from growing hemp are deducted, that's still an obscene monetary gain in profit, especially because most farms in Kentucky are small family farms—not the mammoth corporate farms that typically receive all the USDA subsidies. Over 50 percent of Kentucky (that's 12.8 million acres) is considered farmland, yet 55 percent of its farms (41,800 farms) have had annual sales of less than $10,000.[7] No wonder Kentucky farmers are excited about adding a new crop to their portfolio.

Kentucky farmers are most likely aware of their state's American hemp history legacy. According to a 2002 research article published through the *American Society for Horticultural Science*, "from the end of the Civil War until 1912, virtually all hemp in the US was produced in Kentucky,"[8] and the state was the greatest producer of hemp in the nineteenth and twentieth centuries. Kentucky's current agricultural department (and Kentucky farmers, I'm sure) would want nothing more than to be crowned with that heavyweight title once again.

"Kentucky's success [with hemp] has surpassed all of my expectations," wrote McConnell in an opinion piece published in the *Courier Journal*, days after signing the 2018 Farm Bill.[9] "Hemp can be found in food and clothing, in home insulation and your car dashboard, and in many other Kentucky Proud products. Last year alone, hemp surpassed more than $16 million in product sales and attracted more than $25 million in investments to the commonwealth. Given this remarkable progress, I feel confident that hemp's economic contributions will keep growing with full legalization."

While Mitch McConnell will go down in history as the man who brought back American hemp, and while I'm sure Kentucky farmers are proud of what he's accomplished through the 2018 Farm Bill, I have to say, this legislation is embarrassingly overdue. McConnell isn't the only senator with a state full of prospective hemp farmers, and have you ever heard of a politician running for office that didn't say, "if elected, I promise to create more jobs"? So seriously, Congress, what took so long to legalize hemp, which in turn creates an all-new domestic industry, which in turn creates more jobs?

While the 2018 Farm Bill is a truly historic victory for hemp, and quite frankly could very well be the beginning of the end of cannabis prohibition as a whole—*and I couldn't be happier for the future of our country*—there are some drawbacks to the legislation.

EXCLUSION OF CONTROLLED SUBSTANCE FELONS

For starters, convicted felons can't participate in the program until ten years after their conviction date. I'm not talking about murderers. The 2018 Farm Bill specifically states "any person convicted of a felony relating to a controlled substance" is ineligible to participate until ten years following the date of the conviction. The only exception to this are the felons who are already approved to grow hemp through the 2014 Farm Bill.

So, according to the law, a convicted murderer can grow hemp, but someone convicted of a drug offense cannot? Didn't we already establish that hemp isn't a drug?

Granted, similar parameters involving felons were already in place in some states that participated in the hemp pilot program, plus similar rules currently apply to the marijuana industry, but seriously, I thought we were past this. It's tough enough to make an honest living after being in prison. No one wants to hire a convict. Statistically speaking, we already know that drug offenders are generally nonviolent—and whether or not a convicted felon is violent, that's not even part of the equation here.

In *Jesse Ventura's Marijuana Manifesto,* we covered something known as insourcing. This is essentially the slave labor/sweatshop staff in the United States that is not only legal, but arranged by the prison system. In most instances, prisoners work on farms for less money than undocumented workers. They literally make pennies an hour. While this program also gives tax breaks to the corporations that hire the workers (as a reward for being so generous because they take the risk of hiring criminals *and* paying them next to nothing), the program could be scaled into one that trains convicts in various aspects of the farming industry (sort of like an apprenticeship) if in fact a decent job in agriculture were possible upon their release. Now that hemp-CBD is all the rage, farmers will need to hire more workers, and if convicts have already been trained to work on the farm, finding a stable job in agriculture could be possible.

As a society, if we're looking to reduce recidivism, a convicted felon has already served the time for the crime. Why on earth would Congress stand in the way of allowing nonviolent drug offenders to work in the *agricultural industry* after that's the only job experience they've had, year after year, in prison? This is hemp we're talking about. Not marijuana. A felony conviction should have no bearing on a person's eligibility—or ability—to grow an agricultural crop.

INDUSTRIAL HEMP REGULATIONS AND HEMP LICENSE ELIGIBILITY

The 2018 Farm Bill removed hemp from the Controlled Substances Act, but if a state wants to be the primary regulatory authority over its hemp production, new hemp laws must be submitted to the Department of Agriculture for approval. The Farm Bill instructs states to determine if there should be a limit to how much can be grown, where the crop can be grown, how it should be grown (what chemicals can be used to grow it), what the inspection process is (to ensure THC is below .3 percent), and what products can be made from it. The fees associated with the application to grow it and the license to grow it are also up to the states, but the federal government expects to see a clear outline of parameters from each state.

Since hemp is now federally legal, the FDA and other federal regulatory agencies are now tasked with determining nationwide parameters. While some of these decisions may not be determined immediately, the overall rollout of American hemp will greatly depend on the Secretary of Agriculture, former Georgia governor George Ervin "Sonny" Perdue.

THE ROLE OF THE US SECRETARY OF AGRICULTURE

According to the 2018 Farm Bill, Secretary Perdue must conduct a study of all hemp pilot programs and present the findings to Congress, which is to occur no later than twelve months after the Farm Bill is passed (so the study will be released sometime in December 2019).

In addition to this study, Secretary Perdue will also submit a report to Congress within 120 days to outline the legitimacy of the industrial hemp research to determine "the economic viability of the domestic production and sale of industrial hemp, and hemp products."

The hemp research pilot program will then be repealed a year after the secretary publishes new guidelines for full-scale, nationwide commercial production of hemp. I'm assuming all aspects of hemp will finally be federally legal and treated as a true agricultural crop by that time, but we'll have to wait and see.

Unlike some of President Trump's other appointees, Perdue has had extensive experience in his field. Perdue ran a successful grain and fertilizer business from 2003 to 2011 (prior to becoming governor of Georgia), and he returned to his family agribusiness after leaving office. While he was in favor of a swift passing of the 2018 Farm Bill, and he was quoted as saying Congress's decision would be arrived at prior to Christmas (which it was), he hasn't made any previous pro-hemp statements during his career in public service. Prior to the 2018 Farm Bill, it was illegal to grow hemp in Georgia,[10] so this is all a bit of a cliff-hanger; we'll have to hold on, wait, and see if there are clashes between state and federal regulators.

Regardless, these are exciting times for hemp!

Before getting into the reasons why industrial hemp is so important to America's future, the next couple of chapters are going to explain the difference between hemp and marijuana and how the crop became illegal in the first place.

NOTES

1. Jeff Daniels, "Senate agriculture panel passes farm bill with hemp legalization," Politics, *CNBC,* last modified June 14, 2018, https://www.cnbc.com/2018/06/13/senate-agriculture-panel-passes-farm-bill-with-hemp-legalization.html.
2. Kyle Jaeger, "Read: Here's The Final 2018 Farm Bill That Will Legalize Hemp," Politics, *Marijuana Moment,* Dec. 10, 2018, https://www.marijuanamoment.net/read-heres-the-final-2018-farm-bill-that-will-legalize-hemp/.
3. Jenni Avins and Dan Kopf, "Even farmers are shifting from tobacco to hemp and CBD," *Quartz,* Dec. 10, 2018, https://qz.com/1483381/the-2018-farm-bill-could-make-hemp-the-next-tobacco/.
4. "Hemp State Highlight: Kentucky invests in hemp but wonders if peak has passed," *Hemp Industry Daily*, March 1, 2018, https://hempindustrydaily.com/hemp-state-highlight-kentucky-invests-hemp-wonders-peak-passed/.
5. Jeff Rice, "Hemp can be lucrative, but there are drawbacks," Local News, *Journal Advocate,* Feb. 27, 2018, http://www.journal-advocate.com/sterling-local_news/ci_31699721/hemp-can-be-lucrative-but-there-are-drawbacks.
6. "Quick Facts," *KY Food and Farm*, accessed Jan. 23, 2019, https://www.kyfoodandfarm.com/ky-ag-facts/.
7. Ibid.
8. Small, E & Marcus, D. (2002). *Hemp: A new crop with new uses for North America. Trends in New Crops and New Uses*. 284–326.
9. Mitch McConnell, "McConnell: Hemp legislation will give Kentucky farmers a new cash crop," *Courier Journal,* Dec. 14, 2018, https://www.courier-journal.com/story/opinion/2018/12/14/mitch-mcconnell-hemp-legislation-help-kentucky-farmers/2310577002/.
10. "Georgia Hemp Law," *Vote Hemp,* accessed Jan 23, 2019, https://www.votehemp.com/states/georgia-hemp-law/.

1

HOW TO IDENTIFY HEMP

When Governor Ventura and I were on the *Jesse Ventura's Marijuana Manifesto* book tour during the 2016 presidential election, we found that cannabis legalization is something just about everyone agreed with—and this was during a period of time when our country wasn't so unified. We also noticed something that mainstream media and everyone running for president had missed: People wanted to vote for someone who wanted to legalize cannabis. Gallup polls at the time had indicated that 60 percent of Americans wanted marijuana to be legalized[1] and sure, maybe it's obvious we'd run into the people who felt that way on a pro-cannabis book tour, but what wasn't obvious was whom we were hearing this from: teachers, doctors, lawyers, mothers, fathers, grandparents—even members of law enforcement.

In 1932, presidential candidate Franklin D. Roosevelt promised to end alcohol prohibition if elected (and he did). FDR ran as a Democrat and beat Herbert Hoover in a landslide victory at a time when Republicans were immensely popular. Ever since Abraham Lincoln became the first Republican president in 1860, the party had dominated the executive office almost entirely, yet FDR beat Hoover with 57.3 percent of the popular vote (and there were eight candidates on the ballot at the time). Plus, FDR's election marked the first of five consecutive Democratic presidential wins that amounted to the equivalent of twenty years of Democrats running the country.[2]

Of course, ending alcohol prohibition wasn't the only issue of importance to Roosevelt and to Americans, but this was at a time when it was obvious that the 18th Amendment had failed and was doing more harm than good. People were going to jail simply for drinking alcohol while criminals were getting rich by bootlegging and distributing it to speakeasy establishments all around the country. The topic of ending alcohol prohibition was unifying back then, just as the topic of ending the war on drugs and ending cannabis prohibition is unifying today. Sure, cannabis legalization is technically one issue, but it has a lot of teeth.

As my Texan aunt so cleverly put it: "A remedy for seizures, a plant that removes pollution from the soil, a source for durable *and* lightweight plastics *and* clothing *and* biofuel *and* a building material that's also healthy to eat? Sounds like snake oil!"

That's one phone conversation I'll never forget, and I'm hoping that by writing this book, even more people will come to learn the immense value of hemp. But it does sound too good to be true, doesn't it? Is hemp the elixir of life or twenty-first-century snake oil?

Outside of the United States, there are approximately twenty-nine countries legally cultivating hemp.[3]

Hemp Business Journal and *Vote Hemp* reported that in 2016, the total value of hemp-based products sold in the United States was $688 million.[4] Suffice to say the modern industrial hemp industry isn't a snake-oil scam, but Americans have been getting ripped off nonetheless. Prior to the passage of the 2018 Farm Bill, the federal government stated it was illegal to commercially cultivate and manufacture hemp products, so American industries were importing hemp products from other countries. For American hemp sales to reach nearly $700 million in 2016, clearly the government was willingly ignoring a viable industry. The hemp industry was thriving in other countries—not only for farming and rural communities, but for all industries necessary in the manufacturing process—and now it finally has the potential to thrive in the United States.

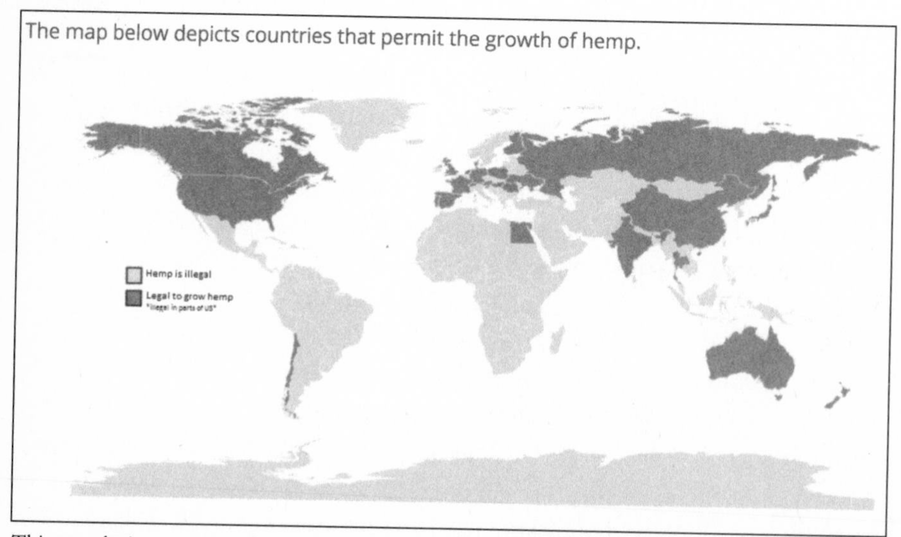

This map depicts countries that permit the growth of hemp, from Hempalta.com. *Source: https://ministryofhemp.com/hemp/countries/*

Although the United States has been lacking when compared to other countries that already have commercial hemp industries, most of these countries haven't been growing hemp to their full potential, either. Many are still in research programs or have only recently allowed the industry to expand. In the majority of these countries, marijuana is still illegal or decriminalized, so it goes without saying there is a recognizable difference between the two cannabis plants, and recognizing this difference is the first step in identifying what hemp is.

Canada has laws that differentiate between hemp and marijuana. Under Canada's Controlled Drugs and Substances Act (CDSA), cannabis was classified as a controlled substance. "All varieties regardless of the tetrahydrocannabinol [THC] content are prohibited" unless the government issued a regulation or exemption.[5] Hemp was included in the CDSA—just like it was in the Controlled Substances Act of the United States—but Canada's Industrial Hemp Regulations (IHR) stepped in to grant Canadian farmers a license to grow industrial hemp for commercial production. The IHR program went into effect on March 2, 1998, and it permitted farmers to grow cannabis containing up to .3 percent THC to be used for any commercial purpose[6] —except for medical-grade CBD extraction, but all that has now changed since Canada legalized recreational marijuana in October 2018 and officially ended Canadian cannabis prohibition.

The Canadian Hemp Trade Alliance states there are currently between 300 and 500 farmers growing hemp in Canada; the country saw an 80 percent increase in production in 2017, and the majority of the country's hemp exports went to the United States.[7] However, that's small potatoes when compared to China—a country that never completely banned cannabis. Chinese media has reported that 309 of the 606 medical patents relating to cannabis have been filed in China,[8] and government-funded scientists have been utilizing the plant for military purposes—such as the hemp-based uniforms the Chinese wore during the Vietnam War.[9] Due to this government backing, the country is now one of the world's top exporters of hemp fibers (for clothing), and is looking to become a go-to source for medical CBD.

However, when it comes to marijuana, if someone is found with "more than 5kg of processed marijuana leaves, 10kg of resin, or 150kg of fresh leaves," get ready to face the death penalty, as that's actually the punishment for possessing marijuana under Chinese criminal law.[10] This is despite the fact that marijuana is used as a traditional Chinese medical remedy (and has been for centuries).

One of the classic excuses given by our federal government or from law enforcement as to why hemp was federally illegal (and listed as a Schedule I narcotic) is because there isn't a way to tell the difference between hemp and marijuana. If hemp is legal but marijuana isn't, then it'll be too difficult to prosecute illegal drug dealers, and hemp farmers might get arrested accidentally.

That's clearly a lazy answer, especially now that the 2018 Farm Bill has defined hemp as an agricultural commodity. Again, in China, farmers grow hemp legally, but that same Chinese hemp farmer gets the death penalty if caught with too much weed. That's an insane law, but obviously there's a significant difference between the two plants. Even though marijuana and hemp are the same plant species—known as *Cannabis sativa*—it is easy to visually tell the difference between the two plants because they have different genetics, different appearances, and prefer different cultivation environments. The genetics, appearance, and cultivation environment of the two plants explains why other countries have been successful in creating separate legislation for marijuana and hemp and why hemp farmers aren't being confused with drug cartels.

GENETICS

The difference between marijuana's and hemp's chemical compositions comes down to the percentages of THC (tetrahydrocannibinol) and CBD (cannabidiol):

- A typical marijuana strain contains anywhere from 5 to 20 percent THC content (premium marijuana strains, of course, can contain more).

- Marijuana's THC is found predominantly in the cannabis flower (or bud).

- Conversely, hemp has 1 percent or less THC—typically .3 percent THC or less—essentially making it impossible to feel any psychoactive or "high" effect when it is smoked.[11]

Hemp's low percentage of THC is what distinguishes it as a non-drug and agricultural crop. This internationally accepted distinction between hemp and marijuana is nothing new; it was developed in 1971 by Canadian researcher Dr. Ernest Small and published in his book *The Species Problem in Cannabis*.[12] In 1976, he published a paper with American taxonomist Arthur Conquist titled "A Practical and Natural Taxonomy for Cannabis" in the *International Association for Plant Taxonomy* journal.[13] Both publications make note of two subspecies of cannabis, one he called sativa (industrial hemp used for seeds and fiber), and the other he referred to as indica (marijuana used for drugs).

According to Dana Larsen, author of *The Illustrated History of Cannabis in*

Canada, Dr. Small acknowledged that he couldn't entirely differentiate the two species, but nonetheless, he "drew an arbitrary line on the continuum of cannabis types, and decided that .3 percent THC in a sifted batch of cannabis flowers was the difference between hemp and marijuana."[14]

Today, there are a variety of marijuana and hemp hybrids that are bred to have a combination of indica and sativa properties, but Dr. Small's interpretation of the difference between hemp and marijuana has become standard around the world, regardless that he noted this percentage wasn't standard among all the strains he experimented on; some hempseed used for birdseed and fiber sometimes did contain "moderate or high amounts of THC."[15]

"The worldwide .3 percent THC standard divider between marijuana and hemp is not based on which strains have the most agricultural benefit, nor is it based on an analysis of the THC level required for psychoactivity," Larsen states in her book. "It's based on an arbitrary decision of a Canadian scientist growing cannabis in Ottawa."

CBD VERSUS THC

Both hemp and marijuana contain CBD, which has been effectively therapeutic for those suffering from seizures, epilepsy, multiple sclerosis pain, and other disorders. Hemp that is grown for CBD extraction is typically preferred in the medical community because CBD accounts for up to 40 percent of the plant's entire extract.[16]

Hypothetically, if a police officer found cannabis in someone's car, it can be taken to a lab and tested for THC content (the same way drugs are tested to determine their potency or if they're synthetic). In fact, in most of today's American hemp laws, the state's agricultural department (or sometimes even the DEA) is given "unfettered access" to conduct surprise inspections of hemp farms.[17] Sample buds are taken randomly from the field and tested to ensure there is .3 percent or less THC.

In Minnesota's industrial hemp pilot program, if a grower's plant sample tests above the acceptable .3 percent THC, then the farmer can pay to have a second test analyzed (the farmer has to pay for the first test as well). If the second test still doesn't pass, then the entire crop must be destroyed.[18] This procedure is fairly common among hemp pilot programs, so evidently the states have already used science to determine one plant from the other without total chaos ensuing.

Just to back this up even further, a 2015 Canadian study entitled "The Genetic Structure of Marijuana and Hemp" officially and scientifically determined that the genetic structures of marijuana and hemp are different. After genotyping eighty-one marijuana and forty-three hemp samples, the study concluded "marijuana and hemp are significantly differentiated at a genome-wide level, demonstrating that the distinction between these populations is not limited to genes underlying

THC production."[19] In laymen's terms, today's botanist can easily tell the difference between these two plants.

From a genetic perspective, the study proved that only the marijuana plant contains "high amounts of the psychoactive cannabinoid delta-9-tetrahydrocannabinol (THC)"—thus making it the choice candidate for recreational drug use. Hemp, on the other hand, contains "low amounts of THC," usually less than .3 percent, cannot be used for psychoactive purposes, and is therefore best to be cultivated as a food and fiber source.[20]

APPEARANCE

Due to this THC/CBD genetic distinction, hemp and marijuana have different physical appearances. Hemp is typically a sturdy, hardy, tall crop. According to the *Encyclopaedia Britannica*, the cultivation of hemp fiber was recorded in China as early as 2800 BC, and the plant is known to grow up to sixteen to twenty feet in height.[21] Hemp has a thick stem, ranging in diameter from ¼ inch to ¾ inch with little to no branching. The reason why hemp can be used for multiple purposes is partly due to its unique stalk. The inner core (referred to as hemp hurds) has a low density and a high absorbency while the outer layer fiber (known as bast) contains the fiber that is processed into textiles.

Hemp's stalks give it an appearance similar to bamboo. Hemp crop in Suffolk, UK, Sept. 2009. *Source: Adrian Cable (https://commons.wikimedia.org/wiki/File:Hemp_Crop_in_Peasenhall_Road,_Walpole_-_geograph.org.uk_-_1470339.jpg)*

In midwestern states such as Nebraska, Indiana, Minnesota, and Iowa, hemp can be found growing along rural roadsides, among wildflowers and weeds.[22] This type of wild hemp is descended from the industrial hemp of the 1940s Hemp for Victory campaign, when hemp was sown to create raw materials for parachute cords, military uniforms, and other World War II necessities. This type of hemp is often referred to as "feral cannabis" or "ditch weed" and it can grow to be eight to ten feet tall.[23]

Since ditch weed is *technically Cannabis sativa* (with less than .3 percent THC content), it was *technically* a Schedule I narcotic, and therefore a previous adversary of the federal government's war on drugs. Prior to hemp's new agricultural classification, local, state, and federal law enforcement agencies tried repeatedly (without success) to eradicate and/or stop ditch weed from spreading. Funnily enough, the seeds can lie dormant for seven to ten years, then sprout again, making the task a losing battle against nature[24]—especially since the seeds are easily spread by the wind or the animals that eat them.

The Northwest Indiana Times reported that in 2000 alone, the DEA spent $13 million to support ninety-six state and local agencies actively trying to get rid of ditch weed.[25] According to StopTheDrugWar.org, from 1984 to 2007, the DEA's Domestic Cannabis Eradication/Suppression Program "seized or destroyed" 4.7 billion feral hemp plants, and the agency spent at least $175 million "in direct spending and grants to the states" in a failed attempt to eradicate ditch weed.[26]

Ditch weed in Buffalo County, Nebraska, June 2017. *Source: Ammodramus (https://commons.wikimedia.org/wiki/File:Marijuana,_Buffalo_County,_Nebraska,_2017-06-15.jpg)*

Since its inception, the war on drugs spent over $175 million in taxpayer money to fight against ditch weed, so presumably law enforcement has always been able to tell the difference between marijuana and hemp. If they're smart enough to know ditch weed when they see it, it goes to reason they know just how different it looks from marijuana.

Two examples of marijuana's bushy appearance. Sativa marijuana strain "Mama Thai" (left) and Indica-based hybrid Pineapple Chunk (right), Lakehead, CA, August 2017. *Source: Hobbs Greenery*

While hemp stalks are tall and skinny, the marijuana plant is short, fat, and stubby. Marijuana has more branches, which allows it to produce more flowers.

Hemp leaf on the left; marijuana leaf on the right. *Source: https://en.wikipedia.org/wiki/File:Cannabis_sativa01.jpg*

Weed can reach tall heights similar to hemp when it's grown outside, but the shape of its leaves also gives it away. Both hemp and marijuana have the same diagnostic venation pattern of at least five symmetrical leaves attached to a single stem (same pattern as Japanese maple tree leaves), but the marijuana leaf is typically broader and wider than hemp's.

The flowering stages of outdoor marijuana from thin, milky white hairs (pistils) to trichome-rich cola buds, Lakehead, CA, September 2017. *Source: Hobbs Greenery*

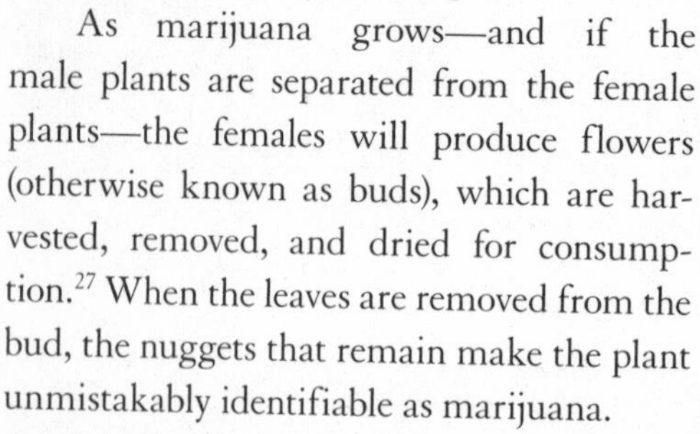

As marijuana grows—and if the male plants are separated from the female plants—the females will produce flowers (otherwise known as buds), which are harvested, removed, and dried for consumption.[27] When the leaves are removed from the bud, the nuggets that remain make the plant unmistakably identifiable as marijuana.

Hemp seeds grow in bundles in the center of hemp leaves in a similar formation to marijuana buds. Hemp crop in Suffolk, UK, Sept. 2009. *Source: Adrian Cable (https://commons.wikimedia.org/wiki/File:Hemp_Crop_in_Peasenhall_Road,_Walpole_-_geograph.org.uk_-_1470340.jpg)*

Conversely, as hemp grows, the plant's skinny leaves remain concentrated at the top of the plant, which gives it a similar appearance to bamboo. The "buds" on the hemp plant contain a high concentration of hempseed.

If the plants are grown outdoors, both hemp and marijuana produce a similar distinctive odor, which is easily carried by the wind, but their cultivation environments again set them apart.

CULTIVATION ENVIRONMENT

When comparing a marijuana farm to a hemp farm, the two look like completely different crops (and technically they are). An industrial hemp field looks similar to a wheat field or cornfield. The seeds are planted closely together, as close as four inches apart. Although the exact distance between each seed depends on what the plant will be used for, the crop is traditionally grown in a large, multiacre plot, and it can be grown on the same plot for several years without being rotated.[28]

Harvesting Kentucky-grown, all-American hemp by hand (left) versus machine (right). *Source: courtesy of the National Archives, photos taken in 1942*

Since it's beneficial for hemp to reach its full height to ensure the maximum yield from the crop, it shouldn't be grown indoors. Nor does it make sense to attempt to grow it indoors because it can be grown in a variety of climates with very little maintenance. It grows heartily in a field environment because pests aren't as attracted to it as other crops, and farmers can grow acres of it without the use of pesticides. In fact, there aren't any herbicides approved for hemp in the United States, so American farmers are typically growing it organically.

When compared to marijuana, hemp requires very little maintenance and care for it to grow. Male and female hemp plants are commonly planted close together to encourage seed production through wind and pollination.[29] Hemp is ready for harvest in approximately four months (about 120 days), and its yield is abundant. Per acre, it producers four times as much paper as trees.[30]

While hemp is an agricultural crop grown for its seeds, fiber, and oil, marijuana is a horticultural plant that is grown specifically for its THC content. Due to the fact that marijuana is highly susceptible to pests and disease, the plants cannot be grown too closely together, and growers typically prefer for it to be grown inside in a controlled environment.

If weed is grown outside, it'll be transferred to a plot of land after the seedlings have reached their vegetative phase. This is when the stem of the seedling grows thicker and taller and starts producing nodes—which produce new branches and leaves.

Once marijuana is strong enough to handle the outside world, the bushy plants are placed in the ground about six feet apart, giving them plenty of room to grow. While sown hemp looks like a wheat field, an outdoor marijuana farm looks similar to an apple orchard.

Marijuana grown in a controlled indoor environment, then transferred to outdoor conditions. *Source: Hobbs Greenery*

If the outdoor marijuana grow is successful, the plants can reach a comparable height to hemp, but the amount of leaves and branches on each plant makes weed look more like young trees.

Freshly planted outdoor marijuana, spreading its roots and its branches accordingly at Hobbs Greenery, Lakehead, CA, 2017

Weed is also pickier about where it is grown and how it is grown—which is why it's typically grown inside under a controlled environment with stable lighting, temperature, and humidity conditions. Marijuana is a thirsty plant and prefers a carefully controlled, warm, and humid environment for proper growth, leaving growers with massive monthly electric and water bills. However, weed has a faster grow cycle than hemp—it's ready to harvest in sixty to ninety days—so growers usually turn a profit by growing multiple indoor harvests each year.[31]

2017 marijuana grow at Hobbs Greenery (Left) versus 1942 Kentucky hemp (Right). Under the right conditions, outdoor marijuana can grow to similar heights as hemp, but the two plants still differ in appearance. *Source: Hobbs Greenery, Lakehead, CA, 2017 and courtesy of the National Archives, photo taken in Georgetown, KY on the farm of Patterson Moore, Sept. 1942*

Also, since marijuana is grown specifically for its THC content (which is found in the female plant), only the females are grown together. After germination (germination is the first step in growing weed when it is determined if a seedling is male or female), the male plants are always removed from female plants to stop the females from becoming fertilized.[32] If the males grow with the females, the males will cross-pollinate and turn the female plants into males. This means that instead of flowering with sticky buds packed with THC, the marijuana plant will produce a lot of seeds, and be virtually useless for recreational marijuana purposes.

When it comes to the ratio of THC and CBD, male marijuana plants are equivalent to hemp. A 2012 study of cannabis plants grown in Northern Thailand determined that the CBD content among male plants was always dominant, thus making them ideal for medical CBD. The study took samples from seized street marijuana, outdoor male marijuana (that was already growing in a marijuana field), and male marijuana purposely grown in trial fields for the study. The seized marijuana contained the most THC content, at 2.068 percent. The male plants that were grown purposely for the study contained anywhere from .722 to .848 percent THC. The THC in male plants that were grown in existing outdoor marijuana fields had a range of .874 to 1.480 percent.[33] Seeds taken from all three groups to grow the next generation produced cannabis with even less THC. What this tells us is that the male plants are dominant, compared to the females, and if the males aren't removed from a field of female marijuana, they will reduce the THC content.

A similar principle applies to corn. If sweet corn is grown in too close of a proximity to yellow farm corn, the farm corn will cross-pollinate with the sweet corn and turn it into farm corn. This is why farmers grow sweet corn in different rotations from farm corn—or at least at a distance far enough apart.

Cross-pollination also affects outdoor marijuana grows if they're too close to or downwind from hemp farms. The finicky female marijuana buds might not even flower if hemp is able to cross-pollinate, and their THC content would definitely be compromised.

When it comes to cultivating marijuana, the greatest threat to growing the dankest sticky icky *isn't* the war on drugs, it's too close of a proximity to hemp. Guess the DEA never stopped to consider that ditch weed was on their side.

So, what's the biggest difference between hemp and marijuana?

Over the centuries, male cannabis and female cannabis were bred for different purposes, so essentially, there are two genetic versions of *Cannabis sativa*—one that gets people high and one that doesn't.

In the United States, hemp is defined as *Cannabis sativa L*, a plant that contains *less than* .3 percent THC, and marijuana is defined as *Cannabis sativa*, a plant that contains *more than* .3 percent THC.

Regardless, the paranoia surrounding this *minuscule* amount of THC has to stop. Sure, industrial hemp will create new jobs and a new source of tax revenue in the US economy. Sure, hemp is a building material, a medication, a healthy snack—even a source for fuel, paper, clothing, and other commodities. But the biggest difference between hemp and marijuana is what hemp can do for our future: it can be grown for environmental purposes to provide us with cleaner air, water, and soil; all we have to do is let it grow.

NOTES

1. Art Swift, "Support for Legal Marijuana Use Up to 60% in U.S.," Social & Policy Issues, *Gallup*, October 19, 2016, https://news.gallup.com/poll/196550/support-legal-marijuana.aspx.
2. "United States presidential election of 1932," *Encyclopaedia Britannica*, accessed Jan 23, 2019, https://www.britannica.com/event/United-States-presidential-election-of-1932.
3. "Legalization," *Hempalta,* accessed Jan 23, 2019, https://www.hempalta.com/legislation/#international.
4. "Market Size: Hemp industry sales grow to $688 Million in 2016," *Hemp Business Journal,* accessed Jan 23, 2019, https://www.hempbizjournal.com/market-size-hemp-industry-sales-grow-to-688-million-in-2016/.
5. "NEW—What is the relationship between the Controlled Drugs and Substances Act (CDSA) and the Industrial Hemp Regulations (IHR)?" *Government of Canada,* accessed Jan 23, 2019, https://www.canada.ca/en/health-canada/services/drugs-medication/cannabis/producing-selling-hemp/about-hemp-canada-hemp-industry/frequently-asked-questions.html#a20.
6. Ibid.
7. Mark Halsall, "Hemp area may stall in 2018," *Country Guide,* March 2, 2018, https://www.country-guide.ca/crops/are-canadas-hemp-acres-poised-for-a-downward-slide-in-2018/52710/.
8. Ian Johnston, "As cannabis is widely legalised, China cashes in on an unprecedented boom," Asia, *Independent,* Jan 5, 2014, https://www.independent.co.uk/news/world/asia/as-cannabis-is-widely-legalised-china-cashes-in-on-an-unprecedented-boom-9039191.html.
9. Alex Halperin, "The hemp revival: why marijuana's cousin could soon be big business," Society, *The Guardian,* June 11, 2018, https://www.theguardian.com/society/2018/jun/11/the-hemp-revival-why-marijuanas-cousin-could-soon-be-big-business.
10. Stephen Chen, "Green gold: how China quietly grew into a cannabis superpower," China science, *South China Morning Post,* last modified Aug, 28, 2017, https://www.scmp.com/news/china/society/article/2108347/green-gold-how-china-quietly-grew-cannabis-superpower.
11. "Hemp vs Marijuana," *Ministry of Hemp,* accessed Jan 23, 2019, https://ministryofhemp.com/hemp/not-marijuana/.
12. Small, E. *The Species Problem in Cannabis: Science & Semantics*. Corpus, 1979. https://books.google.ca/books/about/The_Species_Problem_in_Cannabis.html?id=lb8gAQAAIAAJ.

13. Ernest Small and Arthur Cronquist, "A Practical and Natural Taxonomy for Cannabis," *Taxon*, Vol. 25, No. 4 (Aug., 1976), pp. 405–435, accessed Jan 23, 2019, https://www.jstor.org/stable/1220524?seq=1#page_scan_tab_contents.
14. Angela Bacca, "What's the Difference Between Hemp and Marijuana?" *Alternet,* June 5, 2014, https://www.alternet.org/drugs/whats-difference-between-hemp-and-marijuana.
15. Ibid.
16. Campos, A. C., Moreira, F. A., Gomes, F. V., Del Bel, E. A., Guimarães, F. S., (2012). "Multiple mechanisms involved in the large-spectrum therapeutic potential of cannabidiol in psychiatric disorders." *Philosophical Transactions of the Royal Society B: Biological Sciences.* 367 http://doi.org/10.1098/rstb.2011.0389.
17. "Faqs Regarding Minnesota's Industrial Hemp Pilot Program," *Minnesota Department of Agriculture,* accessed Jan 23, 2019, https://www.mda.state.mn.us/plants/hemp/industhempquestions.
18. Ibid.
19. Sawler J., Stout J., Gardner K., Hudson D., et al. "The Genetic Structure of Marijuana and Hemp." *PLoS One.* 2015; 10(8): e0133292. pmid: 26308334. https://www.ncbi.nlm.nih.gov/pmc/articles/PMC4550350/.
20. Ibid.
21. "Hemp," *Encyclopaedia Britannica,* last updated Dec 6, 2018, https://www.britannica.com/plant/hemp.
22. "Hemp vs Marijuana," *Ministry of Hemp,* accessed Jan 23, 2019, https://ministryofhemp.com/hemp/not-marijuana/.
23. Christian Sheckler, "Wild pot grows unhindered in farmland," *South Bend Tribune,* Sept. 8, 2014, https://www.southbendtribune.com/news/wild-pot-grows-unhindered-in-farmland/article_e58fe274-3743-11e4-a268-0017a43b2370.html.
24. Carmen McCollum, "State cops, county officials fight ditch weed," *NWI Times,* July 6, 2001, https://www.nwitimes.com/uncategorized/state-cops-county-officials-fight-ditch-weed/article_5e15b58b-3e27-5534-bdc7-318ee81f03ce.html.
25. Ibid.
26. P. Smith, "Hemp: DEA Has Spent $175 Million Eradicating 'Ditch Weed' Plants That Don't Get You High," *Stop the Drug War,* Jan 4, 2017, https://stopthedrugwar.org/chronicle/2007/jan/04/hemp_dea_has_spent_175_million_e.
27. "Hemp vs Marijuana," *Ministry of Hemp,* accessed Jan 23, 2019, https://ministryofhemp.com/hemp/not-marijuana/.

28. Purdue Hemp Project, "Hemp Production," *Purdue University,* accessed Jan 23, 2019, https://dev.purduehemp.org/hemp-production/.
29. "Hemp vs Weed: Same Plant, Different Chemical Structure and Uses," *Medical Marijuana, Inc.,* Nov. 28, 2017, https://www.medicalmarijuanainc.com/hemp-vs-weed-plant-different-chemical-structure-uses/.
30. Trevor Hennings, "Tips for Growing Industrial Hemp," Growing, *Leafly,* Oct 4, 2017. https://www.leafly.com/news/growing/tips-for-growing-industrial-hemp.
31. "Hemp vs Weed: Same Plant, Different Chemical Structure and Uses," *Medical Marijuana, Inc.,* Nov. 28, 2017, https://www.medicalmarijuanainc.com/hemp-vs-weed-plant-different-chemical-structure-uses/.
32. Anthony Franciosi, "The 7 Key Stages Of The Marijuana Plant Life Cycle," *Honest Marijuana,* June 22, 2018. https://honestmarijuana.com/marijuana-plant/?age-verified=071f602a26.
33. Tipparat P, Natakankitkul S, Chamnivikaipong P, Chutiwat S., "Characteristics of cannabinoids composition of Cannabis plants grown in Northern Thailand and its forensic application." Forensic Sci Int. 2012 Feb 10; 215(1-3): 164–70. pmid: 21636228. https://www.ncbi.nlm.nih.gov/pubmed/21636228.

2

HISTORY OF HEMP

Here's a likely scenario: a run-of-the-mill, middle-aged, working-class American walks into a grocery store. This is an average individual, one of those people who is not too short, not too tall, with high cholesterol and a doughy physical appearance. One of those people who is still paying off a high-interest student loan on an overpriced college degree. A degree that didn't have any bearing on the cubicle career path from business "assistant" to "associate" to "coordinator" to "specialist" to "liaison" or to the coveted "supervisor" title that everyone says Average Joe is a shoo-in for in the near future.

Every summer, this guy can be heard complaining about pollen allergies and rising gas prices, and every winter—as soon as the first snowflakes stick and accumulate—the tired complaints against the Weather Channel's obscenely exaggerated forecasts for the impending apocalyptic blizzard of the century resume.

So this particular everyday Joe walks into this particular grocery store two days after participating in an epic New Year's Eve debauchery that will never be fully remembered, no matter how many photos from the night are plastered on social media. It took two days to completely get rid of the hangover, but this American is now ready and committed to eat healthier and to be more active. This will be the year. Eat more leafy greens, join the company softball league, get that raise, pay off the credit card, vacation in the Bahamas, return with a dark tan and not a sunburn.

And so today, the security cameras at the non-GMO, 100 percent organic, locally sourced, farm-to-table emporium bore witness to this brave soul's exploratory stroll down the first aisle, where heart-healthy hemp products were soon discovered.

The packaging on the hempseed, hemp oil, hemp milk, hemp waffles, hemp butter, hemp tofu, hemp protein powder, and hemp energy bars all promised that Hemp the Superfood delivered calcium, iron, zinc, and more protein and omega-3s in three tablespoons than an entire fillet of salmon. There were CBD products as well—mainly lotions—and images of pot leaves on the labels, yet there were no rolling papers or bongs or vape pens in the store. This greatly confused the American specimen. A consultation was necessary, but asking a store clerk—and therefore immediately being received as ignorant while in earshot of any marginally attractive fellow shopper—was out of the question.

"Hey, Siri. Is hemp the same as marijuana?"

"No, hemp is not the same as marijuana, but they are both cannabis species."

"Siri, is hemp legal in the United States?"

"The federal government classified hemp as a Schedule I narcotic under the Controlled Substances Act. It became a legal agricultural product with the passing of the 2018 Farm Bill."

"Siri, why was hemp illegal in the United States?"

"Hemp was made illegal due to the decisions of highly motivated people in the federal government."

"Siri, what does that mean?"

"What does *what* mean?"

"Siri, why was hemp illegal in the United States?"

"The Marihuana Tax Act of 1937."

The average American decides to purchase a vacuum-sealed baggie of raw shelled hempseed (a product of Canada), making a mental note to google more on this later.

I can't say how many times a scenario like this has occurred nationwide, but since America imported approximately $688 million in hemp products in 2016 alone, I'm going to assume that the majority of Americans are buying hemp not only frequently, but repeatedly, and on purpose, whether or not the first purchase was out of curiosity.

This is good news because for the first time in generations, the future does look bright for developing and sustaining a robust American hemp industry. Thanks to nonprofit organizations that have been dedicated to educating lawmakers about the benefits of this plant, legislation has been set in motion for American hemp farming to make a comeback on a commercial scale, and for the crop to be processed domestically again as well.

Prior to the 2018 Farm Bill, the 2014 Farm Bill allowed states to participate in a pilot program called "Legitimacy of Industrial Hemp Research." What this did was give states the right to grow hemp, typically for the purpose of research studies, with a possible path to commercialization. Even though this pilot program was extremely regulated in most states, as of December 2018, there were thirty-four states that passed legislation defining industrial hemp as distinct from marijuana: Alabama, Arkansas, California, Colorado, Connecticut, Delaware, Florida,

Hawaii, Illinois, Indiana, Kentucky, Maine, Maryland, Massachusetts, Michigan, Minnesota, Montana, Nebraska, Nevada, New Hampshire, New York, North Carolina, North Dakota, Oregon, Pennsylvania, Rhode Island, South Carolina, Tennessee, Utah, Vermont, Virginia, Washington, West Virginia, and Wyoming.

In 2017 alone—three years after the 2014 Farm Bill was passed—a total of 25,541 acres of hemp were grown in nineteen states; there were 1,424 state hemp licenses issued to farmers; and thirty-two universities conducted research on the crop.[1] Despite this incredible progress, the main drawback for most farmers was that the plant was mainly being grown for research purposes instead of for commercial purposes, and they weren't able to capitalize on the full monetary value of the plant.

But hemp advocates continued to move forward regardless. Since 2010, an annual Hemp History Week (which is now typically celebrated in June) has been bringing together grassroots organizers, hemp companies, researchers, farmers and supporters in an effort to change the federal policy on hemp in the United States. The main focus is to educate and inform the public about the many benefits of hemp, correct misconceptions, and of course get more people involved in the democratic process to push hemp legislation through Congress. Even retail stores that sell hemp products get involved by offering free samples and discounts on everything from hempseed to hemp milk to hemp soaps.

What seemed to initially inspire this annual weeklong event was the discovery of Lyster H. Dewey's diaries—diaries that contained detailed notes on growing hemp for government research, dating from 1896 to 1944. During this time, Dewey worked for the U.S. Department of Agriculture, and he researched and grew hemp for the federal government on a plot of land called Arlington Farms—which is where the Pentagon sits today. Dewey's lost hemp diaries were found in 2010—by complete chance—at a garage sale in upstate New York.[2]

The Hemp Industries Association displayed Dewey's diaries and photos during the first Hemp History Week in 2010. At the time, most hemp activists and American historians were unaware of Uncle Sam's hemp farm, and here was a firsthand account, complete with photos, from the cultivator of that hemp.

What's interesting about Lyster Dewey's contributions to hemp research is the time frame in which that research took place. Dewey began his research with the USDA when hemp was an active part of the American economy. His research proved the economic value of a thriving American hemp industry, yet this didn't seem to matter to the next generation of lawmakers who lumped it in with marijuana in the Marihuana Tax Act.

One example of Dewey's contributions to hemp research was a paper titled "Hemp Hurds as Paper-Making Material." As the Botanist in Charge of Fiber-Plant

Investigations, Dewey collaborated with Jason L. Merrill, the Paper-Plant Chemist in Paper-Plant Investigations to conduct the research. The USDA published their findings on October 14, 1916. This means that the federal government knew way back in 1916 that among many other uses, hemp was a more environmentally friendly method of manufacturing paper.

Interesting to note that in the early 1900s there were government-employed scientists and researchers who were concerned with the environment and aware of the impact that major industries were beginning to have upon it. In the conclusion of their report, Dewey and Merrill stated:[3]

> *There appears to be little doubt that under the present system of forest use and consumption the present supply can not withstand the demands placed upon it. After several trials, under conditions of treatment and manufacture which are regarded as favorable in comparison with those used with pulp wood, paper was produced [with hemp hurds] which received very favorable comment from both investigators and for the trade and which according to official tests would be classified as a No. 1 machine-finish printing paper.*

So what exactly are "hemp hurds"? They're essentially the waste created in the industrial hemp manufacturing process.

Dewey and Merrill defined hemp hurds as "the woody inner portion of the hemp stalk [that is] broken into pieces and separated from the fiber"[4] as hemp stalks are being cut and processed by a machine. In the early 1900s, there wasn't much purpose for hemp hurds; farmers were using them for barnyard litter and animal bedding because they're extremely absorbent.

Prior to the early 1900s, hemp fiber was removed from the plant by hand. The hemp stalks had to be physically broken to access the fiber and this was a time-consuming, laborious task, to say the least. However, a new manufacturing process using machine brakes (instead of hand brakes) was on the rise. The machines separated the fiber from the plant at a much faster rate, and the new process also left farmers with the waste by-product, hemp hurds—essentially the equivalent of wood chips—in large, neat piles.

At the time, machine brakes were being used in Wisconsin, Indiana, Ohio, California, and to a limited extent in Kentucky; America was not as technologically advanced as Europe, where machine brakes were common. Dewey and Merrill noted "all of the best hemp in Italy, commanding the highest market price paid for any hemp, is broken by machines."[5]

In their paper, Dewey and Merrill also noted that Europeans were developing even newer models of the hemp-breaking machines, and farmers in the United

States were considering upgrading their equipment because the new technology provided greater monetary benefits from hemp farming not previously possible.

If farmers upgraded to the new hemp-breaking machines, Dewey and Merrill estimated hemp hurds could be sold for $4–$6 per ton.[6] That amount of money might not seem all that significant today, but when adjusting for inflation, in 1916, a mere $4 had the same buying power as $95.74—and $6 in 1916 was worth the equivalent of $143.61—meaning farmers could make between $95.74 and $143.61 per ton of hemp hurds. The two researchers broke down the math as follows:[7]

> *During the last season, 1915, about 1,500 acres of hemp have been harvested outside of Kentucky and in regions where machine brakes are used. Estimating the yield of hurds at 2 ½ tons per acre, this should give a total quantity of about 3,750 tons.*

In addition to that number, Dewey and Merrill also noted that there was approximately seven thousand tons of hemp hurds still available from the 1914 harvest, as well as an additional seven thousand acres in Kentucky that was still being harvested by hand in 1915 (they didn't have an accurate statistic for the total amount of hemp hurds collected from that acreage at the time the paper was published because it wasn't reported yet). Going with their conservative number of 3,750 tons of hemp hurds, that's approximately $15,000 to $22,500 total (worth $259,015.53 to $538,523.30 in 2018). Not a bad for something that was being used in animal stalls to absorb poop.

Theoretically, if American hemp farmers modernized their production, they'd find additional income from selling hemp hurds to paper manufacturers. That monetary incentive was the beginning of a bigger picture. With modern equipment, the United States could become a true contender in exporting hemp fiber and hemp products.

Nothing in Dewey's notes indicated that he was trying to put American industries out of business by advocating for the hemp industry, but his position at the USDA proved what we know to be true today: if you want to find alternative ways to grow the economy, then expand hemp farming. And yet, at the time, American industries seemed threatened by the competition. As motivated as Dewey was to research this phenomenal plant and its many uses, he wasn't one of the highly motivated people who ultimately caused it to become illegal.

In the words of George Carlin:[8] "Motivation is bullshit. If you ask me, this country could use a little less motivation. The people who are motivated are the ones who are causing all the trouble. Stock swindlers, serial killers, child molesters, Christian conservatives. These people are highly motivated."

More specifically, and as it pertains to hemp, two highly motivated troublemakers in the 1930s were Herman Oliphant and Harry Anslinger. If you've read

Jesse Ventura's Marijuana Manifesto, then these names should be familiar to you. Regardless, this chapter expands on who they were, how they were successful in prohibiting hemp in America, and why they did it.

Harry Anslinger was the first commissioner of the Treasury Department's Federal Bureau of Narcotics, what is now the DEA. He is a prime example of someone in an authoritative position within the federal government who lacked a supervisor or a conscience to rein him in, so instead, the authoritarian reigned—he bullied and threatened and ruled by intimidation—and he did so without considering the long-term effects of his thirst for power. Anslinger and J. Edgar Hoover—the FBI director who put George Carlin on a watchlist in 1969 after the comedian did an impression of him on the *Jackie Gleason Show*[9] —would've been besties if they ever met in real life. Or, like the snakes they both were, one would've eaten the other alive.

Herman Oliphant was the general counsel for the U.S. Department of the Treasury during the 1930s. He figured out that the federal government could outlaw something if it was taxed to the point where it was not affordable to purchase. Oliphant cooked up the legal parameters of the National Firearms Act (1934) and the Marihuana Tax Act (1937). Both bills were successful because they were passed as additions to the Internal Revenue Code—which had never been done before—and this revolutionary taxation even held up in court.

The Marihuana Tax Act was the legislation that successfully made Anslinger the nation's first drug czar, with cannabis as his first adversary. This bill was successful in eliminating marijuana and hemp because it used Oliphant's 1934 National Firearms Act (NFA) as a foundation. Although guns and plants don't share much in common, the NFA used taxation and fines to *remove* automatic machine guns from American society, and that same taxation technique was used to *remove* cannabis from American soil.

1934 NATIONAL FIREARMS ACT (NFA)

If you're unfamiliar with the NFA, it was passed because the federal government wanted to outlaw machine guns, but Congress couldn't figure out a way to ban them without infringing upon commerce laws that were in effect at the time.

Machine guns were the first fully automatic weapon, and they were the weapon of choice for organized crime during and after Prohibition. After the 1929 Saint Valentine's Day Massacre in Chicago and the 1933 assassination attempt on President Roosevelt, there was widespread support for the bill—even the NRA (National Rifle Association) openly supported it.[10]

What makes the NFA interesting in a historical context is that it was the first exorbitant excise tax created by the U.S. Department of the Treasury and passed by Congress. And it worked. It greatly prohibited the buying and selling of machine guns, and the act still remains in effect today.

When the NFA was passed, if a US citizen wanted to buy a machine gun, that person had to pay a $200 transfer tax on top of the cost of the gun.[11] Keep in mind, this was during the Great Depression, and when adjusting for inflation, that tax today would be approximately $3,500. Anyone who purchased a machine gun and paid the tax was put on a registry, so there was a record of the sale and the payment.[12] If a citizen was a criminal who didn't legally purchase the gun, or if someone bought the gun without paying the tax, that person would face jail time and a much bigger monetary fine of $2,000—which would equate to a $35,218.00 fine today.[13]

This bill was so effective that machine guns essentially ceased to exist. No one could afford the gun, so the demand ceased. No one manufactured them, so the supply ceased.

Interestingly, the prohibition of machine guns was made even stronger in the generations that followed. By the 1960s, it was illegal to import machine guns, and by 1986, it was illegal for anyone to manufacture new ones for civilian use.[14]

As of February 2016, the Bureau of Alcohol, Tobacco, Firearms, and Explosives database lists that there are 490,664 fully automatic machine guns in circulation.[15] Today, if someone wanted to purchase a machine gun legally, the price point is estimated to start around $20,000 due to the limited supply, and the $200 tax is still applied.[16]

If someone does purchase a machine gun today, regardless of who that person is, the buyer "must pass an extensive background check—including submission of a photograph and fingerprints—as well as fully register the firearm, and receive written permission by the Bureau of Alcohol, Tobacco, Firearms and Explosives before moving the firearm across state lines."[17] Plus, if someone is found in possession of an unregistered machine gun today, that person can face fines up to $250,000 and prison sentences up to ten years.[18]

What does all of this have to do with hemp?

For lawmakers back in the 1930s, the beauty of the National Firearms Act was that it proved the government can be successful in essentially banning a product without passing legislation that specifically calls for a ban. The excise tax worked for machine guns, so in theory, a tax could be applied to cannabis, as well as any other product the government wanted to control.

Oliphant's excise tax held up in court during the 1937 *United States v. Miller* decision when the Supreme Court stated the law did not violate or infringe upon Second Amendment rights.[19] Two weeks after this decision, a decision which proved Oliphant's experimental excise tax passed the test of constitutionality, Herman Oliphant and Harry Anslinger introduced the Marihuana Tax Act to the House Means and Ways Committee. President Roosevelt signed it into law in August 1937, and the bill took effect in October 1937. This bill paved the way for the Controlled Substances Act of 1971, and thus from one generation to the next, the legal consequences of marijuana and hemp became more and more serious.

1937 MARIHUANA TAX ACT

The reason why both marijuana and hemp were classified as Schedule I narcotics by the Controlled Substances Act is because the Marihuana Tax Act didn't completely differentiate between hemp and marijuana when it laid out the taxation system. This was done on purpose. Harry Anslinger was more interested in prosecuting marijuana use than dealing with hemp farmers. It was easy to convince Congress that the marijuana plant was dangerous. The movie *Reefer Madness* had been released in 1936, and William Randolph Hearst's newspapers carried "yellow journalism" articles (the 1930s terminology for "fake news," or creating obscenely inaccurate stories and headlines to sell newspapers) about how smoking marijuana caused violent crimes. But Anslinger and Hearst's real problem child was hemp.

William Randolph Hearst built the largest newspaper chain in the country, founded the media company Hearst Communications, *and* owned the forests that were used to produce all the paper for his empire. Hearst helped Anslinger's cause immensely by writing sensationally false stories that demonized cannabis and people of color simultaneously. According to Hearst, the greatest domestic threat in every small town and big city across America were these "marijuana-crazed negroes" raping white women.[20] Just to round out this image of Hearst (because the tour guides at his extravagant California estate known as "Hearst Castle" won't bring it up), in 1934, he visited Berlin to interview Adolf Hitler, which helped to "legitimize Hitler's leadership in Germany."[21] After the interview, Hearst wrote articles stating the Germans regarded Hitler "as a Moses leading them out of their bondage, and their bondage since the war [World War I] has been utter and bitter" while simultaneously claiming that the Nazis' anti-Semitism "is already well on the way to abandonment."[22] (Even though the first concentration camps in Germany were already established in January 1933, prior to Hearst's visit.[23])

By 1937, Hearst was facing financial ruin—the stock market crash and Great Depression meant not many Americans could afford to buy his newspapers. He had to shut down several of his publications, his corporation faced a court-ordered reorganization, and he was forced to sell many of his antiques to pay his debts.[24] Don't worry—he bounced back. In 2016, Forbes listed the Hearst family among

America's richest families with a net worth of $28 billion.[25] But in 1937, lashing out against a revival of the hemp industry vis-à-vis a marijuana propaganda campaign was most likely Hearst's self-preservation strategy.

Why hemp was of any concern to Harry Anslinger was the same reason he was appointed as the Director of the Bureau of Narcotics and Dangerous Drugs: nepotism. **Andrew Mellon**, the director of the Mellon Bank and secretary of the U.S. Department of the Treasury—who also happened to be the uncle of Anslinger's wife—hired Anslinger. Remember at this time, the Federal Bureau of Narcotics was established as part of the U.S. Department of the Treasury, which meant Andrew Mellon was Herman Oliphant and Harry Anslinger's boss,[26] and the boss had his own financial incentive for terminating the American hemp industry.

As secretary of the treasury, Mellon managed the federal government's revenue, including the nation's financial assets and debts, and he looked for new ways to collect revenue for the federal government. *Why else would the Federal Bureau of Narcotics be placed under the U.S. Department of the Treasury if it couldn't turn a profit by taxing drugs?*

As a private banker, Mellon was financing DuPont, a rising petrochemical and pulp-paper company. A commercialized hemp industry would jeopardize DuPont's lucrative chemical patents,[27] so naturally the secretary of the treasury of all people had a vested interest in passing the Marihuana Tax Act. Luckily for Mellon, Oliphant, Anslinger, and Hearst, hemp and marijuana looked similar. The seeds were identical. The leaves were fairly identical. How would anyone aside from a seasoned botanist be able to tell the difference? So, a seasoned botanist was called upon to answer these questions.

Lyster Dewey retired from the USDA on September 30, 1935, but he testified at the Marihuana Tax Act hearings on April 28, 1937.[28] He was seventy-two at the time. He was also called upon to consult in drafting the legal terminology in the Marihuana Tax Act so that all forms of the cannabis plant were included in the legislation—including *Cannabis sativa* L (hemp).[29] His input was only required to verify what was already known:

- Even to a layman, the cannabis seed was distinctly identifiable from all other seeds.
- Hemp seeds and marijuana seeds were virtually identical.
- The leaves of the hemp plant and the marijuana plant were "very characteristic" when compared to any other plant.

- The unmistakable look of hemp and marijuana leaves made them easily identifiable and not so easily distinguishable—unless closely examined by a professional.

- Although the two plants had different genetics, had different growing cycles, and grew to different heights and widths, their leaves made them look fairly identical to a layman.

- The term "cannabis" should be used whenever referring to marijuana and hemp and all properties of the plant to ensure all varieties were included under the new law.

During the hearings, no one discussed the fact that if hemp was grown in close proximity to marijuana, it would turn the females into males, thus sterilizing its *Reefer Madness* potential, yet there's documented proof that America's founding fathers were aware of this because they grew hemp for economic purposes. George Washington's diaries and farm reports indicated he cultivated hemp at all of his five farms (Mansion House, River Farm, Dogue Run Farm, Muddy Hole Farm, and Union Farm).[30] In his diaries, he even mentioned what happened when the male and the female plants cross-pollinated.

On August 7, 1765, Washington wrote: "began to separate the Male from the Female hemp at Do. [Do is his abbreviation for Dogue Run Farm]—rather too late."[31] Why was he too late? Most marijuana enthusiasts interpret this to mean that the female had already turned male. Maybe Washington was hoping to harvest female buds for his pipe, but there isn't much in the diary to indicate he was pissed to miss out on the opportunity to smoke his own weed, so who really knows.

A year later, on August 29, 1766, Washington wrote: "Began to pull Hemp at the Mill and at Muddy hole—too late for the blossom Hemp by three Weeks or a Month."[32] The "blossom" he is referring to is the marijuana bud, which again indicates his female plants had already turned male.

Although the Marihuana Tax Act did not criminalize cannabis, the freedom to grow, cultivate, and manufacture hemp came with a price. The law stated "every person who sells, deals in, dispenses, or gives away marihuana must register with the Internal Revenue Service and pay a special occupational tax."[33] The occupational tax on cannabis took the form of a yearly "license," which was less costly than the tax imposed on machine guns, but also came with stipulations. Here are some examples of the taxation tiers, paraphrased from the text of the 1937 Marihuana Tax Act:[34]

- Importers, manufacturers, and compounders of marihuana pay a tax of $24 per year (equates to approximately $423 per year with inflation).

- Producers of marihuana pay $1 per year—whether they grow it all year or for a few months. (One dollar equates to about $17.61 per year with inflation.)

- Anyone in the medical field who distributes, dispenses, prescribes, gives away, or administers marijuana to patients pays $1 per year.

- Anyone who obtains and uses marihuana in a laboratory setting for research, instruction, or analysis, or anyone who produces marihuana for any such purpose pays $1 per year.

- Anyone who is not in the medical field who deals in, dispenses, or gives away marihuana pays $3 per year (approximately $54 per year today).

Remember, marihuana in this case refers to cannabis as a whole—including hemp—and all properties of the cannabis plant are subject to this law—from seeds to oil to fiber.

This tax was paid in advance. Once it was paid, the federal government issued a "marijuana stamp" —which served as the license—to prove the person paid the tax and had the legal right to the cannabis. This stamp system is why the Marihuana Tax Act is sometimes referred to the Marihuana Stamp Act.

On the surface, this occupational tax might seem reasonable, and even in the 1930s, a yearly tax of $1 was something most businesses could afford. However, what deterred people from continuing in the cannabis business were the stipulations that followed once the stamp was received.

A hemp farmer's taxes *increased* once the annual $1 tax was paid under something called a "transfer tax."[36]

- If a farmer paid the yearly $1 tax, and was already registered in the federal system, and then wants to transfer the cannabis to another person, a tax of $1 per ounce must be paid.

- If the $1 per ounce tax is not paid, then that person must pay a fine of $100 per ounce.

- All transactions are recorded and submitted to the IRS.

What this means is if a farmer is looking to sell hemp to a manufacturing facility, that farmer must pay $1 per ounce once it is "transferred" to the manufacturer.

At the time, the US economy was still recovering from the Great Depression. Farmers, who were already struggling, had to then consider the ongoing fees involved to acquire and renew the permit to grow cannabis. When one crop starts being taxed at a higher rate than other crops, any reasonable person would do the math to figure out the return on the investment of planting that crop, so hemp farmers considered planting something else.

The funny thing is (actually, it's really not all that funny at all), four years after the Marihuana Tax Act was passed, America desperately needed hemp for World War II, so the industry was rebooted as a wartime crop.

NOTES

1. "State Hemp Legislation," Vote Hemp, accessed Jan 23, 2019, http://www.votehemp.com/PR/PDF/Vote-Hemp-2017-US-Hemp-Crop-Report.pdf.
2. Manuel Roig-Franzia, "Hemp fans look toward Lyster Dewey's past, and the Pentagon, for higher ground," Style, *Washington Post,* May 13, 2010, http://www.washingtonpost.com/wp-dyn/content/article/2010/05/12/AR2010051204933.html.
3. Bulletin 404, U.S. Department of Agriculture: Hemp Hurds as Paper-Making Material; Dewey and Merrill, p. 25.
4. Bulletin 404, U.S. Department of Agriculture: Hemp Hurds as Paper-Making Material; Dewey and Merrill, p. 1.
5. Bulletin 404, U.S. Department of Agriculture: Hemp Hurds as Paper-Making Material; Dewey and Merrill, p. 4.
6. Bulletin 404, U.S. Department of Agriculture: Hemp Hurds as Paper-Making Material; Dewey and Merrill, p. 6.
7. Bulletin 404, U.S. Department of Agriculture: Hemp Hurds as Paper-Making Material; Dewey and Merrill, p. 5.
8. George Carlin's "A List of People Who Ought to be Killed," *Complaints and Grievances*, 2001, https://www.youtube.com/watch?v=tVlkxrNlp10.
9. Carly Sitrin, "George Carlin's J. Edgar Hoover impression got him put on the FBI's watchlist," *Muck Rock,* Aug 13, 2015, https://www.muckrock.com/news/archives/2015/aug/13/george-carlins-fbi-file/.
10. "When Capone's gang massacred people, Congress wasted no time banning Tommy guns," News, *Chicago Sun Times,* March 19, 2018, https://chicago.suntimes.com/columnists/when-capones-gang-massacred-people-congress-wasted-no-time-banning-tommy-guns/.

11. "National Firearms Act of 1934," *The Internet Archive,* accessed Jan 23, 2019, https://archive.org/stream/NationalFirearmsActOf1934/National_Firearms_Act_of_1934_djvu.txt.
12. "National Firearms Act," *Bureau of Alcohol, Tobacco, Firearms and Explosives,* April 26, 2018, https://www.atf.gov/rules-and-regulations/national-firearms-act.
13. "National Firearms Act of 1934," Legal Dictionary, *The Free Dictionary by Farlex,* accessed Jan 23, 2019, https://legal-dictionary.thefreedictionary.com/National+Firearms+Act+of+1934.
14. "Key Federal Acts Regulating Firearms." *Giffords Law Center,* accessed Jan 23, 2019, http://lawcenter.giffords.org/gun-laws/federal-law/background-resources/key-federal-acts-regulating-firearms/.
15. Boucher, Stephanie. Letter to Jeffrey E. Folloder. Feb 24, 2016. U.S. Department of Justice, Bureau of Alcohol, Tobacco, Firearms and Explosives. Retrieved from http://www.nfatca.org/pubs/MG_Count_FOIA_2016.pdf.
16. Christian Lowe, "Did Las Vegas Shooter Stephen Paddock Use a Fully Automatic Rifle?" *Weekly Standard,* Oct 2, 2017, http://www.weeklystandard.com/did-las-vegas-shooter-stephen-paddock-use-a-fully-automatic-rifle/article/2009907.
17. "Prohibition-Era Gang Violence Spurred Congress To Pass First Gun Law," All Things Considered, *NPR*, June 30, 2016, https://www.npr.org/2016/06/30/484215890/prohibition-era-gang-violence-spurred-congress-to-pass-first-gun-law.
18. M. L. Nestel & Andrea Miller, "What to know about machine gun laws in the US," *ABC News,* Oct 4, 2017, http://abcnews.go.com/US/machine-gun-laws-us/story?id=50256580.
19. "United States v. Miller, 307 U.S. 174 (1939)," *Justia US Supreme Court,* accessed Jan 23, 2019, https://supreme.justia.com/cases/federal/us/307/174/case.html.
20. Jack Herer, "Chapter 4," *The Emperor Wears No Clothes,* accessed Jan 23, 2019, https://jackherer.com/emperor-3/chapter-4/.
21. "William Randolph Hearst Biography," *Biography,* last updated April 27, 2017, https://www.biography.com/people/william-randolph-hearst-9332973.
22. David Nasaw, *The Chief: The Life of William Randolph Hearst*, p. 497–498 .
23. Holocaust Encyclopedia, "Concentration Camps, 1933–39," *United States Holocaust Memorial Museum,* accessed Jan 23, 2019, https://www.ushmm.org/wlc/en/article.php?ModuleId=10005263.
24. "William Randolph Hearst Biography," *Biography,* last updated April 27, 2017, https://www.biography.com/people/william-randolph-hearst-9332973.
25. "America's Richest Families Net Worth," *Forbes.com*, June 29, 2016, https://www.forbes.com/profile/hearst/.

26. Joseph W. Jacob, "Medical Uses of Marijuana," Trafford Publishing, 2009, p. 83.
27. Joseph W. Jacob, "Medical Uses of Marijuana," Trafford Publishing, 2009, p. 84.
28. "Timeline," *Lyster Dewey Diaries,* accessed Jan 23, 2019, http://www.lysterdewey.com/timeline/.
29. "Conference on Cannabis Sativa L.," January 14, 1937—Room 81 Treasury Building, 10:30 AM, *Drug Library,* accessed Jan 23, 2019, http://www.druglibrary.org/schaffer/hemp/taxact/canncon.htm.
30. "Did George Washington Grow Hemp?," *George Washington's Mount Vernon,* Mount Vernon Ladies' Association, accessed Jan 23, 2019, http://www.mountvernon.org/george-washington/the-man-the-myth/george-washington-grew-hemp.
31. George Washington, "Diary Entry: 7 August 1765," *Founders Online,* National Archives, accessed Jan 23, 2019, https://founders.archives.gov/documents/Washington/01-01-02-0011-0006-0004.
32. George Washington, "Diary Entry: 29 August 1766," *Founders Online,* National Archives, accessed Jan 23, 2019, https://founders.archives.gov/documents/Washington/01-02-02-0001-0006-0006.
33. "Marijuana Tax Act Law and Legal Definition," *US Legal,* accessed Jan 23, 2019, https://definitions.uslegal.com/m/marijuana-tax-act%20/.
34. "The Marihuana Tax Act of 1937," The Schaffer Library of Drug Policy, Jan 23, 2019, http://www.druglibrary.org/schaffer/hemp/taxact/mjtaxact.htm.
35. Ibid.
36. Ibid.

3

HEMP AS A WARTIME CROP

In *Jesse Ventura's Marijuana Manifesto,* we documented many of the cultures that incorporated hemp into paper, rope, and clothing. Centuries prior to World War II, hemp's demand was tied into wartime necessities. The British colonized the New World in part due to the need for cheap access to hemp for the British Royal Navy. British flax and hemp were major industries in medieval times, and by 1533, Henry VIII ordered farmers to grow a quarter acre of flax or hemp for every sixty acres of land[1] to keep up with supply demands for the naval fleet. Each ship required more than two hundred tons of rope and sails, all made from hemp fiber.[2] Thirty years later, the British navy had significantly increased, so Queen Elizabeth I increased the amount of hemp that farmers were required to produce and increased penalties for not growing it.

Due to the fact that England is an island with very limited agricultural space, the country was running out of room to produce both food and hemp.[3] At the time, the British were also purchasing hemp from allies including Russia, Poland, France, Holland, Italy, and Prussia (Germany). Queen Elizabeth I soon found a way for her island nation to sprout more viable farmland for hemp by colonizing Ireland, and by encouraging British citizens to colonize America. By colonizing other countries, England was able to grow its own hemp at a lesser cost than purchasing it outright.

In 1619, King James I ordered every American colonist to grow one hundred hemp plants for export. The crop was so valuable colonial governments including Virginia, Maryland, and Pennsylvania allowed farmers to pay one-fourth of their taxes in hemp.[4] Thomas Jefferson actually received the first US patent for inventing a machine to break hemp to extract the fibers.[5] George Washington, Thomas Jefferson, and Ben Franklin's wealth was intertwined with either growing hemp or producing hemp products, which makes sense because 95 percent of Colonial America was involved in agriculture.[6]

Fast-forward to the early 1900s and there were other means of manufacturing products that were originally derived from hemp. Twenty-three years prior to the Marihuana Tax Act passing, Lyster Dewey listed five reasons why American hemp farming was declining. In the 1913 Yearbook of the USDA, Dewey writes:[7]

> *The cultivation of hemp is declining in the United States because of the (1) increasing difficulty in securing sufficient labor for handling the crop with present methods, (2) lack of labor-saving machinery as compared with machinery for handling other crops, (3) increasing profits in other crops, (4) competition of other fibers, especially jute, and (5) lack of knowledge of the crop outside of a limited area in Kentucky.*

Although Dewey openly admitted the challenges to hemp farming in 1913 without prejudice, he made another claim in the 1913 USDA Yearbook: "The two fiber-producing plants most promising for cultivation in the central United States and most certain to yield satisfactory profits are hemp and flax."

Seems a bit contradictory, but in the late 1930s, technology was beginning to resolve the five issues Dewey had addressed. American hemp farming was predicted to make a big comeback, and this pitted it against already established American businesses, such as synthetic fibers.

On February 26, 1937, a presentation about the hemp industry was given by the Process Industries Division of the American Society of Mechanical Engineers at an agricultural processing meeting in New Brunswick, New York.[8] This resulted in the publication of two articles: one was featured in *Popular Mechanics* and titled "Billion-Dollar Crop";[9] the other, titled "The Most Profitable and Desirable Crop that Can Be Grown," was published in *Mechanical Engineering.*

Both articles spoke to new hemp-processing machines developed in Europe that had the potential to revolutionize the hemp industry. The authors of the articles also estimated how much additional money the crop could bring into the American economy if the industry was successfully modernized. Unfortunately, due to printing deadlines in the 1930s, the articles were published too late to make a difference. The newly developing American hemp industry was no longer on a fast track due to the Marihuana Tax Act.

Here is the text from those two articles.[10] Note how similar they are to articles written about hemp today: the excitement for the "new" hemp industry and the need to explain why hemp and marijuana may look the same but are completely different. There's a sense of awe when the authors list the vast number of products that can be made from hemp (including plastics!), especially after calculating the astounding amount of money that can be made from this crop that essentially founded the American colonies.

NEW BILLION-DOLLAR CROP, *POPULAR MECHANICS*, FEBRUARY 1938

AMERICAN farmers are promised a new cash crop with an annual value of several hundred million dollars, all because a machine has been invented which solves a problem more than 6,000 years old. It is hemp, a crop that will not compete with other American products.

Instead, it will displace imports of raw material and manufactured products produced by underpaid coolie and peasant labor and it will provide thousands of jobs for American workers throughout the land.

The machine which makes this possible is designed for removing the fiber-bearing cortex from the rest of the stalk, making hemp fiber available for use without a prohibitive amount of human labor. Hemp is the standard fiber of the world. It has great tensile strength and durability. It is used to produce more than 5,000 textile products, ranging from rope to fine laces, and the woody "hurds" remaining after the fiber has been removed contain more than seventy-seven per cent cellulose, and can be used to produce more than 25,000 products, ranging from dynamite to Cellophane.

Machines now in service in Texas, Illinois, Minnesota and other states are producing fiber at a manufacturing cost of half a cent a pound, and are finding a profitable market for the rest of the stalk. Machine operators are making a good profit in competition with coolie-produced foreign fiber while paying farmers fifteen dollars a ton for hemp as it comes from the field.

From the farmers' point of view, hemp is an easy crop to grow and will yield from three to six tons per acre on any land that will grow corn, wheat, or oats. It has a short growing season, so that it can be planted after other crops are in. It can be grown in any state of the union. The long roots penetrate and break the soil to leave it in perfect condition for the next year's crop. The dense shock of leaves, eight to twelve feet above the ground, chokes out weeds. Two successive crops are enough to reclaim land that has been abandoned because of Canadian thistles or quack grass.

Under old methods, hemp was cut and allowed to lie in the fields for weeks until it "retted" enough so the fibers could be pulled off by hand. Retting is simply rotting as a result of dew, rain and bacterial action. Machines were developed to separate the fibers mechanically after retting was complete, but the cost was high, the loss of fiber great, and the quality of fiber comparatively low.

With the new machine, known as a decorticator, hemp is cut with a slightly modified grain binder. It is delivered to the machine where an automatic chain conveyor feeds it to the breaking arms at the rate of two or three tons per hour. The hurds are broken into fine pieces which drop into the hopper, from where they are delivered by blower to a baler or to truck or freight car for loose shipment. The fiber comes from the other end of the machine, ready for baling.

From this point on almost anything can happen. The raw fiber can be used to produce strong twine or rope, woven into burlap, used for carpet warp or linoleum backing or it may be bleached and refined, with resinous by-products of high commercial value. It can, in fact, be used to replace the foreign fibers which now flood our markets.

Thousands of tons of hemp hurds are used every year by one large powder company for the manufacture of dynamite and TNT. A large paper company, which has been paying more than a million dollars a year in duties on foreign-made cigarette papers, now is manufacturing these papers from American hemp grown in Minnesota. A new factory in Illinois is producing fine bond papers from hemp. The natural materials in hemp make it an economical source of pulp for any grade of paper manufactured, and the high percentage of alpha cellulose promises an unlimited supply of raw material for the thousands of cellulose products our chemists have developed.

It is generally believed that all linen is produced from flax. Actually, the majority comes from hemp—authorities estimate that more than half of our imported linen fabrics are manufactured from hemp fiber. Another misconception is that burlap is made from hemp. Actually, its source is usually jute, and practically all of the burlap we use is woven by laborers in India who receive only four cents a day. Binder twine is usually made from sisal which comes from Yucatan and East Africa.

All of these products, now imported, can be produced from home-grown hemp. Fish nets, bow strings, canvas, strong rope, overalls, damask tablecloths, fine linen garments, towels, bed linen and thousands of other everyday items can be grown on American farms.

Our imports of foreign fabrics and fibers average about $200,000,000 per year; in raw fibers alone we imported over $50,000,000 in the first six months of 1937. All of this income can be made available for Americans.

The paper industry offers even greater possibilities. As an industry it amounts to over $1,000,000,000 a year, and of that eighty percent is imported. But hemp will produce every grade of paper, and government figures estimate that 10,000 acres devoted to hemp will produce as much paper as 40,000 acres of average pulp land.

One obstacle in the onward march of hemp is the reluctance of farmers to try new crops. The problem is complicated by the need for proper equipment a reasonable distance from the farm. The machine cannot be operated profitably unless there is enough acreage within driving range and farmers cannot find a profitable market unless there is machinery to handle the crop. Another obstacle is that the blossom of the female hemp plant contains marijuana, a narcotic, and it is impossible to grow hemp without producing the blossom. Federal regulations now being drawn up require registration of hemp growers, and tentative proposals for preventing narcotic production are rather stringent.

However, the connection of hemp as a crop and marijuana seems to be exaggerated. The drug is usually produced from wild hemp or locoweed which can be found on vacant lots and along railroad tracks in every state. If federal regulations can be drawn to protect the public without preventing the legitimate culture of hemp, this new crop can add immeasurably to American agriculture and industry.

THE MOST PROFITABLE AND DESIRABLE CROP THAT CAN BE GROWN, *MECHANICAL ENGINEERING*

Note: "Flax and Hemp: From the Seed to the Loom" was published in the February 1938 issue of Mechanical Engineering magazine. It was originally presented at the Agricultural Processing Meeting of the American Society of Mechanical Engineers in New Brunswick, New York, on February 26, 1937, by the Process Industries Division.

Flax and Hemp: From the Seed to the Loom, by George A. Lower

This country imports practically all of its fibers except cotton. The Whitney gin, combined with improved spinning methods, enabled this country to produce cotton goods so far below the cost of linen that linen manufacture practically ceased in the United States. We cannot produce our fibers at less cost than can other farmers of the world. Aside from the higher cost of labor, we do not get as large production. For instance, Yugoslavia, which has the greatest fiber production per acre in Europe, recently had a yield of 883 lbs. Comparable figures for other countries are Argentina, 749 lbs.; Egypt 616 lbs.; and India, 393 lbs.; while the average yield in this country is 383 lbs.

To meet world competition profitably, we must improve our methods all the way from the field to the loom.

Flax is still pulled up by the roots, retted in a pond, dried in the sun, broken until the fibers separate from the wood, then spun, and finally bleached with lye from wood ashes, potash from burned seaweed, or lime. Improvements in tilling, planting, and harvesting mechanisms have materially helped the large farmers and, to a certain degree, the smaller ones, but the processes from the crop to the yarn are crude, wasteful and land injurious. Hemp, the strongest of the vegetable fibers, gives the greatest production per acre and requires the least attention. It not only requires no weeding, but also kills off all the weeds and leaves the soil in splendid condition for the following crop. This, irrespective of its own monetary value, makes it a desirable crop to grow.

In climate and cultivation, its requisites are similar to flax and like flax, should be harvested before it is too ripe. The best time is when the lower leaves on the stalk wither and the flowers shed their pollen.

Like flax, the fibers run out where leaf stems are on the stalks and are made up of laminated fibers that are held together by pectose gums. When chemically treated like flax, hemp yields a beautiful fiber so closely resembling flax that a high-power microscope is needed to tell the difference—and only then because in hemp, some of the ends are split. Wetting a few strands of fiber and holding them suspended will definitely identify the two because, upon drying, flax will be found to turn to the right or clockwise, and hemp to the left or counterclockwise.

Before [World War I], Russia produced 400,000 tons of hemp, all of which is still hand-broken and hand-scutched. They now produce half that quantity and use most of it themselves, as also does Italy from whom we had large importations.

In this country, hemp, when planted one bu. per acre, yields about three tons of dry straw per acre. From 15 to 20 percent of this is fiber, and 80 to 85 percent is woody material. The rapidly growing market for cellulose and wood flower for plastics gives good reason to believe that this hitherto wasted material may prove sufficiently profitable to pay for the crop, leaving the cost of the fiber sufficiently low to compete with 500,000 tons of hard fiber now imported annually.

Hemp being from two to three times as strong as any of the hard fibers, much less weight is required to give the same yardage. For instance, sisal binder twine of 40-lb. tensile strength runs 450 ft. to the lb. A better twine made of hemp would run 1280 ft. to the lb. Hemp is not subject to as many kinds of deterioration as are the tropical fibers, and none of them lasts as long in either fresh or salt water.

While the theory in the past has been that straw should be cut when the pollen starts to fly, some of the best fiber handled by Minnesota hemp people was heavy with seed. This point should be proved as soon as possible by planting a few acres and then harvesting the first quarter when the pollen is flying, the second and third a week or 10 days apart, and the last when the seed is fully matured. These four lots should be kept separate and scutched and processed separately to detect any difference in the quality and quantity of the fiber and seed.

Several types of machines are available in this country for harvesting hemp. One of these was brought out several years ago by the International Harvester Company. Recently, growers of hemp in the Middle West have rebuilt regular grain binders for this work. This rebuilding is not particularly expensive and the machines are reported to give satisfactory service.

Degumming of hemp is analogous to the treatment given flax. The shards probably offer slightly more resistance to digestion. On the other hand, they break down readily upon completion of the digestion process. And excellent fiber can, therefore, be obtained from hemp also. Hemp, when treated by a known chemical process, can be spun on cotton, wool, and worsted machinery, and has as much absorbency and wearing quality as linen.

Several types of machines for scutching the hemp stalks are also on the market. Scutch mills formerly operating in Illinois and Wisconsin used the system that consisted of a set of eight pairs of fluted rollers, through which the dried straw was passed to break up the woody portion. From there, the fiber with adhering shards—or hurds, as they are called—was transferred by an operator to an endless chain conveyer. This carries the fiber past two revolving single drums in tandem, all having beating blades on their periphery, which beat off most of the hurds as well as the fibers that do not run the full length of the stalks. The proportion of line fiber to tow is 50% each. Tow or short tangled fibers then go to a vibrating cleaner that shakes out some of the hurds. In Minnesota and Illinois, another type has been tried out. This machine consists of a feeding table upon which the stalks are placed horizontally. Conveyor chains carry the stalks along until they are grasped by a clamping chain that grips them and carries them through half of the machine.

A pair of intermeshing lawnmower-type beaters are placed at a 45-degree angle to the feeding chain and break the hemp stalks over the sharp edge of a steel plate, the object being to break the woody portion of the straw and whip the hurds from the fiber. On the other side and slightly beyond the first set of lawnmower beaters is another set, which is placed 90-degrees from the first pair and whips out the hurds.

The first clamping chain transfers the stalks to another to scutch the fiber that was under the clamp at the beginning. Unfortunately, this type of scutcher makes even more tow than the so-called Wisconsin type. This tow is difficult to re-clean because the hurds are broken into long slivers that tenaciously adhere to the fiber.

Another type passes the stalks through a series of graduated fluted rollers. This breaks up the woody portion into hurds about ¾ inch long, and the fiber then passes on through a series of reciprocating slotted plates working between stationary slotted plates.

Adhering hurds are removed from the fiber which continues on a conveyer to the baling press. Because no beating of the fiber against the grain occurs, this type of scutcher make only line fiber. This is then processed by the same methods as those for flax.

Paint and lacquer manufacturers are interested in hempseed oil which is a good drying agent. When markets have been developed for the products now being wasted, seed and hurds, hemp will prove, both for the farmer and the public, the most profitable and desirable crop that can be grown, and one that can make American mills independent of importations.

Recent floods and dust storms have given warnings against the destruction of timber. Possibly, the hitherto waste products of flax and hemp may yet meet a good part of that need, especially in the plastic field which is growing by leaps and bounds.

* * *

It's a bit sickening, isn't it? To realize America has been in this same cyclical loop since the 1930s: recognizing hemp's value but completely ignoring it at the same time. Until we need it for a war effort, that is. World War II proved beyond any doubt the value of domestic American hemp, yet shortly after the war, the federal government went back to eradicating it.

HEMP FOR VICTORY!

War disrupts global supply chains, and the supply chain to the United States was disrupted in World War II when Japan cut off US access to hemp from the Philippines.[11] This forced the federal government to turn to its own farmers and ask them to produce hemp for the needs of the army, navy, and air force.

In 1942, the Department of Agriculture and the US Army urged farmers to grow hemp for fiber, which was needed for uniforms, parachutes, rope, and other necessities of war. They launched a pro-hemp campaign called Hemp for Victory, which included a newsreel film by the same title and the distribution of 400,000 pounds of seeds to farmers.[12] The regulations previously set forth in the Marihuana Tax Act were more or less revoked. Tax-free permits were being issued to farmers, and the government created pamphlets explaining the growing cycle and optimal harvesting methods for hemp—stating the plant wasn't a drug and had no psychoactive effects, like many people at the time were led to believe. The government also subsidized the costs for industrial hemp farming if a farmer grew more than a million acres.[13]

Keep in mind, America was in a race against time to ramp up hemp production. Many farmers were producing zero acres of hemp; now they were being asked to grow as much as possible as quickly as possible. Agricultural communities were now job creators. New manufacturing buildings had to be constructed, and more laborers were necessary to harvest and process the hemp.

One example of the government stepping in to completely revitalize the hemp industry to meet supply demands for World War II took place in Jackson County, Minnesota. *The Jackson County Pilot* chronicled how the federal government paid millions to revive "the century-old hemp industry"—and then much to everyone's surprise, the government closed the factories soon after the war—including the one that was constructed in Jackson County.

Here are some bullet points taken from the *Jackson County Pilot* articles that speak to how many people were employed, how much federal money was spent to build hemp factories, and how much hemp was cultivated between 1942 and 1944 for war efforts in Minnesota and other states:[14]

- The War Production Board's multimillion-dollar program for reviving the hemp industry was focused in Kentucky, Minnesota, Wisconsin, Illinois, Iowa, and Indiana.

- A total of seventy-one hemp factories (referred to as breaking plants) were constructed and 3,500 harvesters were built.

- Each plant cost approximately $275,000 to build and employed seventy-five to one hundred men in the breaking season (harvest season).

- The goal was to plant 300,000 acres in Kentucky to be harvested in 1943 because the hemp fiber the United States had in reserves from the Philippines was expected to be completely depleted by the end of 1943.

- Jackson County, Minnesota, was designated as one of fifteen locations in Minnesota where the government established a hemp factory. (The government spent $330,000–$335,000 to erect each factory in Minnesota.)

- The plant in Jackson County cost $335,000, and forty workers were needed to construct it within 160 days.

- A well was also built at the site to account for the need for water for the crop.

- Each hemp mill was estimated to produce a minimum of four thousand acres of hemp; Minnesota had a total of fifteen mills and harvested sixty thousand acres in 1943. The state was approved for eleven new plants to be built later that year.

- On January 21, 1943, the *Jackson County Pilot* published an article titled "Hemp, the New Million Dollar Cash Crop for Farmers," which republished many of the same ideas from the 1938 *Popular Mechanics* article. The local newspaper's article stated the many uses for this newly established, modern hemp industry, touting that hemp can easily transition to a "peacetime crop."

- Farmers in Jackson County received a total of $57,000 worth of hempseed to be grown in 1943; the hempseed was worth $14 per bushel (three bushels were required per acre); the government purchased the entire crop—which was estimated to be 460,000 tons of fiber.

- At harvest time, between 120 and 150 people were needed to keep the mill operating around the clock in two ten-hour shifts. Twenty-five to thirty employees were also needed to haul the hemp to the mill (there was a surplus of people who qualified for these jobs).

- In 1943, farmers saw a net return of $125 per acre of hemp in Wisconsin and a net return of $75–$100 per acre of hemp in Minnesota.

- After all expenses were deducted, farmers in Jackson County (all of whom had allotted 4,000 acres or more to growing hemp) were receiving between $48.44 and $1,155.74 per acre, with an average net income of $82.55 per acre.

A total of approximately 42,000 tons of hemp fiber was cultivated annually between 1942 and 1945 for war efforts.[15] Considering how profitable the one hemp facility in Jackson County was, everyone assumed the hemp industry would continue to flourish after the war. But when the war ended, so did the support of the US Army and the Department of Agriculture. The processing plants were built with federal money, and the federal government decided to shut them down. Forget the fact that federal money is taxpayer money, and technically those factories belonged to those communities.

On December 30, 1943, an article published in the *Jackson County Pilot* was titled "Revival of Hemp Industry Promised in United States," but on January 20, 1944, there was a significant update: "Manager Hislop Is Told No 1944 Crop to Be Put in Here."[16] The newspaper reported that Jackson County and Sherburn County plants were told to cease operations. The facilities that cost taxpayers $330,000 each were to close in Iowa and Minnesota. Why? The US global supply chain was restored! Even though the government saw that the supply chain of hemp from Central Asia could be compromised due to war, the decision was made to cut the cord with domestic hemp and the American farming community.

The announcement came from Fred E. Butcher, the president of War Hemp Industries, Inc. This private corporation was created specifically for the Hemp for Victory campaign to oversee the planting and harvesting of hempseed and fiber.[17] What was missing in all of the newspaper articles at the time was the strategic way in which the Hemp for Victory campaign was constructed to ensure that hemp would stay a "wartime crop" and not transition into becoming a "peacetime crop."

During the Hemp for Victory campaign, any farmer who did not have hempseed or the necessary equipment to sow it had to purchase USDA-approved hempseed and rent machines from War Hemp Industries, Inc.—a private company that acted on behalf of the Commodity Credit Corporation, a corporation that essentially served as the financing institution for the USDA.

The Commodity Credit Corporation (or Commodity as it is referred to in financial documents from the 1930s) is a corporation made up of USDA officials that is owned by the federal government—it is actually still in existence today (hence the relevance).

The Commodity was created in 1933 to "stabilize, support, and protect farm income and prices"; it also "helps maintain balanced and adequate supplies of agricultural commodities and aids in their orderly distribution."[18] Today, it provides income support and disaster assistance as well as other financial services to the agricultural community. In 2018, the Commodity was tasked with providing payments up to $12 billion to producers of soybean, sorghum, corn, wheat, cotton, dairy, shelled almonds, sweet cherries, and hogs as part of a Trade Retaliation Mitigation Aid Package.[19] Trade disputes with China, Mexico, Canada, and other countries financially impacted American farmers, and the Trump administration agreed to have the Commodity provide direct payments to farmers as "a short-term relief strategy" while the Trump administration worked on "free, fair, and reciprocal trade deals" in global markets.[20]

During World War II, the Commodity Credit Corporation ensured there was enough hempseed available to satisfy the military's fiber demands and enough money available to pay for all the equipment and manufacturing needs. It worked with the Defense Plant Corporation to build the necessary hemp facilities—such as the one built in Jackson County, Minnesota— that were then leased to War Hemp Industries, Inc., and then War Hemp Industries, Inc. leased them to the farmers to utilize.

THE DEFENSE PLANT CORPORATION

Another government corporation founded during World War II worked with the Commodity to fund the building projects and blueprints for the hemp facilities. The Defense Plant Corporation was part of the Reconstruction Finance Corporation (RFC), which was also owned and operated by the federal government between 1932 and 1957. The RFC was designed to provide financial support and government loans to banks, railroads, mortgage associations, and other businesses to boost the country after the start of the Great Depression. Under the New Deal (and President FDR), the RFC was expanded to provide loans for agriculture. The loans were then distributed to the Defense Plant Corporation, and then to the Commodity.

When the Defense Plant Corp. was founded in June 1940, its main priority was to expand the industrial base of the country to meet military demands during

World War II. It did this by contracting with private businesses to operate defense-related industries for the war. These private businesses then restructured themselves to create products intentionally for war efforts. If local businesses didn't have the necessary factory or machines, the private companies would ask the Defense Plant Corp. to build or supply what was necessary, and then those manufacturing plants or machines were leased to the local businesses. By the time the Defense Plant Corp. was dissolved (at the end of World War II), it owned approximately 10–12 percent of the industrial capacity of the United States.[21]

Since the hemp industry was created specifically for war purposes, it was deemed no longer necessary after the war had come to an end. The same held true for other industries; clothing companies could go back to producing sweaters, jackets, sneakers, socks, and all other essential products for their usual client base rather than producing uniforms, boots, and other gear for the soldiers in active duty.

This symbiotic relationship between government and corporation during World War II still exists today—private military contractors such as Halliburton, Academi, and DynCorp receive millions of dollars in federal funding to carry out all kinds of projects on behalf of the United States of America. Today's private wartime corporations have been known to act with very little government oversight or even accountability; however, according to the master contract between Commodity and War Hemp Industries, the USDA actually pulled the strings in War Hemp Industries (at least on paper):[22]

- War Hemp Industries had to wait for approval from the Commodity for nearly every decision, but once the approval was granted, the private corporation was "an independent operating agent."

- War Hemp Industries was entitled to purchase the hempseed from the Commodity, sell it on the Commodity's behalf, and deposit money into the Commodity's bank account.

- War Hemp Industries acted on behalf of the Commodity, as if it was the Commodity, and the Commodity allowed it to do this.

- Although Hemp for Victory was a government campaign, hemp farmers interacted directly with War Hemp Industries instead of the Commodity.

Since hemp was being rebranded as a wartime crop, a wartime corporation was put in charge of it. This made sense from a logistical standpoint as War Hemp

Industries, Inc. was based in Chicago—much closer to the agricultural heartland than Washington, DC—and after the Marihuana Tax Act, the federal government couldn't be involved in any scandals involving *selling and distributing* cannabis to farmers, so by proxy, War Hemp Industries was the perfect solution.

Also, in the contract between Commodity and War Hemp Industries, the Commodity actually put the private corporation on the hook for any potential legal battles with Harry Anslinger or law enforcement. The contract states that War Hemp Industries, Inc. "agrees to indemnify and hold Commodity harmless from any liability or penalty with may be lawfully imposed by local or state authorities or any department or bureau thereof."[23] Which is humorous, considering that the contract also shows that the private corporation is being fully funded and controlled by the Commodity. In fact, the Commodity loaned War Hemp Industries the funds for employee salaries and all operating costs—while at the same time forbidding the private corporation and its employees from borrowing funds from any other source.

In the contract, the loan for War Hemp Industries was described as "an amount sufficient for operating expenses at the rate of three percent per annum on the actual cash balance" that accrued until the contract between the Commodity and War Hemp Industries expired.[24] In other words, when the war was over, the contract was over, and at that time, War Hemp Industries was to pay back the amount that was spent, plus 3 percent interest. However, this would be easy to repay as War Hemp Industries also received a "compensating operating fee"—or a commission—on everything it did on behalf of Commodity. The Commodity also had control over the prices of everything War Hemp Industries sold or leased to farmers, so (in theory) the private company couldn't make a large profit from the war efforts. The private corporation had to provide monthly budgeting statements, including all costs and all transactions, for the Commodity's approval, and the Commodity had the authority to "change the operating budget at any time" for any reason.[25]

When it came to profits, the Commodity's contract also outlined that if a significant profit was made by War Hemp Industries, the money would be transferred over to the Commodity when the contract had expired (at the conclusion of World War II), and the Commodity would determine if that money was to be kept by the USDA or if any of it would be returned and distributed to War Hemp Industries.

The Commodity's contract also outlined salary caps: "No salaries shall be paid by War Hemp in excess of $9,000 per year, and the entire salary . . . is subject to approval by the Commodity. Any salary to be paid by War Hemp to any officer or employee in excess of $5,000 per annum shall be subject to the specific written approval of the Commodity."[27]

In return for its service, War Hemp Industries was allowed to operate research units—subject to the approval of the Commodity, of course—to develop "the most efficient machinery and processes for the production of the highest quality" of hemp as well as investigate and research "new uses of hemp fiber and by-products."[28]

Upon studying these documents—the scanned copies are easily accessible on Hempology.org—it becomes clear why the wartime crop never stood a chance to become a peacetime crop. The Defense Plant Corp. fronted the money for all the hemp facilities and supplied the money for all the equipment, but the source of the funds soon disappeared when the war ended because that's when the corporation's contract with the federal government expired. When the Defense Plant Corp. dissolved, the Commodity also ended its contract with War Hemp Industries, and then War Hemp Industries had to break off its relationship with the local hemp farmers by essentially laying them off.

The only hemp farms that stayed in business after the war were the ones the farmers owned and operated directly. Since the hemp farmers who used their own land and their own equipment couldn't be systematically shut down, this threatened Harry Anslinger's authoritative grasp on cannabis. After World War II, the Marihuana Tax Act was reinstated, but Andrew Mellon was no longer secretary of treasury, nor was Herman Oliphant the chief counsel. They were both deceased. The current secretary of the treasury, Henry Morgenthau, Jr., was on to bigger and better issues than hemp, such as serving as the chairman at the conference that established the International Monetary Fund and the World Bank.[29]

AMENDMENTS TO THE MARIHUANA TAX ACT AND IRS TAX CODE

Since the Marihuana Tax Act was part of the IRS code, there was a hearing on May 24, 1945 before the Congressional Committee on Finance to determine if hemp and marijuana should in fact be separated to appease hemp farmers. Congressman Joseph P. O'Hara from Minnesota provided the opening statements at the hearing:[30]

> *I want to say first that I am not an expert on hemp matters, but because one of the industries in my district is in the business of the manufacture of hemp and has been for a considerable time, and because of the trouble we have been having . . . in regard to some of the regulations which they [the Narcotics Division of the U.S. Department of the Treasury] have imposed up on this industry in my district—I feel that an amendment should be considered which would give the growers of hemp in this country some consideration . . .*
>
> *Permit me to say that since the beginning of the war the Government has installed out in our country some 11 [11 is a typo in the transcript—it was 71] hemp plants. . . . I merely want to say that following the complaints that I received from the industry in my section of the country I took the matter up*

> *with Mr. Anslinger and some of his assistants from the Division of Narcotics and told them of the complaint . . . which the industry felt was a very harsh burden, namely, to place guards around the plants to guard the hemp which was there for processing.*

The Chairman of the Committee on Finance, Georgia Congressman Walter F. George, responded to Mr. O'Hara's opening remarks with words of support and historical reference:

> *Hemp has been grown in the United States since the earliest colonial times. Records of the town of Oxford, MD show that the farmers the Eastern Shore of Maryland produced hemp and sold it to the square-rigged ships for rope making before the Revolutionary War. Kentucky has produced hemp for more than a century and Wisconsin has been producing it continuously since 1916. . . . When the Marihuana Act was passed in 1937, the two representatives of the United States Treasury testifying before the committee stated that the legislation would not in any way interfere with the legitimate production of hemp in the United States. . . . There has been no change in the method of producing hemp fiber since 1917. . . . The method used today is identical with the process that was in general use when the Marihuana Act was passed in 1937.*

According to the hearing transcripts, testimony from farmers who grew hemp *before* the Marihuana Tax Act *and during* World War II backed up the chairman's statements: their methods of growing and processing hemp hadn't changed, but the law was now inexplicably requiring them to employ guards to watch the hemp as it grew, and to pay a tax of $1 for each ounce of leaves on every stalk of harvested hemp.

Hemp farmers also testified that they weren't growing marijuana within their hemp fields, and they also had to assure lawmakers that no one had ever tried to steal hemp either before, during, or after the war for "illicit purposes."

The chairman and members of the committee had already devoured the original transcripts from the first Marihuana Tax Act hearings and read Harry Anslinger's initial testimony aloud. In the initial testimony, Anslinger assured the House Ways and Means Committee that "the production and sale of hemp and its products for industrial purposes will not be adversely affected by this bill." Naturally, legislatures now wanted to know why Anslinger was requiring guards to be hired to oversee hemp fields and hemp-processing facilities—especially since this was not required when the bill was initially passed, nor was this action required during an actual world war.

Congressman George also stated during the hearing that he was aware of only five hemp mills that existed in Wisconsin and Kentucky when the Marihuana Tax Act was passed in 1937, but the federal government had now built forty-two hemp mills in that same area to supply fiber for military purposes. He admitted that "we could not have successfully carried on our tremendous naval expansion program if this [expansion of hemp mills] had not been possible." Whether or not all of the committee members agreed with their chairman, Harry Anslinger now had a problem. What was he going to do with these farmers who wanted to continue to grow hemp without paying for guards and without paying the taxes he outlined in the Marihuana Tax Act?

The hearings also included testimony from farmers looking to educate lawmakers on how ridiculous and inconsistent Anslinger's taxation system was. In one instance, a Wisconsin hemp farmer named Matt Rens—whose family had been in the business for thirty years and had supplied hemp for both world wars—explained that the fees involved with Anslinger's transfer taxation system would ensure "that the hemp business will be killed." Mr. Rens's "transfer" of hempseed involved three transfers—"from the grower to the dealer in Kentucky, from the dealer to us as the mill owner, from us to the farmer"—and each transfer required a $1 tax to be paid per ounce. Mr. Rens stated if this taxation had still been in effect during the war, he would've paid a total transfer tax of $8,997,120 in one year, obviously a sum of money that no farmer in 1945 could possibly afford.

In another instance, a farmer representing the Minnesota Hemp Co.—an organization that grew hemp during World War II—had his crops examined in 1945 by the Narcotics Bureau. The inspection determined the farmer must remove all the leaves from the hemp stalks or the farmer would be taxed $1 an ounce for each ounce of leaves. The inspector then allowed between 10 and 20 percent of the leaves to remain, as removing all the leaves would clearly affect the plant's ability to grow properly, but the cost of going into a field of 4,000 acres and removing that many leaves from each and every stalk of hemp made the inspector's slight compromise impossible to comply with.

Samuel H. McCrory, the Director of the Hemp Division for the Commodity Credit Corporation, also testified in favor of continuing the hemp industry beyond a wartime crop. War Hemp Industries, Inc. was now dissolved, but the Commodity backed up the farmers who stated the impracticality of complying with Anslinger's demands. McCrory added that he knew of no hemp mill that "turned out any marihuana for drug purposes," and his closing statement really put it all in perspective:

Hemp fiber is in a highly competitive situation as in respect to the other fibers that are grown in this country or imported, particularly the soft fibers, flax and jute, and if we are to keep a hemp industry, and I believe that it is in our country's interest to do so, we should not put any more obstacles than we must in the way of the people who are going to grow hemp, requirements that increase their cost and place them at a disadvantage.

George E. Farrell, an Agricultural Specialist from the Bureau of Agricultural Economics at the USDA, also backed up his colleague's statements. He noted that the government had already invested $12,000,000 to construct hemp mills—all of which were to be shut down once the remaining hemp was collected and sold. As a comparison, he noted that corn production ramped up to 30 percent during the war, but those farmers now did not need to grow that much for domestic consumption, so what were corn farmers going to grow on all that additional land?

"It will be necessary to find new crops that can be grown in the Corn Belt," Farrell stated, "and hemp is one of them. It will provide an income for farmers in the Corn Belt, it will increase employment in these small towns where the mills are, it will provide a sizable amount of employment in the spinning and weaving mills. It seems to be a very satisfactory industry that can be developed extensively."

He went on to list how many other countries were currently growing hemp for commercial purposes to theorize that the United States would be losing out on an expanding market as Italy, Yugoslavia, Romania, Bulgaria, Poland, and Russia had taken advantage of utilizing the crop.

The committee also heard from manufacturing mills about the limited supply of cotton and other fibers. Manufacturers now preferred hemp fibers because the fiber was superior to cotton; it was easier to work with, and it was guaranteed to last longer for the consumer.

The hearings took an interesting turn when Harry Anslinger couldn't attend to testify due to an illness. Will S. Wood, the Deputy Commissioner of the Bureau of Narcotics, as well as B. T. Mitchell, the Assistant Chief Counsel of the Bureau of Narcotics, gave testimony in Anslinger's place, and remarkably, common sense prevailed: everyone came to an agreement to amend the Marihuana Tax Act so that the taxation system wasn't open to new interpretations. Specifically, the amended version of the Marihuana Tax Act was to end the transfer tax and end the confusing taxation based on how many leaves were left on the hemp stalk, but the millers were required to register and pay a $1 per year tax, same as the growers.[31]

However, just because the committee agreed on these changes didn't mean they took effect immediately. The Commodity Credit Corporation's Samuel McCrory was tasked with being the liaison between the hemp industry and Anslinger so farmers could avoid any future conflicts with the Federal Bureau of Narcotics, but the speed at which it took Congress to formally amend the preexisting law was not swift.

On July 21, 1945—two months after the initial Committee on Finance hearings—the Senate voted to adopt these amendments, and requested the House of

Representatives also accept the changes to the Marihuana Tax Act. It took the House of Representatives eleven years, but they eventually passed the Senate Amendments and Public Law 320 on March 8, 1956, and the previsions were added to IRS US Code 26 USC 4742 (c), 26 USC 4751 (6), and 26 USC 4753 (b) to officially specify the difference between hemp and marijuana.

According to documents available on the National Hemp Association's website, the changes to the law didn't do much to stimulate a postwar American hemp industry. By the time the IRS amendments took effect, hemp farming was being overshadowed by the increasing popularity of synthetic fibers such as nylon, polyester, spandex, and acrylic.[32]

THE DEFENSE PRODUCTION ACT OF 1950

Nonetheless, the federal government's view of hemp as a wartime crop continued under the Defense Production Act of 1950, and President Clinton and President Obama reiterated the definition through executive orders.

The Defense Production Act gives the president the authority to instruct private businesses to sign contracts or fulfill orders deemed necessary for national defense. Under this law, the president can establish wartime agencies (such as the Defense Plant Corporation), and allocate materials, services, and facilities all in the name of national defense. The president is also granted control over the civilian economy so that scarce or critical materials necessary to national defense are available for defense purposes.

Essentially, the Defense Production Act takes power away from Congress (and all other aspects of government) and gives it to the president. According to the law, the president can "requisition property, force industry to expand production, impose controls over wages, change product prices, settle labor disputes, control consumer and real estate credit, and allocate raw materials towards national defense."[33]

Hemp is included in Part VIII of the Defense Production Act, under "General Provisions." It is categorized as "food resources."[34] Only in times of war, and only if our national defense depended upon it, the president can decree that Americans grow (or eat) hemp.

In 1994, President Clinton issued Executive Order 12919, "National Defense Industrial Resources Preparedness," and in 2012, President Obama issued Executive Order 13603, "National Defense Resources Preparedness"—both included modern updates to the original Defense Protection Act, and both list hemp among the nation's essential agricultural products.[35] So evidently, the federal government has recognized hemp as an agricultural crop for decades prior to the 2018 Farm Bill, but only in times of war.

NOTES

1. John Blair, Nigel Ramsay, "English Medieval Industries," Hambledon Press, 1991, p. 322.
2. Robert Deitch, "Hemp: American History Revisited: The Plant with a Divided History," Algora Publishing, 2003, p. 12.
3. John Blair, Nigel Ramsay, "English Medieval Industries," Hambledon Press, 1991, p. 322.
4. PBS, "Marijuana Timeline," *Frontline,* accessed Jan 23, 2019, https://www.pbs.org/wgbh/pages/frontline/shows/dope/etc/cron.html.
5. Robert Deitch, "Hemp: American History Revisited: The Plant with a Divided History," Algora Publishing, 2003, p. 19.
6. Robert Deitch, "Hemp: American History Revisited: The Plant with a Divided History," Algora Publishing, 2003, p. 20.
7. Lyster H. Dewey, "1913 Yearbook of the United States Department of Agriculture," *Bureau of Plant Industry,* Jan 1, 1913, http://www.globalhemp.com/1913/01/1913-yearbook-of-the-united-states-department-of-agriculture.html.
8. Jack Herer, "Chapter 3," *The Emperor Wears No Clothes,* accessed Jan 23, 2019, https://jackherer.com/emperor-3/chapter-3/.
9. "Billion-Dollar Crop," *Popular Mechanics,* Feb 1938, https://books.google.com/books?id=e9sDAAAAMBAJ.
10. Jack Herer, "Chapter 3," *The Emperor Wears No Clothes,* accessed Jan 23, 2019, https://jackherer.com/emperor-3/chapter-3/.
11. "The History of Hemp in America," *Medical Marijuana Inc,* accessed Jan 23, 2019, https://news.medicalmarijuanainc.com/history-hemp-america.
12. Ibid.
13. "Before the Coming of the Anti-Medical Marihuana Laws," *Antique Cannabis Book,* accessed Jan 23, 2019, http://antiquecannabisbook.com/chap04/Illinois/IL_IHIndustrialHemp.htm.
14. Craig D. Putnam, "War on Hemp," *Hempology.org: The Study of Hemp,* accessed Jan 23, 2019, http://hempology.org/ALL%20HISTORY%20ARTICLES.HTML/1942-44%20HEMP%20FOR%20VICTORY.HTML/1942-44MINNESOTA.html.
15. "The History of Hemp in America," *Medical Marijuana Inc,* accessed Jan 23, 2019, https://news.medicalmarijuanainc.com/history-hemp-america.
16. Craig D. Putnam, "War on Hemp," *Hempology.org: The Study of Hemp,* accessed Jan 23, 2019, http://hempology.org/ALL%20HISTORY%20ARTICLES.HTML/1942-44%20HEMP%20FOR%20VICTORY.HTML/1942-44MINNESOTA.html.

17. "Industrial Hemp in the United States," USDA, p. 3 https://www.ers.usda.gov/webdocs/publications/41740/15853_ages001ec_1_.pdf?v=42087.
18. "Commodity Credit Corporation," United States Department of Agriculture, Farm Service Agency, accessed Jan 23, 2019, https://www.fsa.usda.gov/about-fsa/structure-and-organization/commodity-credit-corporation/index.
19. Kent Thiesse, "USDA sets up avenues to get aid to farmers," AgriNews, *Post Bulletin,* Sept 6, 2018, https://www.postbulletin.com/agrinews/news/midwest/usda-sets-up-avenues-to-get-aid-to-farmers/article_b42cc818-aba3-11e8-97cd-b72df5479088.html.
20. USDA News Release, "USDA adds shelled almonds and fresh sweet cherries to Market Facilitation Program," *Rushville Republican,* Sept 28, 2018, http://www.rushvillerepublican.com/news/lifestyles/usda-adds-shelled-almonds-and-fresh-sweet-cherries-to-market/article_a15b0faf-c5b0-5c99-ab75-42c9ddd7455c.html.
21. "Defense Plant Corporation," Social Networks and Archival Context, accessed Jan 23, 2019, http://snaccooperative.org/ark:/99166/w6tj3qmp.
22. "Master Contract between Commodity Credit Corporation and War Hemp Industries," p. 1, *Hempology.org: The Study of Hemp,* accessed Jan 23, 2019 http://hempology.org/ALL%20HISTORY%20ARTICLES.HTML/1942-44%20HEMP%20FOR%20VICTORY.HTML/1943%20WAR%20HEMP%20INDUSTRIES/MASTER%20CONTRACT%20PG1.html.
23. "Master Contract between Commodity Credit Corporation and War Hemp Industries," p. 11, *Hempology.org: The Study of Hemp,* accessed Jan 23, 2019, http://hempology.org/ALL%20HISTORY%20ARTICLES.HTML/1942-44%20HEMP%20FOR%20VICTORY.HTML/1943%20WAR%20HEMP%20INDUSTRIES/MASTER%20CONTRACT%20PG11.html.
24. "Master Contract between Commodity Credit Corporation and War Hemp Industries," p. 2, *Hempology.org: The Study of Hemp,* accessed Jan 23, 2019, http://hempology.org/ALL%20HISTORY%20ARTICLES.HTML/1942-44%20HEMP%20FOR%20VICTORY.HTML/1943%20WAR%20HEMP%20INDUSTRIES/MASTER%20CONTRACT%20PG2.html.
25. "Master Contract between Commodity Credit Corporation and War Hemp Industries," p. 10, *Hempology.org: The Study of Hemp,* accessed Jan 23, 2019, http://hempology.org/ALL%20HISTORY%20ARTICLES.HTML/1942-44%20HEMP%20FOR%20VICTORY.HTML/1943%20WAR%20HEMP%20INDUSTRIES/MASTER%20CONTRACT%20PG10.html.
26. "Certificate of Stock of the War Hemp Industries," *Hempology.org: The Study of Hemp,* accessed Jan 23, 2019, http://hempology.org/ALL%20HISTORY%20ARTICLES.HTML/1942-44%20HEMP%20FOR%20VICTORY

.HTML/1943%20WAR%20HEMP%20INDUSTRIES/CERTIF%20OF%20 STOCK.html.

27. "Master Contract between Commodity Credit Corporation and War Hemp Industries," p. 9, *Hempology.org: The Study of Hemp,* accessed Jan 23, 2019, http://hempology.org/ALL%20HISTORY%20ARTICLES.HTML/1942-44%20 HEMP%20FOR%20VICTORY.HTML/1943%20WAR%20HEMP%20 INDUSTRIES/MASTER%20CONTRACT%20PG9.html.
28. "Master Contract between Commodity Credit Corporation and War Hemp Industries," p. 12, *Hempology.org: The Study of Hemp,* accessed Jan 23, 2019, http://hempology.org/ALL%20HISTORY%20ARTICLES.HTML/1942-44%20 HEMP%20FOR%20VICTORY.HTML/1943%20WAR%20HEMP%20 INDUSTRIES/MASTER%20CONTRACT%20PG12.html.
29. "History of the Treasury," U.S. Department of the Treasury, accessed Jan 23, 2019, https://www.treasury.gov/about/history/Pages/edu_history_brochure.aspx.
30. "Hearing Before the Committee on Finance United States Senate Seventy-Ninth Congress First Session on H.R. 2348," Committee of Finance, May 24, 1945, https://www.finance.senate.gov/imo/media/doc/79Hrg.Hemp.pdf.
31. "Hemp in the Post-War Era: The Hemp Millers' Exemption from Marihuana Taxes," *Drug Science,* accessed Jan 23, 2019, http://www.drugscience.org /Archive/bcr3/n3_Gettman_Hemp.html.
32. "Hemp History," *National Hemp Association,* accessed Jan 23, 2019, https://nationalhempassociation.org/study-at-academica/.
33. "The Defense Production Act: Choice as to Allocations," Columbia Law Review, March 1951; Lockwood, Defense Production Act: Purpose and Scope, June 22, 2001
34. Office of the Press Secretary, "Executive Order—National Defense Resources Preparedness," *The White House,* March 16, 2012, https://obamawhitehouse.archives .gov/the-press-office/2012/03/16/executive-order-national-defense-resources -preparedness.
35. USDA, "Hemp For Victory," *Hemp Business Journal,* 1942. https://www.hemp bizjournal.com/video-hemp-for-victory/.

4

THE RETURN OF AMERICAN HEMP

One final American hemp history lesson that I was surprised to learn: once upon a time, hemp was as American as the $10 bill because it was on the $10 bill.

The year was 1914, and the Federal Reserve had just started issuing the first printings of official US currency. The $10 bills had President Andrew Jackson's portrait on the front, and on the back was a representation of the cultural divide already taking place in America: On the left-hand side was an illustration of farmers harvesting hemp (representing American agriculture); on the right-hand side was an illustration of a factory (representing American industrialization). There was a blank space between the two images.

For the next fifteen years, every $10 bill issued by the Federal Reserve depicted this image. Then in 1929, the bill was updated. Alexander Hamilton was on the front and the U.S. Department of the Treasury building was on the back. Although the bill has had additional changes throughout the years, the rendering of the U.S. Department of the Treasury building has remained on the back.

The front and back of the $10 bill, as issued by the Federal Reserve in 1914. *(https://en.wikipedia.org/wiki/Federal_Reserve_Note#/media/File:US-$10-FRN-1914-Fr.898a.jpg)*

The adage "a picture is worth a thousand words" certainly describes this symbolic revision of the $10 bill. While I really don't think the 2018 Farm Bill will usher in a new version of the $10 bill, I also didn't think Congress would actually legalize hemp when I started writing this book.

This chapter highlights some of the more notable obstacles that have been overcome by activists and the challenges we still need to overcome ensure the return of the American hemp industry is successful.

The front and back of the $10 bill, as updated in 1929. *(https://en.wikipedia.org/wiki/United_States_ten-dollar_bill#/media/File:US-$10-GC-1928-Fr-2400.jpg)*

WHEN STATES FIRST LEGALIZED HEMP

Looking back at everything the federal government has done to block our access to hemp, the industrial hemp pilot program outlined in the 2014 Farm Bill seems like sudden about-face. However, as many hemp advocates know, the momentum started in 1998, when the United States allowed Canadian food-grade hemp products including hemp oil and hempseed to be imported. Then the 2004 Hemp Industries Association vs DEA decision allowed hemp foods and body care products to be sold in the United States legally. The next logical step was for hemp to be grown and developed into a domestic industry, rather than importing all these products, so the states made strides to legalize it. By 2007, farmers in North Dakota were feeling helpless. Since 1993, a crop fungus had repeatedly diminished wheat returns, which were the primary agricultural commodity in the state, so North Dakota's agricultural department decided to grant hemp licenses to two farmers in an attempt to get more crops in the rotation. Unfortunately, they weren't able to obtain any seed until after the 2014 Farm Bill because of the DEA. The agency stood in the way by denying North Dakota farmers the right to obtain a federal permit and access to hempseed, even after they had paid the non-refundable fees. Lawsuits against the DEA ensued, and it could've very well become a losing battle until the 2014 Farm Bill began to put the DEA in it's place.

While activist group Vote Hemp, North Dakota's Agriculture Commissioner, and North Dakota's farmers fought the DEA, something interesting took place in

2012 in Colorado. In 2012, Colorado and Washington states were the first to legalize marijuana for recreational purposes. The major difference between these two states is that Colorado simultaneously legalized hemp production when it legalized recreational marijuana (Washington didn't do so until 2016).

By spring 2013, the crop was already being grown in Colorado—even before state regulators had the opportunity to write the rules for the industry—and the Centennial State's hemp producers have been successfully doing their own thing ever since. By the time the 2014 Farm Bill ushered in terms for the hemp pilot program, Colorado's state legislators and agricultural department opted to continue to allow growers the complete freedom to acquire their hempseed from whatever source they prefer. There's also no application deadline for interested hemp producers, no acreage limit, and no requirements for processing a hemp license. According to *Hemp Industry Daily*, Colorado currently has seven varieties of certified hemp, has more market opportunities for selling hemp than any other state, and in 2017, it produced more than half of the nation's total hemp.[1] The pioneering spirit of Colorado certainly paid off; hemp producers are ahead of the curve when compared to other states, but Vermont also defied the Controlled Substances Act by legalizing industrial hemp prior to the 2014 Farm Bill.

According to *Hemp Industry Daily*'s Top 10 States for Hemp,[2] Vermont also passed its own laws to legalize hemp in 2013 without any stipulations. As of December 2018, farmers don't have to participate in a pilot project or research collaboration, they don't have to register with state agricultural authorities, and they aren't subject to background checks or acreage limits. Agricultural officials don't keep track of how many licensed acres are in production, how much hemp the farmers are producing, or where/how they're selling it. The state doesn't actively research or monitor or test the plants to ensure the THC content is below .3 percent. Only at the request of law enforcement will Vermont test the THC content (which has only happened once, and the plants in question passed inspection). While Colorado and Vermont have some of the most lenient hemp laws in the nation, it will be interesting to see how their policies change, now that the 2018 Farm Bill has passed.

THE GROWING WARRIORS PROJECT

One of my favorite success stories of the modern American hemp industry is the Growing Warriors Project, which operated the first federally permitted industrial hemp farm since the 1940s Hemp for Victory campaign.

In 2016, the Growing Warriors' farm, located in eastern Kentucky, produced two hundred American flags, made in America, from American hemp.[3] But when they initially started the farm in 2014, the DEA told them they didn't have the right to grow hemp.[4] This is after Mike Lewis, the group's founder, secured a $30,000 grant from Grow Appalachia to begin the program.[5] Lewis took the DEA to court, the Louisville judge sided with the Growing Warriors Project, and they've been growing hemp, expanding their operations, and raising awareness about hemp ever since.

The Growing Warriors Project produced American flags made from American hemp by American veterans . . . has this ever occurred before in American history? In 1793, Eli Whitney invented the cotton gin, and cotton soon became the go-to source for all textiles, so it's tough to say when the last time an all-American hemp flag was produced. Today's historians and vexillologists (those who study the history, symbolism, and usage of flags) assert that early American flags were made from wool, cotton, linen, or silk depending on what was available and where the flag was to be hung.[6] American folklore maintains that Betsy Ross's flag—the first flag—was made from homespun hemp, and that perception has certainly helped intertwine the significance of the plant with American patriotism.

So what's the benefit of making American flags from hemp when most flags are now made from polyester or rayon? Hemp fiber takes to fabric dyes well, resists mildew, blocks ultraviolet light, and has natural antibacterial properties.[8] Plus, the process to create hemp fiber versus synthetic fibers is much less harmful to the environment. Hemp fiber is actually one of the strongest natural fibers in the world, so when it comes to symbolism, wouldn't Americans want the national symbol of the country made from this enduring, homegrown fiber?

Aside from the cultural significance of veterans producing American flags by hand, the location of the Growing Warriors' hemp farm is also noteworthy. Aside from large-scale logging and coal-mining firms—which brought many short-term, low-wage paying jobs to the area—many of the small towns in eastern Kentucky are now the headquarters of American opioid addiction.[9] In 2012, the unemployment rate for Kentucky-based US veterans aged eighteen to twenty-five was nearly 30 percent and the average age of the Kentucky farmer was fifty-seven.[10] Putting these statistics together, Kentucky has a lot of younger people out of work, many addicted to opioids, and a vulnerable agricultural industry with the potential to collapse or die of old age.

According to *Kentucky Monthly*, Growing Warriors Project works with Kentucky Proud Jobs for Vets and Homegrown by Heroes to enable Kentucky veterans to pursue careers in agriculture. The hemp farm is just one of many agricultural projects—the veterans receive training in growing vegetables, livestock, and sustainable agriculture. There are also eight community garden agriculture programs across the state of Kentucky where veterans receive a stipend for producing food for homeless shelters and food pantries.

As Mike Lewis said when he was asked what his biggest success was in hemp farming to date:[11]

> *The relationships have really been meaningful. It is important that people learn to work together—for the future of all humanity. That's the reason I remain involved. People from all walks of life are working together to create solutions.*

The Kentucky Department of Agriculture's Homegrown by Heroes and Kentucky Proud continue to help military veterans find jobs in agriculture, train veterans who are considering a career in agriculture, and provide mentorship opportunities between veterans and veteran-run farms. Participating farms are also given a distinctive label for their products so consumers can easily identify them as grown locally by a veteran-farmer.

HEMP FARMING EQUIPMENT

Of course, when adding a new crop to the rotation, a significant expense for farmers is the cost of specialized equipment.

In the 1942 "Hemp for Victory" film, the USDA instructed farmers:[12]

> *Soil that will grow good corn will usually grow hemp. For fiber, hemp should be sown closely, the closer the rows, the better. . . . These rows are spaced about four inches. . . . Hemp for fiber is ready to harvest when the pollen is shedding and the leaves are falling. In Kentucky, hemp harvest comes in August. Here the old standby has been the self-rake reaper, which has been used for a generation or more. . . . A modified rice binder has been used to some extent. This machine works well on average hemp. Recently, the improved hemp harvester, used for many years in Wisconsin, has been introduced in Kentucky. This machine spreads the hemp in a continuous swath. It is a far cry from this fast and efficient modern harvester, that doesn't stall in the heaviest hemp.*

Farmers were able to sow and harvest the crop using much of the same equipment they already had on hand, and even today's start-up hemp farm operations utilize machines that are used for other crops.

"I have probably rolled more cannabis than Willie Nelson," industrial hemp farmer Glenn Rodes told *Lancaster Farming* in October 2018. Rodes has a ten-acre hemp farm in Virginia that partners with James Madison University as part of the state's industrial hemp pilot program. "Part of our research with JMU was to determine if industrial hemp can be grown and harvested with existing equipment on the farm," Rodes explained. "It can be."[13]

Since most agricultural crops aren't nearly as tall or as strong as hemp stalks, in certain situations, updated equipment is necessary.[14] An article in *LandTechnik Agricultural Engineering* explains: "Even if standard machines at farm or contractors level are used, certain modifications or device couplings are still necessary due to the specific characteristics of the crop."[15] Plus, farmers may prefer to use different equipment depending if the crop is harvested for CBD or for fiber.[16]

Sharing resources among farms could also be a way to mitigate costs for those just starting out in the hemp industry. In October 2018, county officials in Wetaskiwin, Canada, began brainstorming to determine how to best support the hemp farming industry in Alberta. Since new equipment is expensive, the councilors are considering adopting a co-op model,[17] which could greatly help all farmers in the industry to expand.

As the hemp industry expands, technology will advance, and farming and processing methods will certainly become more and more specialized. It's also good to know that hemp cultivators don't have to spend a lot of money on the latest toys, gizmos, and gadgets to make money. Prior to the 2018 Farm Bill, hemp farmers couldn't rely on bank loans to purchase new equipment; most banks wouldn't work with the hemp industry at all because the crop was in the Controlled Substances Act. Now that this is no longer an issue, hemp farmers are able to operate as any other agricultural business.

TROUBLESHOOTING THE BANKING INDUSTRY

Banks frequently decline to work with any cannabis-related businesses. Because banks abide by federal law, and federal law states cannabis is illegal, they can deny loans and refuse to open a bank account for businesses simply for having the word *marijuana* in their name—regardless if the business sells it or grows it or merely promotes it. Prior to the 2018 Farm Bill, hemp growers were also affected.

For instance, hemp growers in Colorado harvested approximately nine thousand acres of hemp in 2017, yet they found it is nearly impossible to get a bank account or approval for crop insurance through the Federal Crop Insurance Program.[18] Without crop insurance, there's nothing to protect a farmer from unforeseen circumstances, such as an early frost that wipes out acres of crops before they can be harvested.

Dani Billings, owner of LoCo Farms and cofounder of the Colorado Hemp Project, has been growing hemp since 2014 and is one of those farmers who was denied federal crop insurance, banking services, and credit card processing.[19] When she did find a bank to deposit funds in, she had to pay deposit insurance because her business was considered high risk.

Thankfully, these stresses and expenses are now obsolete, and the 2018 Farm Bill also extends hemp investment opportunities to the stock market.

According to a September 2018 article on Forbes.com, the legal marijuana industry as a whole is estimated at $9 billion in sales (which is equivalent to the entire stock market), but only 30 percent of businesses have bank accounts.[20] Now that hemp companies can obtain bank accounts, they can also start qualifying to list their businesses on NASDAQ, NYSE, or publicly traded stock exchanges in other countries, such as the Toronto Stock Exchange or Canadian Ventura Exchange.

However, we may see more Canadian cannabis businesses on the country's stock exchange within the next year. When Canada passed its Cannabis Act in 2018, the law specifically stated that all of the country's banks can conduct business with the cannabis industry, just as they would with any other industry. In June 2018, the Bank of Montreal extended a $200 million loan to Aurora Cannabis, one of the world's largest marijuana companies, based in Edmonton, Alberta.[21] Not that I'd ever advocate for Wall Street and the banking industry to pressure Capitol Hill, but it's certainly possible for lobbying efforts to now focus on the obvious multibillion-dollar reason why all US-based cannabis companies should have access to banking services—especially since the Canadian stock exchange welcomed Canadian cannabis companies.

HEMP OVERCOMES IRS CODE 280E

In *Jesse Ventura's Marijuana Manifesto,* we discussed section 280E of the IRS tax code, which essentially states that any business involved in "trafficking" controlled substances is prohibited from taking any deductions or credits.[22] The controlled substances are those listed in the Controlled Substances Act under Schedule I and Schedule II narcotics (marijuana is listed as Schedule I, and up until recently, so was hemp).

Therefore, if a business grows or sells marijuana legally according to state law, that business cannot take any tax deductions. When states legalize marijuana, they're nullifying federal law, but states can't seem to nullify IRS laws, and the "punishment" for being in the marijuana industry is to pay about 80 to 90 percent in taxes from the revenue that the business generated. The biggest hypocrisy here is that pharmaceutical companies producing the controlled substances applicable to tax law 280E are able to take massive deductions.

Here's what the Controlled Substances Act (CSA) classifies as Schedule I (not currently accepted for medical use and a high potential for abuse): heroin, lysergic acid diethylamide (LSD), marijuana, 3,4-methylenedioxymethamphetamine (ecstasy), methaqualone, and peyote.

Here's what the CSA classifies as Schedule II (a high potential for abuse, with use potentially leading to severe psychological or physical dependence): Combination products with less than 15 milligrams of hydrocodone per dosage unit (Vicodin), cocaine, methamphetamine, methadone, hydromorphone (Dilaudid), meperidine (Demerol), oxycodone (OxyContin), fentanyl, Dexedrine, Adderall, and Ritalin.[23]

So why aren't pharmaceutical companies that produce these Schedule II drugs or the pharmacies that sell them constrained by 280E the way marijuana growers and marijuana dispensaries are? The DEA has determined those particular drugs have "accepted medical use" and this means that:[24]

1. The drug's chemistry is known and reproducible.
2. There are adequate safety studies.
3. There are adequate and well-controlled studies proving efficacy.
4. The drug is accepted by qualified experts.
5. The scientific evidence is widely available.

Now that hemp is no longer subject to 280E, marjiuana industries might be looking to a certain medical marijuana study to prove it has accepted medical use. The first and only study right now with FDA approval for smoked marijuana as a treatment for severe cases of PTSD in veterans took seven years to get approved. If this trial is successful, it would be the first step to make smokable marijuana acceptable for medical use under the DEA and FDA's criteria.

As of May 2018, Dr. Sue Sisley's study, which is funded by a $2.5 million grant from the University of Colorado, is in its third year; she's now in Phase 2 and hoping the FDA will approve a Phase-3 trial (which is typically the final step before a drug can be considered for FDA approval). Because the study is FDA approved and because marijuana is federally illegal, the federal government must supply Sisley with the weed for the study and she obtains it from the University of Mississippi, where it is grown under the guidance of the National Institute on Drug Abuse and the DEA for federally approved studies. She has stated publicly that "she has concerns regarding the potency of the cannabis supplied by the government, which contains extraneous plant material like sticks and seeds."[25]

"If half [of the marijuana] is extraneous non-medical plant material, it could be sabotaging efficacy studies," Dr. Sisley explained at the American Psychiatric Association Annual Meeting in May 2018. "If you're trying to measure effectiveness of cannabis and you're forced to use a very sub-optimal plant material, this could harm the outcome of tests looking at how effective cannabis is at treating a certain illness."[26]

Meanwhile, in June 2018, the FDA approved Epidiolex, a drug produced by GW Pharmaceuticals to treat seizures, which is made from a highly purified cannabidiol (CBD) synthesized in a lab. Anyone selling or distributing Epidiolex won't be subject to 280E; however, CBD oil taken directly from the marijuana plant (which has also proven to be effective in treating seizures) still isn't considered as "acceptable medical use" by the federal government.

In fact, in December 2016, the DEA updated its regulations to include CBD oil in the definition of marijuana, which means businesses "trafficking" CBD oil also weren't permitted to make tax deductions. According to the drug code, any "extract containing one or more cannabinoids that have been derived from any plant of the genus Cannabis"[27]—including CBD oil—is illegal and listed as a Schedule I narcotic under the Controlled Substances Act; therefore, 280E technically applies to marijuana-CBD businesses.

Luckily, hemp (and hemp-CBD) is now exempt from 280E because the federal government did make some distinctions between hemp and marijuana through the Agricultural Act of 2014,[28] and of course the 2018 Farm Bill took hemp off the Controlled Substances Act entirely.

HEMP-CBD OIL LEGALIZATION

Industrial hemp is typically lumped together as a specific plant that can be used for three things:

1. Produce hempseed and hemp oil for food
2. Produce fiber for products (clothing, plastics, etc.)
3. Produce CBD oil (medication)

However, there are *varieties* of industrial hemp:

- Industrial hemp specifically grown for fiber and textiles;
- Industrial hemp specifically grown for food;
- Industrial hemp considered "dual" purpose, which can be used for both fiber and food production (but experts state the quality isn't as good); and
- A phytocannabinoid-rich hemp (PCR-Hemp) used for CBD production.

There are many different *strains* of certified hempseed, and each of these strains are grown for at least one of the above purposes. When a farmer purchases certified industrial hempseed, the seed's origin, kind, and many other classifications are listed, all the way down to the name and address of the producer or grower. There are different ways to grow the crop, depending on what it is going to be used for:

- Hemp strains used to produce fiber are harvested before they reach maturity because this produces the strongest fibers. The farmer wouldn't have a reason to try to extract CBD from these plants. Since they're harvested early, the quality of CBD would be subpar (and there wouldn't be enough of it to really justify the expense).

- When hemp is being grown for food such as the hulled hempseed sold in stores, the focus is on producing hempseed. To produce enough of the seeds, both male and female plants are grown together. Once the flowers are fertilized, they stop producing cannabinoids and put all energy into producing seeds. Typically, the industrial hemp won't have much (if any) THC content or CBD content when harvested because the cannabinoids have simply been bred out of them.

- PCR-hemp or CBD-rich hemp strains are technically a modified breed of industrial hemp. They started off as marijuana strains that had the THC bred out of them, or they could be a cross-breed of hemp and marijuana (like Charlotte's Web). Farmers looking to grow hemp for CBD extraction will typically choose a PCR-hemp strain. Medical-grade CBD oils are extracted from the plant using CO_2 (which kills the terpenes in THC).

Sometimes articles will refer to CBD oil and hemp oil as if they're the same thing, but the major difference is that hemp oil doesn't contain any CBD or THC. Hemp oil is extracted from the hempseed that is produced by the industrial hemp grown for food. There are aren't any cannabinoids in hemp oil, just as there aren't any in hemp milk, hemp butter, hemp ice cream, and so on because the seeds used to produce these food products do not contain CBD or THC.

However, hemp-CBD oil is extracted from all other parts of the hemp plant (leaves, stems, etc.) and those parts of the plant contain cannabinoids.[29] Since cannabinoids (and CBD) aren't legal in all fifty states, some legal hemp farmers in the pilot program were restricted from producing CBD oil for medical research or for retail opportunities.

As of May 2018, there are seventeen states that allow the possession of CBD oil: Alabama, Georgia, Indiana, Iowa, Kentucky, Mississippi, Missouri, North

Carolina, Oklahoma, South Carolina, South Dakota, Tennessee, Texas, Utah, Virginia, Wisconsin, and Wyoming. While it may seem unlikely to be arrested for CBD oil, it certainly has happened.

In August 2017, a forty-one-year-old photographer and filmmaker was actually charged with possessing marijuana during a traffic stop in Hamilton County, Indiana for having CBD oil in his car. Mamadou Ndiaye was facing jail time and a $1,000 fine.[30] An independent lab report discovered there wasn't any THC in his CBD oil,[31] and eventually all charges were dropped. The case led to Indiana updating state laws to legalize CBD oil that contains .3 percent THC or less.

We'll have to wait and see if the 2018 Farm Bill (which made hemp-CBD legal) will push states to legalize CBD oil in general to avoid confusion and complications, or if it will cause CBD to be removed from the Controlled Substances Act entirely.

Regardless if CBD came from a marijuana plant or a hemp plant, it's still CBD, so this may present an interesting legal argument for CBD in general.

FOCUSES OF THE HEMP PILOT PROGRAM

Research has shown there is relatively little correlation between the name of the marijuana strain and its true origin/genetic structure. When it comes to hemp, the name of the strain and the genetics do match up.

Industrial hemp is grown in many different countries and bred for different purposes; there are many varieties—they don't have cool names like Purple Urkle or Laughing Buddha or Ghost Train Haze—but there are subtle differences that set them apart. For example, when an American hemp farmer purchases certified seeds for Futura75 (a Spanish strain), the farmer is purchasing that strain for a specific purpose, and those seeds have guidelines for when to plant, ideal climate, soil content, and so on to maximize the yield.

For the past five years, the hemp pilot and research programs mainly focused on developing the best growing practices for specific regions and specific varieties of hemp, including how much water the plants should receive, how far apart they should be grown, and what month the seeds should be planted. Most research collected under the program makes note of a particular strain's aspects: how strong the fibers are, how many seeds are produced, most effective pest-control measures, the best commercial use based on the quality and quantity of the yield, etc. Here are some of the varieties of industrial hemp that have been imported for research purposes under the 2014 Farm Bill's hemp pilot program:

- **Spanish Industrial Hemp**
  Futura75 is one of the more predominant varieties of hemp imported under the industrial hemp pilot program. Futura75 is currently used to produce hemp oils and extracts in Europe.

- **French Industrial Hemp**
  Fedora17 is currently grown in the European Union for hemp fiber or CBD oil;[33] Santhica27 and Ferimon used primarily for fiber.

- **Canadian Industrial Hemp**
  Canda, from Canada, is used for multiple purposes including fiber, biomass production, and food consumption.[34]

- **Finnish Industrial Hemp**
  Finola is used primarily for consumption, both as a grain and as a food source.[35]

POLITICAL SUPPORT FOR HEMP

The hemp industry is largely being jump-started by both Republicans and Democrats who listened to their constituents and saw there was a need to bring another industry to their states. The good news is hemp pilot programs are making great progress and expanding operations from year to year because the initiative is being supported by elected officials.

One example of this is New York state. In 2017, New York's Industrial Hemp Research Initiative took a huge leap forward. Governor Cuomo announced up to $10 million in grant funding to advance hemp research and economic development opportunities for industrial hemp businesses.[36] As of April 2018, Governor Cuomo stated there were more than sixty new industrial hemp research partners—farms and businesses—that received research permits to conduct trials in food and fiber. According to the governor's website:

- In 2018, approximately 3,500 acres of New York farmland were approved for industrial hemp research trials. The studies are to focus on utilizing the crop for a food source, fiber and grain for the production of animal bedding, insulation, pellets for heating, and many other consumer products.

- Researchers are also exploring the "potential cosmetic and wellness benefits of CBDs."[37] Cornell University, SUNY Sullivan,

Binghamton University, and SUNY Morrisville are part of the academic research division.

- New York state also launched a $5 million Industrial Hemp Processors Grant Fund in 2017 to help businesses cover costs related to hemp processing, including new construction and the purchase of equipment.

- In January 2018, the state also invested $650,000 to establish a $3.17 million industrial hemp-processing facility in the greater Binghamton area.[38] Assuming all goes well, this facility should bring in major job growth for an area struggling with unemployment. In July 2018, the New York Department of Labor noted that the Binghamton labor area (including Broome and Tioga counties) was the third-worst-performing region in the entire state.[39] From 2017 to 2018, there was a 0.1 percent gain in employment (which equates to only 100 new job openings within a calendar year),[40] so suffice to say, if New York state is any indication, hemp production has the potential to bring much needed jobs to struggling regions.

- According to *Hemp Industry Daily*'s 2018 Hemp Report on the top ten hemp-growing states, New York's investment in hemp cultivation *and* processing "represented a national first."[41] While most states have established industrial hemp programs, growers don't get much support from their state's agriculture department. The state doesn't promote the plant or work to find buyers or processors for the crop. All aspects of the business are put on the grower's shoulders. Farmers even provide the funding for their state's hemp program through the application fees and yearly license cost they pay to the state for the privilege to grow the crop.

Since hemp is now legal, the burden placed on the growers should be examined. The 2018 Farm Bill has made it possible for hemp growers to apply for loans, grants, and crop insurance. They're now eligible for tax exemptions and reimbursements from the costs associated in academic studies (research that advances our understanding of hemp, but doesn't pay much to the hemp farmer for providing the research material). The success of the industry as a whole is going to depend on continued political support beyond the legislation, especially now that hemp seeds are a commodity. Now that there's a higher demand for them, the price per hemp seed will certainly increase, thus affecting the current price-points of the entire industry.

The 2018 Farm Bill also reauthorized $400 billion in agriculture subsidies, which are given to farms to offset operating costs, and will provide funding to the agricultural community for the next five years. (In addition, the bill covers SNAP, the food stamp program that feeds 40 million low-income Americans.)

According to a database compiled by the Environmental Working Group (EWG), the larger the farm, the larger the subsidy received. In 2015 and 2016, the USDA gave more than $14 billion in taxpayer-funded subsidies. The majority went to corn and soybean farming facilities,[42] with mega-farms receiving the bulk of the money. USDA data confirms that from 1995 to 2016, the top 1 percent of large commercial farms received 26 percent of all subsidies, which equated to $1.7 million per recipient. The top recipient of USDA subsidies was Deline Farms Partnership, which received more than $4 million in 2016, for its farms based in Kentucky, Arkansas, Illinois, Missouri, and Mississippi.[43] EWG also reported that in 2016, the median household incomes for large commercial farms was $1.1 million. Redistributing the subsidy program so that it benefits those most in need of financial assistance, including newly emerging hemp farms, will help guarantee the successful return of the American hemp industry.

HEMP ADVOCACY

Without citizens lobbying for change, there would be no change. Industrial hemp might seem to be set up for success right now and moving in the right direction, but that doesn't mean it's time for us to be complacent. There is still a long road ahead to make hemp completely accessible to every American, including those convicted of a felony related to a controlled substance, who are still ineligible to grow hemp until ten years *after* their conviction date, an equivalent of another prison sentence after time served.

Keep in mind: hemp used to be on the back of the $10 bill. As of 2019, that was ninety years ago. We've come a long way since the Marihuana Tax Act, but we seem to always be taking two steps forward and one step back. Government officials didn't just wake up one day and decide to legalize hemp; this is the result of decades of advocacy efforts.

One such advocacy group with an extensive website is VoteHemp.com. Their website is particularly valuable because they are focused on activism, registering people to vote, lobbying for hemp, and getting the word out on hemp legislation to enable anyone to take action on a grassroots level, where the pro-hemp organizations originated.

NOTES

1. "Hemp Report: Top 10 U.S. States," Hemp Industry Daily, last accessed Jan 23. 2019, https://mjbizdaily.com/wp-content/uploads/2018/04/Hemp-Report_Top-10-US-States.pdf.
2. Ibid.
3. Tim Devaney, "Hemp-made flag flies over the US Capitol," *The Hill,* Nov 11, 2015, http://thehill.com/regulation/defense/259857-flying-high-hemp-made-flag-adorns-us-capitol.
4. The Growing Warriors Project, accessed October 20, 2018, http://www.growingwarriors.org/media/.
5. Kristy Robinson Horine, "Growing Hope, One Hero At A Time," People, *Kentucky Monthly,* Oct 21, 2013, http://www.kentuckymonthly.com/culture/people/homegrown-by-heroes/.
6. Anthony Iasso, "Fabrics," Rare Flags, last modified 2014, http://www.rareflags.com/rareflags_collecting_fabrics.htm.
7. "Flag Materials," *Flags Unlimited,* last accessed Jan 23, 2019, http://www.usflags.com/flagmaterials.aspx.
8. International Year of Natural Fibres 2009, "Profiles of 15 of the world's major plant and animal fibres," Food and Agriculture Organization of the United Nations, accessed Jan 23, 2019, http://www.fao.org/natural-fibres-2009/about/15-natural-fibres/en/.
9. Kristy Robinson Horine, "Growing Hope, One Hero At A Time," People, *Kentucky Monthly,* Oct 21, 2013, http://www.kentuckymonthly.com/culture/people/homegrown-by-heroes/.
10. Ibid.
11. "Mike Lewis, Hemp Farmer, Kentucky," Hemp History Week, last modified June 4, 2018, https://hemphistoryweek.com/about/u-s-hemp-farmers/mike-lewis/.
12. USDA, "Hemp For Victory" transcript (Jan 1, 1942), *Global Hemp,* accessed Jan 23, 2019, http://www.globalhemp.com/1942/01/hemp-for-victory.html.
13. Rick Hemphill, "Virginia Hemp Company Seeks Growers," *Lancaster Farming*, Oct 5, 2018, https://www.lancasterfarming.com/farming/field_crops/virginia-hemp-company-seeks-growers/article_39ebb86c-4e44-523c-844b-805181c053bd.html.
14. Roth, G., Harper, J., Manzo, H., et. al, "Industrial Hemp Production," *PennState Extension*, July 2, 2018, https://extension.psu.edu/industrial-hemp-production.
15. Gusovius, H., Hoffmann, T., Budde, J., Lühr, C., "Still special? Harvesting procedures for industrial hemp," *Landtechnik* 71(1): 14-24, Jan 29, 2016, https://www.researchgate.net/publication/292155140_Still_special_Harvesting_procedures_for_industrial_hemp.

16. "Hemp Equipment by Crop Type," *Hemp Harvest Works,* accessed Jan 23, 2019, https://www.hempharvestworks.com/cbd-hemp-equipment/.
17. Stu Salkeld, "County of Wetaskiwin council discusses hemp co-op," Local News, *Pipestone Flyer*, Oct 31, 2018, https://www.pipestoneflyer.ca/news/county-of-wetaskiwin-council-discusses-hemp-co-op.
18. Libby Rainey, "Going into its fourth harvest, industrial hemp industry still facing growing pains," Business, *Denver Post,* Sept 10, 2017, https://www.denverpost.com/2017/09/10/colorado-culvitars-hemp-farm-harvest.
19. Jake Altinger, "Hemp almost legal as Big Pharma moves in on CBD," News Cover Story, *Colorado Springs Independent,* July 25, 2018, https://www.csindy.com/coloradosprings/hemp-almost-legal-as-big-pharma-moves-in-on-cbd/Content?oid=13994816.
20. Kevin Murphy, "Legal Marijuana: The $9 Billion Industry That Most Banks Won't Touch," Markets, *Forbes.com,* Sept 6, 2018, https://www.forbes.com/sites/kevinmurphy/2018/09/06/legal-marijuana-the-9-billion-industry-that-most-banks-wont-touch/#4704e7033c68.
21. NCV Newswire, "Aurora Cannabis Gains Access to $200 Million in Loans from Leading Canadian Bank," *New Cannabis Ventures,* June 26, 2018, https://www.newcannabisventures.com/aurora-cannabis-gains-access-to-200-million-in-loans-from-leading-canadian-bank/.
22. Legal Information Institute, "26 U.S. Code § 280E—Expenditures in connection with the illegal sale of drugs," *Cornell Law School,* accessed Jan 23, 2019, https://www.law.cornell.edu/uscode/text/26/280E.
23. "Drug Scheduling," *United States Drug Enforcement Administration,* accessed Jan 23, 2019, https://www.dea.gov/drug-scheduling.
24. Jola Mehmeti, "DEA Considers Rescheduling Cannabis," *Pharmacy Times,* May 30, 2016, https://www.pharmacytimes.com/contributor/jola-mehmeti-pharmd-mba-candidate-2018/2016/05/dea-considers-rescheduling-cannabis.
25. Healio, "First FDA-approved study of cannabis for PTSD in veterans underway," *Psychiatric Annals,* May 6, 2018, https://www.healio.com/psychiatry/ptsd/news/online/%7Bac35ff1a-3729-416e-bdfa-f0a6d68d5f43%7D/first-fda-approved-study-of-cannabis-for-ptsd-in-veterans-underway.
26. Ibid.
27. "Establishment of a New Drug Code for Marihuana Extract," *Federal Register Rules and Regulations, Vol. 81, No. 240*, Dec 14, 2016, https://www.gpo.gov/fdsys/pkg/FR-2016-12-14/pdf/2016-29941.pdf.
28. "Agricultural Act of 2014: Highlights and Implications," *USDA Economic Research Service*, accessed Jan 23, 2019, https://www.ers.usda.gov/agricultural-act-of-2014-highlights-and-implications.aspx.

29. "Clarification of the New Drug Code (7350) for Marijuana Extract," *US Department of Justice Drug Enforcement Administration*, accessed Jan 23, 2019, https://www.deadiversion.usdoj.gov/schedules/marijuana/m_extract_7350.html.
30. Mona Zhang, "No, CBD Is Not 'Legal In All 50 States,'" Editor's Pick, *Forbes.com*, April 5, 2018, https://www.forbes.com/sites/monazhang/2018/04/05/no-cbd-is-not-legal-in-all-50-states/#69c5adfa762c.
31. PSI Labs, "Certificate of Analysis, CBD Oil Sample," Sept 18, 2017, https://www.wthr.com/sites/wthr.com/files/CBDoil.LabResults1.pdf.
32. Reuters, "Woody Harrelson's Hemp Woes Over," *ABC News,* Aug 25, 2007, https://abcnews.go.com/Entertainment/story?id=2969864&page=1.
33. "Fedora 17 Cannabis sativa," Canna Pot, accessed Jan 23, 2019, https://www.cannapot.com/shop/cannabis-seeds/nativcanna/fedora-17-cannabis-sativa_seeds.html.
34. "Canda," *Parkland Industrial Hemp Growers,* accessed Jan 23, 2019, https://www.pihg.net/currently-available.
35. Hanson, B., Johnson, B., Hermann, A., et. al, "Industrial Hemp Performance in North Dakota—2015," NDSU Langdon Research Extension Center, last modified Feb 2, 2016, https://industrialhemp.ces.ncsu.edu/wp-content/uploads/2016/10/Industrial-Hemp-Variety-Trials-FR-revised-Feb2-2016-1.pdf?fwd=no.
36. "Industrial Hemp Research Initiative in New York State," *Empire State Development,* accessed Jan 23, 2019, https://esd.ny.gov/industrial-hemp.
37. Governor Cuomo's Press Office, "Governor Cuomo Announces More Than 60 New Industrial Hemp Research Partners Join New York State Pilot Program," Agriculture, *Governor Andrew M. Cuomo,* April 2, 2018, https://www.governor.ny.gov/news/governor-cuomo-announces-more-60-new-industrial-hemp-research-partners-join-new-york-state.
38. Governor Cuomo's Press Office, "Governor Cuomo Announces Industrial Hemp Research Forum to be Held in the Southern Tier," Agriculture, *Governor Andrew M. Cuomo,* Feb 26, 2018, https://www.governor.ny.gov/news/governor-cuomo-announces-industrial-hemp-research-forum-be-held-southern-tier.
39. Jeff Platsky, "Ithaca job growth exceeds NYC's; Elmira, Binghamton lag," Binghamton Press & Sun-Bulletin, July 20, 2018, https://www.pressconnects.com/story/money/2018/07/20/ithaca-elmira-binghamton-job-counts/802447002/.
40. US Census bureau, "Population Demographics for Binghamton, New York in 2017, 2018," *Suburban Stats,* accessed Jan 23, 2019, https://suburbanstats.org/population/new-york/how-many-people-live-in-binghamton.

41. "Hemp Report: Top 10 U.S. States," Hemp Industry Daily, last accessed Jan 23. 2019, https://mjbizdaily.com/wp-content/uploads/2018/04/Hemp-Report_Top-10-US-States.pdf.
42. Alex Formuzis, "EWG: Mega-Farms Reap Billions from Taxpayers in Farm Subsidies," News Releases, *EGW.org,* Nov 2, 2017, https://www.ewg.org/release/ewg-mega-farms-reap-billions-taxpayers-farm-subsidies.

5

HEMP DISRUPTS AMERICAN FARMING

If hemp history has taught me anything, it's that the struggle to make hemp legal is due to the fault of the plant. Hemp is a disrupter. It mocks modern ingenuity. The more we research it, the more we find it offers yet another natural way to improve modern industry. Time and time again, hemp is the more eco-friendly and safer solution when compared to what humans have invented. One such example that pertains to American farming is the weed- and pest-killing industry. This is an industry our agriculture depends on, but it's simultaneously harming us due to the consequences of synthetic chemicals. While we've pushed away the naturally effective solution for so long, we now know current methods are not sustainable, nor are they healthy.

On August 10, 2018 a jury ruled that former school groundskeeper Dewayne Johnson was entitled to $289 million in damages from Monsanto.[1] The San Francisco Supreme Court made this determination because Johnson is suffering from terminal lymphoma at forty-six years of age as a result of overexposure to glyphosate, a key ingredient in the weed killer Roundup. Glyphosate happens to be carcinogenic (a chemical that can cause cancer), a fact that Monsanto has publicly denied for generations.

"The jury found Monsanto acted with malice and oppression because they knew what they were doing was wrong and doing it with reckless disregard for human life," said Robert F. Kennedy Jr., a member of Johnson's legal team.[2]

While Monsanto still maintains that its products do not cause cancer, evidence revealed in Johnson's trial include internal corporate records and emails that discuss how executives ghost-wrote scientific papers and colluded with the EPA to manipulate and deceive the public. Monsanto memos also include plans to discredit international health agencies for declaring the main ingredient in Roundup (glyphosate) to be "a probable human carcinogen."[3]

"Monsanto, for forty years, has been taking the playbook from the tobacco industry by ghostwriting science, buying science, using all the different PR strategies and legal strategies to confuse the science," Kennedy explained during a video press conference after the verdict. "I'm so glad that this jury has held them accountable."[4]

The global chemical corporation was quick to appeal the verdict, but Johnson's case was solid enough to prevail a second time, and he was awarded $78 million on October 22, 2018. When Monsanto attempted to appeal Johnson's case yet again on November 20, the judge refused to grant a new trial, and Johnson suddenly found himself the victor of a historic lawsuit over one of the most powerful and controversial companies in the world. Monsanto now faces eight thousand US lawsuits—and 5,200 of those are regarding glyphosate in particular.[5] Whether or not Monsanto takes all these cases to trial or settles with the plaintiffs, hopefully Dewayne Johnson's case has sent a message to the company's boardroom.

A mere six days after Johnson's initial $289 million award was announced, General Mills found itself facing a lawsuit against three consumer advocacy groups: Beyond Pesticides, Moms Across America, and the Organic Consumers Association.[6] This is after the nonprofit Environmental Working Group (EWG) tested General Mills products to find they contained levels of glyphosate. General Mills was able to settle the lawsuit by agreeing to remove the phrase "Made with 100% Natural Whole Grain Oats," since the oats were the likely source of the glyphosate.[7] Whether the oats were grown 100 percent naturally or not, the fact that they could be laced with glyphosate made the claim of "all natural" ingredients pretty misleading.

Since 2008, EWG has been running tests and researching the safety of everything from consumer products to tap water to determine what toxins can be lurking in the everyday lives of Americans. When EWG tested General Mills' oat cereals (such as Cheerios), oatmeal, granola, and snack bars, researchers found that the products contained trace amounts of glyphosate. Even "all natural" products were found to contain glyphosate. Since General Mills wasn't adding glyphosate during the manufacturing process, EWG's research proved that the integrity of a farmer's crops can be compromised. Water, soil, a nearby stream or water source, even the wind and the process of cross-pollination can take particles of glyphosate and deposit them in a farm that isn't using pesticides. Plus, if processing facilities are handling both organic and nonorganic produce, the likelihood of contaminating organic products with herbicide debris increases.

While this is obviously problematic, it shouldn't come as a shock. Crops grown on outdoor farms are subject to all kinds of threats. In Oregon, there are marijuana growers who've been quoted in the media as referring to hemp as "biological terrorism" because the cross-pollination from a hemp farm can diminish marijuana's THC content.[8] But in this case, the "biological terrorism" for an organic farmer are the pesticides sprayed by nonorganic farms, which in turn compromise and contaminate organic produce.

Since glyphosate is the most widely used herbicide in US agriculture, and organic farms are typically smaller operations when compared to large-scale commercial farming operations, are we doomed to find glyphosate residue in just about everything we consume? Not necessarily. EWG did find some General Mills products were devoid of glyphosate. Of the sixteen organic samples, five had trace amounts of glyphosate; out of the forty-five conventional/nonorganic samples tested, forty-three contained glyphosate.

Here are the results:[9]

Glyphosate Was Found on Most Samples of Oat-Based Foods

Type of Food	Product Name	Glyphosate (ppb)*		
		Sample 1	Sample 2	Sample 3
Granola	Nature's Path Organic Honey Almond granola	ND**	ND	
	Back to Nature Classic Granola	620	170	
	Quaker Simply Granola Oats, Honey, Raisins & Almonds	430	400	
	Back to Nature Banana Walnut Granola Clusters	30	30	340
	Nature Valley Granola Protein Oats 'n Honey	220	170	
	KIND Vanilla, Blueberry Clusters with Flax Seeds	50	60	
Instant Oats	Giant Instant Oatmeal, Original Flavor	760		
	Simple Truth Organic Instant Oatmeal, Original	ND	ND	
	Quaker Dinosaur Eggs, Brown Sugar, Instant Oatmeal	620	780	
	Great Value Original Instant Oatmeal	450		
	Umpqua Oats, Maple Pecan	220	220	
	Market Pantry Instant Oatmeal, Strawberries & Cream	120	520	

Oat Breakfast Cereal	Kashi Heart to Heart Organic Honey Toasted cereal	ND	ND	
	Cheerios Toasted Whole Grain Oat Cereal	490	470	530
	Lucky Charms****	400	230	
	Barbara's Multigrain Spoonfuls, Original, Cereal	340	300	
	Kellogg's Cracklin' Oat Bran oat cereal	250	120	
Snack Bar	Cascadian Farm Organic Harvest Berry, granola bar	ND	ND	
	KIND Oats & Honey with Toasted Coconut	ND	120	
	Nature Valley Crunchy Granola Bars, Oats 'n Honey	340	120	
	Quaker Chewy Chocolate Chip granola bar	120	160	
	Kellogg's Nutrigrain Soft Baked Breakfast Bars, Strawberry	30	80	
Whole Oats	365 Organic Old-Fashioned Rolled Oats	ND	ND	
	Quaker Steel Cut Oats	530	290	
	Quaker Old Fashioned Oats	390	1100	1300
	Bob's Red Mill Steel Cut Oats	300	ND	
	Nature's Path Organic Old Fashioned Organic Oats	30	20	
	Whole Foods Bulk Bin conventional rolled oats	10	40	
	Bob's Red Mill Organic Old Fashioned Rolled Oats (4 samples tested)	ND	10	20, 20***

Source: EWG, from tests by Eurofin Analytical Laboratories
* *EWG's child-protective health benchmark for daily exposure to glyphosate in food is 160 ppb.*
** *ND = none detected*
*** *Two product samples tested both had 20 ppb glyphosate concentration.*
**** *Lucky Charms Frosted Toasted Oat Cereal with Marshmallows. Marshmallows were manually removed from the samples prior to shipping to the lab and testing for glyphosate.*

So how do we know what's really organic and what's not? *Who is going to test every single product on the shelf?* What we do know is that more than 250 million pounds of glyphosate are sprayed on American crops every year, according to the U.S. Department of the Interior.[10] Genetically modified corn and soybeans that are resistant to the herbicide are sprayed most frequently, but the highest glyphosate levels are not found in products made with GMO corn or soybeans—they're found in grain products.

Farmers also use glyphosate on wheat, barley, oats, and beans that are not genetically engineered, and yes, it's in these crops where the largest deposits of the herbicide are found. This is because the chemical is used right before it's time to harvest the crops to accelerate the ripening process. This practice originated in Scotland in the 1980s, and has become fairly common since that time. Farmers apply glyphosate to a crop at the end of the growing season to essentially kill it and dry it out instead of waiting for it to naturally die and dry in the field. This allows the crop to be harvested sooner (allowing another crop rotation sooner) and ensures the less mature plants "catch up" to their companions, so farmers can maximize their yield.[11] An organic farmer wouldn't practice this method, but if the wind is a-blowin', the herbicide can make its way onto their crops during harvest season.

Sure, the levels of glyphosate that were found in General Mills products were lower than the EPA's safety threshold, so the lawsuit against the corporation can appear frivolous, but the EPA's standard is significantly different from those of other environmental groups, including the EWG. When food is tested for glyphosate, the chemical is measured in parts per billion (ppb), and the EPA states up to 30,000 ppb is safe.[12] However, the EWG's threshold for safe consumption is 160 ppb. Cheerios, for example, had between 470 to 530 ppm of glyphosate, so the findings were particularly significant by the EWG's standards.

In any case, there's a huge discrepancy of safety here between what the EPA states is safe and what the EWG states is safe— so what are these safety standards based on?

The EPA calculates that one- to two-year-olds have the highest exposure to glyphosate,[13] while at the same time, the agency states the chemical isn't carcinogenic, so there's no need for concern. In a December 2017 press release, the EPA concludes:[14]

> *Glyphosate is not likely to be carcinogenic to humans. The Agency's assessment found no other meaningful risks to human health when the product is used according to the pesticide label. The Agency's scientific findings are consistent with the conclusions of science reviews by a number of other countries as well as the 2017 National Institute of Health Agricultural Health Survey.*

However, when it comes to other scientific findings, in 2015, the World Health Organization classified glyphosate as "probably carcinogenic"[15]—as did the International Agency for Research on Cancer[16]—and as we now know from Dewayne Johnson's trial against Monsanto, that didn't sit very well with executives.

Meanwhile, in July 2017, the state of California listed glyphosate as a "probable carcinogen"[17] and added it to the list of Prop 65 chemicals that are known to cause cancer.[18] California's Office of Environmental Health Hazard Assessment (OEHHA) set their cap for glyphosate exposure to 1.1 milligrams per day for an adult of about 154 pounds. In comparison, the EPA's safe level is 2 milligrams per kilogram of body weight (or 140 milligrams per day for an adult weighing 154 pounds).[19] The EWG states adults should ingest no more than .01 milligrams per day, and if General Mills products are any indication, rest assured the average American is consuming much more glyphosate than that per day, and there's no way to "calorie count" the amount of glyphosate we're consuming, since it isn't listed as an ingredient.

Although there is yet to be a universally accepted safe level for glyphosate, what we do know from recently released Monsanto emails is that there is a level of harm that the company has been aware of and continues to cover up:[20]

- Monsanto emails discussed "what to do" to discredit or lessen the impact of a 2008 European study found that glyphosate is toxic to rats if inhaled (the findings led European scientists to believe that it should be classified as a human carcinogen due to respiration toxicity).

- Also discussed: Studies from the late 1990s to the early 2000s that show glyphosate can be absorbed through the skin; Monsanto executives were aware that studies showed the chemical is absorbed at a higher, more toxic rate than what the company claimed. Emails from 2011 also show that Monsanto executives were aware Roundup would fail the UK and EU's dermal absorption tests.

- Wallace Hayes, the editor of *Food and Chemical Toxicology*, also worked as a consultant to Monsanto. As editor, he was able to retract a study from the publication in 2012 that found glyphosate to be a possible carcinogen in rats.

- Monsanto emails reference scientist Dr. David Saltmiras and how he was involved with ghostwriting scientific studies on glyphosate (without disclosing his conflict of interest as a Monsanto employee)—including cancer review papers from 2015, 2013, and 2012.

- In 2008, Monsanto executive Dean Nasser emailed Dr. Donna Farmer a study published in *Beyond Pesticides* that found a link between glyphosate and Hodgkin's disease. She said she "knew it would only be a matter of time before the activists pick it up" and, since she is one of Monsanto's primary expert witnesses, she asked Nasser "how do we combat this?"

- In an email from Dr. Farmer to Dr. Acquavella in 2014, Dr. Farmer states that the IARC (International Agency for Research on Cancer) is planning to test glyphosate in March 2015—"what we have long been concerned about has happened," she stated.

- In 2015, Dr. Larry Kier, a consultant for Monsanto, was paid to be on an expert panel (the Intertek Expert Panel) for the IARC. The panel was supposed to be composed of independent scientists to determine if glyphosate is a carcinogen.

- Monsanto also paid $20,000 to expert panelist Dr. John Acquavella in August 2015, but Acquavella eventually had a conflict with his conscience. He wrote to Monsanto in November 2015 to state he "can't be a part of deceptive authorship on a presentation or publication . . . we call that ghost writing and it is unethical."

- In 2015, EPA official Jess Rowland, a dear friend of Monsanto, visited the IARC to participate in meetings as an observer. Rowland also served as cochair of the EPA's CARC report, a report that concludes it is "biologically improbable for glyphosate to act as a human carcinogen." He also helped Roundup pass the EPA's regulatory process, which proved that the EPA's safety evaluation of glyphosate wasn't ever independent.

- Emails also show that EPA officials who left the EPA started working for Monsanto, including Mary Manibusan, who worked in the EPA's Office of Pesticide Programs and wrote the CARC report with Rowland.

- Further collusion with the EPA is revealed in text messages from 2013 to 2016. Monsanto execs ask "is there anyone we can get to in the EPA?" after the IARC revealed glyphosate was a "probable carcinogen." Texts reveal that Monsanto's head of regulatory affairs, Mr. Daniel Jenkins, spoke to the EPA, and the agency is "going to conclude that IARC is wrong."

- In October 2015, the company tried to influence the state of California, asking the California Office of Environmental Health Hazard Assessment to exempt glyphosate and Roundup from its Proposition 65 rating, but was ultimately unsuccessful.

- In 2016, Monsanto used its EPA connections to delay the Scientific Advisory Panel's review of glyphosate, which was to be conducted after the IARC's classification.

While it's clear that the company was successful in controlling the narrative from the EPA, which still lists glyphosate as "not likely" to be carcinogenic, a study titled "Major Pesticides Are More Toxic to Human Cells than Their Declared Active Principles" that was published in 2014 in *BioMed Research International* summed up what the Monsanto emails used in Dewayne Johnson's lawsuit now prove:[21]

> *It is commonly believed that Roundup is among the safest pesticides. This idea is spread by manufacturers, mostly in the reviews they promote, which are often cited in toxicological evaluations of glyphosate-based herbicides. However, Roundup was found in this experiment to be 125 times more toxic than glyphosate. Moreover, despite its reputation, Roundup was by far the most toxic among the herbicides and insecticides tested. This inconsistency between scientific fact and industrial claim may be attributed to huge economic interests, which have been found to falsify health risk assessments and delay health policy decisions.*

So if glyphosate-based herbicides are toxic enough to cause cancer through prolonged exposure, would farmers continue to use them, and if not, what would be a suitable replacement?

Roundup was first brought to market in 1974, and in 1994, the herbicide was named one of the "Top Ten Products that Changed the Face of Agriculture" by *Farm Chemicals Magazine*.[22] Wouldn't we have to reinvent modern farming entirely if the product was suddenly banned? Yes and no.

SELF-CONTAINED FARMING METHODS

Right now, there is a movement to reinvent the way we think about farming. Marijuana is being grown indoors, under controlled conditions—everything from the amount of sunlight the plants receive to the air and water purification systems are utilized to ensure the entire process is secure. The facilities used are typically repurposed, vacant warehouses left over from the days when America manufactured products.

Farmers outside the marijuana industry are now buying these empty buildings to grow what is actually 100 percent all-natural, organic produce—food that has never known pesticide-ridden wind or soil or compromised water. Here are just a few examples of self-contained, eco-conscious farming methods:

- In 2016, the town of Pagosa Springs, Colorado, started building dome-shaped geothermal greenhouses that are being used to grow vegetables year-round.[23] Pagosa Springs has owned and operated a geothermal system since December 1982, which provides a geothermal heating system to the entire town, and this system was modified to allow produce to grow under controlled conditions in the greenhouses.[24]

- Scottish company Intelligent Growth Solutions has created a vertical farm where plants are hydroponically grown from the floor to the ceiling of a climate-controlled building. Because produce is grown vertically, their facilities produce yields up to 200 percent more than that of a traditional horizontal greenhouse.[25]

- Gotham Greens in New York and Lufa Farms in Montreal grow food on the roofs of skyscrapers. Although they can't grow as much food as artificially lit vertical farms, they're at least 70 percent less expensive because they utilize naturally lit hydroponic greenhouses.[26]

- Rooftop and vertical farming projects are also utilizing aquaponic systems, which grow food with the help of fish, so that there's a contained ecosystem, including fish poop used as fertilizer.

- There are also home robots like Seedo and Leaf that can incubate and grow plants—including cannabis—in a small, self-sustaining unit about the size of a mini-fridge. The automatic hydroponic system is controlled by a smartphone app. The unit remains sealed until it is unlocked by the app (to keep odors in and possible pests out). Aside from creating pesticide-free, organic produce, these plug-'n'-plant systems offer the capability to have consistency of product. Since the growth cycle will be the same each time, the results will also be the same, which is a huge advantage for the home cannabis grower.[27]

According to research by Jelle Bruinsma of the Food and Agriculture Organization of the United Nations, by 2050 global food production will need to increase in developed countries by an estimated 70 percent and 100 percent in developed countries to match current trends in population growth (this is based on production information from 2005 to 2007).[28]

In countries that already use the majority of their land for farming, it's not really possible to rely on field/outdoor farming. For example, the UK already uses 72 percent of its landmass for agriculture,[29] yet the country still imports nearly *half* of the food being consumed.[30] The good news is solutions like urban, vertical, and indoor farming are being embraced. Companies such as American Express are also getting on board. AmEx purchased six indoor farm kits from the Brooklyn-based start-up Farmshelf to supply vegetables for its corporate cafeteria. According to Farmshelf's CEO, the kits produce about 140 heads of lettuce per month. The lettuce springs up faster than field farmed lettuce due to the controlled conditions inside the growing kit (lettuce will reach maturity in 20 to 28 days instead of 60 days), and the plants use 90 percent less water.[32]

While these advances are exciting for organic food lovers and nonprofit watchdog groups like EWG that would like to see zero ppb of glyphosate in our breakfast cereals, this phenomena is benefiting crops that take up the least amount of room to grow, such as strawberries, tomatoes, lettuce, basil, and other herbs. Sure, lemon trees can be planted on rooftop greenhouses in theory, but to get maximum yields (and maximize profits), plants with smaller root systems that take up less space are obviously preferable.

And what about plants that need to cross-pollinate to bear fruit, such as blueberries? Safe to say field farming isn't going to be 100 percent replaced, and while it's here, we're going to have to deal with glyphosate, right? Well, maybe not. At least for the bees' sake, let's hope not.

GLYPHOSATE AND BEES (AND WEEDS AND TADPOLES . . . OH MY!)

Going back to the EPA's December 2017 press release on the effects of glyphosate, the agency stated "the ecological risk assessment indicates that there is potential for effects on birds, mammals, and terrestrial and aquatic plants."[33]

Wait. Aren't humans mammals? So the EPA is stating there potential risk for humans?

Forget about humans for a minute, let's focus on the bees.

A study conducted by the University of Texas discovered that bees are vulnerable to glyphosate poisoning. The health of a honeybee is directly tied to its microbiota, which are the helpful organisms in their gut that regulate the bee's metabolism, weight gain, and immune system functions.[34] When bees are exposed to glyphosate, they have significant decreases in their gut bacteria, and the lifespan of a bee is significantly diminished if the amount of microbiota in the gut decreases.

The study, which was published in the *Proceedings of the National Academy of Sciences* on September 24, 2018, involved collecting hundreds of honeybees from a single hive and exposing them to the levels of glyphosate that they'd commonly encounter in the environment. The bees were then returned to the hive and compared to the ones not exposed to the herbicide. After three days, the gut bacteria in the exposed bees had decreased, leaving them more susceptible to death. This experiment was carried out in other hives with the same results.

According to the U.S. Department of Agriculture, a third of our food relies on pollinators like bees, and they're significantly decreasing in numbers. In a January 2015 report prepared for Congress, the USDA states:[35]

> *In the United States alone, the value of insect pollination to U.S. agricultural production is estimated at $16 billion annually, of which about three-fourths is attributable to honey bees. Worldwide, the contribution of bees and other insects to global crop production for human food is valued at about $190 billion. Given the importance of honey bees and other bee species to food production, many have expressed concern about whether a "pollinator crisis" has been occurring in recent decades. In the United States, USDA estimates of overwinter colony losses from all causes have averaged more than 30 percent annually since 2006.*

Is glyphosate the primary cause for bee populations to decrease by 30 percent every year? Well, a June 2018 study conducted by the Royal Society B determined that bumblebees are healthier in urban areas. Apparently (for British bees anyway) the air pollution in cities is *less problematic* than the fresh country air. The study found urban bees have larger colonies, live longer, are better fed, and are less prone to disease.[36] Perhaps because they aren't getting sprayed to death by herbicides in cities?

So where does this leave American field farming when . . .

- The majority of our farms use chemical herbicides with glyphosate to protect their crops from pests and/or to ensure a maximum harvest.

- Exposure to these herbicides is killing off the bee population, which in turn affects how many crops can be pollinated, which in turn affects how large the yield of a harvest can be.

- Organic, all-natural, non-GMO products are popular with consumers, but those products can still carry trace amounts of chemicals.

Just as a side note, obviously farmers aren't the only ones using glyphosate—Roundup is readily available to anyone for weed control. As of September 2018, Portland's Parks and Recreational Department is still using Monsanto's Roundup Pro to "manage vegetation" in city parks,[37] so next time you go for a jog, maybe call the Parks & Rec department first to ask the last time they sprayed? Otherwise, the exercise routine could be counterintuitive if you're breathing in the very chemicals Dewayne Johnson claims caused his extremely aggressive and terminal lymphoma.

Of course, there are many other legitimate studies showing the impact of the chemical on the environment. For example, there are studies indicating it is harmful to wildlife, including frogs. In the early 1990s, the Australian government discovered Roundup "was much more toxic to amphibians than the active ingredient [glyphosate] alone"[38]—which is why the product was required to carry a specific label in Australia: "Do NOT apply to weeds growing in or over water. Do NOT spray across open water bodies, and do NOT allow spray to enter the water." A 2012 study conducted by the biological sciences department at the University of Pittsburgh discovered Roundup "induced morphological changes" in tadpoles to change the shape of their tails and activates the tadpoles' "developmental pathways used for anti predator responses."[39] Monsanto states that glyphosate degrades in water within seven to seventy days, "depending on environmental conditions," but the University of Pittsburgh noted that when wetland wildlife is exposed to Roundup, tadpoles died within one or two days of exposure.[40]

MODERN FARMING VERSUS THE ENVIRONMENT

Thanks to Dewayne Johnson's lawsuit and the subsequent lawsuits against Monsanto, we now know that Monsanto paid "think tank" organizations such as the Genetic Literacy Project and the American Council on Science and Health to write articles debasing scientists and scientific studies that show glyphosate is harmful to human health and wildlife.[41] The company also hired internet trolls to write positive comments on news articles and Facebook posts to defend Monsanto, its chemicals, and its GMOs.[42] As far as modern farming is concerned, utilizing GMO crops to deter pests and remove weeds is also problematic:

- Good in theory, GMO crops require much less chemicals to thrive because the chemicals to repel pests are already in their genes, but bugs become resistant to the chemicals over time, which leads to more glyphosate spraying.

- Cross-pollination between GMO plants leads to "superweeds" that can be just as resistant as the crops, making weeds even more difficult to get rid of.

- Although GMO crops clear bugs from the area, if the GMO pollen gets on organic crops, it can contaminate them and their seeds, and thus turn organic produce into GMO produce.

- GMO crops also negatively impact the microbes and other nutrients in the soil, and essentially the crops can "kill" the soil health, which in turn makes our food less nutritious.

Plus there are even more concerns with the environmental effects of modern farming—ones that don't involve Monsanto—when animals are taken into account. For instance, dairy cows can suffer from a stomach condition known as hardware disease because they swallow stray pieces of metal whether they graze in open spaces or at a feedlot. Nails, screws, wire, you name it, they swallow it. Farmers have to feed the cows a magnet so that the metal scraps will collect, pass through the cow's four stomachs, and eventually all the way through the cow. If this isn't done, the metal can stay inside the cow's body and can eventually kill the cow. A 1993 study conducted by the University of Missouri found that 55 to 75 percent of the cattle slaughtered in the eastern United States contained metal hardware stuck in their stomachs and rectum.[43]

Cow poop is also contributing to global warming because it contains methane, a greenhouse gas roughly 30 percent more potent than carbon dioxide,[44] making our love of cow meat and dairy products at odds with the Paris/UN Climate Agreement (the global initiative to lessen greenhouse gas emissions). In 2011, livestock pooped an estimated 119.1 million tons of methane into the air;[45] 55 percent of that came from cattle raised for beef. When we recycle cow poop into fertilizer, it produces twice as much methane as what was present when it came out of the cow,[46] so needless to say, yes, farming in general has more than one carbon footprint. We need food to sustain life, but the way in which we're going about producing food is essentially destroying life.

But, hey, don't worry—this is a book about simple solutions to prevent global catastrophe, remember?

HEMP AS A "COMPANION PLANT"

The main use for glyphosate and GMO crops is to repel weeds and pests that impact the crop's growth and overall harvest yield. Most pro-hemp websites make reference to hemp as being a natural repellent, and it can be grown with crops to keep pests away. The source for the majority of these claims comes from research that druglibrary.net compiled from the 1990s, dating back to the 1800s. To summarize:[47]

- As far back as 1885, hemp has been used as a "companion plant" to institute a form of biological control in cotton fields against cotton worms. It also safeguards vegetables from cabbage caterpillars and potato fields from the potato beetle.

- Hemp suppresses the growth of natural weeds among beet, maize, potato, lupine, and brassicas crops.

- There are studies proving hemp can suppress soybean cyst nematodes, and when planted among tomatoes and cucumbers, hemp rids the soil of root knot nematodes.

- Dried leaves or juice from fresh hemp leaves have been known to drive off bedbugs and kill the larvae of ticks.

As fascinating as this research is, some of these studies contradict each other. Some pests controlled by cannabis are also known to attack hemp and marijuana fields, and it's not clear:

1. What genetic ingredients in cannabis actually deter bugs consistently;
2. If the deterrent only works during certain periods of time within the plant's grow cycle;
3. If the results are limited to certain climates;
4. If some of the studies are referring to hemp's abilities or marijuana's, as the term "cannabis" is used interchangeably.

The good news is that because of America's industrial hemp pilot programs, these studies are being reexamined and reevaluated.

For example, a 2014 research paper from Washington State University titled "Industrial Hemp: Opportunities and Challenges for Washington" notes that when industrial hemp is grown for fiber—and when it is grown in rotation with other crops—hemp "reduces weed and insect pressure in other crops." The paper goes on to state:[48]

- Farmers in China use industrial hemp "as a barrier to repel insects from vegetable crops."

- Canadian farmers grow hemp with a rotation of soybeans "to reduce cyst nematodes."

Selecting plants that "get along" with each other is actually part of the planting process for most organic farmers because they can't use glyphosate to control weeds or pests. Without going into every single vegetable's needs, let's look at red peppers as one example. There are over thirty different kinds of flowers and herbs that can be grown with red peppers to accomplish what chemical pesticides can do. These companion plants also *attract* bees and other pollinators instead of killing them. For instance:[49]

- If red peppers (hot peppers or sweet peppers) are grown around basil, the herb will boost the flavor of the peppers, plus repel mosquitoes, files, thrips, and spider mites.

- Dill is also a great companion plant because its aroma attracts insects that will eat the pests.

- Growing buckwheat around peppers attracts pollinators and beneficial insects that only eat the pesky bugs.

The best part of these companion plants—including hemp—is that many of them can be used as mulch to regenerate the soil (what is also known as "green manure") to cut down on all that greenhouse gas that animal manure creates. If a plant removes certain nutrients from the soil, then that plant's stems, leaves, and "waste" by-products all contain levels of those nutrients. When plants are laid back onto the soil as green manure, the nutrients return to the earth.

Since hemp is biodegradable, microorganisms such as earthworms love it, but predators such as snails and slugs avoid it. When hemp is used as mulch, it deposits humus to the soil, which significantly increases the soil's density and helps the soil retain moisture and nutrients.

The process of regenerating the soil with companion plants dates back to the "three sisters" farming methods of the Native Americans. When corn, climbing beans, and squash are planted together, the corn serves as a natural trellis for the beans to grow on. In return, the beans add nitrogen (a must-have nutrient for plants) to the soil. Simultaneously, the corn and climbing beans provided shade for the squash to grow. The shade also kept the ground moist, which is ideal for agricultural soil (especially prior to modern irrigation systems), and the small spikes on squash vines deterred pests from accessing all three crops.[50]

Although the Native Americans weren't microbiologists, their methods have been proven effective by today's scientists. In March 2017, the peer-reviewed scientific journal *Frontiers in Plant Science* published a Kenyan study that determined climbing beans add nitrogen to agricultural soil that has been deprived of the nutrient due to overuse.[51] Of course, most farmers rotate their crops so that climbing beans and other nitrogen-fixing crops such as alfalfa or legumes don't overexpose the soil. But if a farmer plants hemp immediately following a nitrogen-fixing crop, rotating the crops won't be necessary. Cannabis loves nitrogen and absorbs enough of it to ensure that by the time hemp is ready for harvest, the soil is depleted of nitrogen and ready for the next batch of climbing beans. Not growing climbing beans again? Don't worry—nitrogen can still be returned to the soil if the "waste" from hemp—such as the leaves—are used as mulch, which is what organic hemp farmers in Canada are already doing to manage soil content.[52]

There are plenty of other modern scientific research studies to prove the efficiency of companion plants. The ATTRA Sustainable Agriculture Program, funded by the National Center for Appropriate Technology (NCAT) and the USDA, is just one example of researchers who supply alternative agriculture measures for farmers.

By *sustainable* measures, we're talking about agricultural methods that don't further the risk of polluting the soil and wetlands as well as other means that don't involve killing bees and tadpoles to feed humankind. We're already moving in the right direction—even if we take one step forward and two steps back—the research is mounting to prove there are successful farming methods that don't include bathing vegetables in chemicals.

Part of ATTRA's focus is the Scientific Foundation for Companion Planting. Here is some of the research the foundation compiled and areas of focus (if I listed and explained them all, we'd never get out of chapter 5):[53]

- The insect department (entomology) determines what the best plants are for "trap cropping" (a method that uses a neighboring crop to attract and "trap" pests from the main crop) as well as other modified forms of trap cropping.

- Another focus is symbiotic nitrogen fixation—how legumes such as peas should be grown with potatoes because potatoes will gobble up all the nitrogen the peas produce.

- And when it comes to animals? Apparently if peas are used as a green manure crop, the nitrogen levels in the soil will increase. If animals (such as sheep) are grazing in a field that has rich nitrogen deposits, the animals will naturally gain weight—without the use of any steroids or antibiotics or growth-stimulating hormones or whatever other drugs commercial farms are feeding the animals to fatten them up these days.

One of the most precious resources in farming is water. On average, here's how many inches per year outdoor crops need:[54]

- Corn: twenty to twenty-five inches of water per year
- Alfalfa: thirty to forty inches of water per year
- Tomato: fifteen to twenty-five inches of water per year
- Hemp: twelve to twenty-five inches of water per year

If hemp is used as a neighboring crop, designed to trap the pests and keep them away from the main crop (i.e., corn), then hemp can be given less water than the main crop (and therefore uses less resources). If used as a companion crop, grown within the same field as the main crop to deter pests, then hemp can consume just as much (or as little) water as the main crop needs, therefore avoiding root rot and other risks of overwatering.

In 2017, the industrial hemp pilot program in North Dakota had thirty-four growers plant a total of 3,020 acres of industrial hemp. The growers were able to plant, maintain, and harvest hemp without any significant modifications to their current farming equipment and practices. That year, North Dakota experienced drought conditions, but since the crop grows best in warmer soil, two of the four regions still harvested between 785 to 1,800 pounds per acre, which is in line with Canadian expectations. In the areas most affected by the drought, yields were between 215 to 1,621 pounds per acre.[55] Farmers noted that their typical pests—cutworm, bertha armyworm, corn borers, lygus bugs, aphids, and grasshoppers—weren't seen "at any site" where hemp was being grown. They also noted that "bees were prevalent during flowering time at several field locations."

Isn't that fascinating? One way to ensure bee colonies continue to thrive in rural America is to grow hemp!

HEMP AS COVER CROP TO KILL WEEDS

In much the same way Native Americans used corn and climbing beans to create the opportune conditions for growing squash, hemp can be used to kill weeds.

Cover crops and canopy cover are used in organic farming; these are tall companion plants used to create shade among main crops to cut off sunlight to weeds (thus starving and killing them).

"Hemp is very competitive against weeds under ideal conditions," states 2018 documentation from the Plant Industries Division of the North Dakota Department of Agriculture. "It emerges very rapidly (three to four days) in warm/moist soils and can quickly shade and outcompete weeds. Growers commented on how quickly hemp develops."[56]

The University of Vermont conducted a weed control trial with industrial hemp in 2016 at Borderview Research Farm. Researchers found "when the hemp was 8–10 inches tall, it grew rapidly past the weeds and became far more competitive and clearly could grow past the weed pressure."[57]

Aside from killing weeds naturally, cover crops are used to:[58]

- Reduce soil erosion from wind and water
- Increase organic matter in the soil
- Increase biodiversity, which can decrease the likelihood of disease
- Capture excess nutrients in the soil
- Reduce energy use and promote biological nitrogen fixation
- Maintain soil health for vegetables when repurposed as "green manure"

In California, farmers use cover crops for more than just weed control. There are more than 400,000 acres of agricultural land affected by poor water penetration. The soil is dense, so farmers have to use more water than other parts of the country to keep fruit and nut crops nurtured. Even so, according to research by University of California's Agriculture and Natural Resources department, the soil's poor water penetration resulted in economic losses estimated as high as $486 per acre for orchards in 1992.[59] Water usage is always a concern in a state on high drought alert, and studies by the University of California have noted that cover crops can lower soil surface strength and improve soil permeability.

Currently, the Rodale Institute is conducting a four-year research project into hemp's ability to address weed issues and enhance soil health in organic agriculture.

"Hemp fits well into a diverse organic crop rotation," states the website for the Rodale Institute Industrial Hemp Research Project. "Its high biomass and canopy production has the potential to shade out weeds resulting in lower weed seed germination and growth."[60]

So again, hemp the disrupter has shown there are natural ways to achieve the same effects as synthetic chemical pesticides and herbicides.

BIOPESTICIDES AND BOTANICAL PESTICIDES

Here are the biopesticide terms I want to clarify before moving forward with this section:[61]

- Botanical insecticides (also known as natural insecticides) are naturally occurring chemicals that are extracted from plants or minerals. These natural extracts—such as citrus oil—are toxic to insects, and if sprayed on crops, they ward off insects without exposing the crops to synthetic chemicals.

- Biochemicals (used for insect repellents or insect attractants) are also derived from naturally occurring substances, such as plant extracts.

- Microbial products contain microorganisms that can fertilize plants and soil or repel pests, such as caterpillars.

- Plant-incorporated protectants (PIP) are plants that have had genes inserted into them to produce a pesticide inside its own tissues. Although this means the plants are genetically modified, the genes inserted are pesticidal proteins already found in plants, which is why they are considered to be a safer alternative to synthetic chemicals. However, they aren't supposed to be used in greenhouses, and it's important to know how they interact with other plants before using them as a biopesticide. Plus, if an insect already possesses a high-level resistance to the PIP proteins, they'll continue to feed on the plant, which is the same problem experienced with other genetically modified plants.

Some farms might use a combination of biopesticides with chemical pesticides to cut back on the amount of chemicals they use. Glyphosate is a chemical that generations of farmers have been using—one could say it's an integral part of American farming at this point—but what if there was a plant that the FDA hasn't cleared for chemical pesticide use? What if a farmer wanted to grow that crop?

As it turns out, hemp *and marijuana* growers aren't legally able to use the most popular synthetic chemicals on their crops because there aren't any federally approved for cannabis (yet).

Since marijuana has been legalized in more than half the states in the United States, growers all over the country have turned to the biological pesticide industry to keep pests under control. Forbes.com has gone so far as state that the marijuana industry is creating a "boom" in the bio-pest industry, and this boom has helped reinforce the science behind biological pesticides.[62]

So why do American farmers (and most farmers worldwide) use Roundup and other chemical products? Is it because glyphosate is the most reliable and efficient means to repel pests? Or is this the result of the power of corporate advertising and years of self-fulfilling research that is bought and paid for by Monsanto?

According to science—and by science, I mean studies conducted without the bias that comes along with corporate payoffs—some all-natural pesticides have been found to be *more effective* in controlling pests than chemical means. Which is quite remarkable, considering there are botanical pesticides that can be made at home in a blender with common ingredients. Of course, if someone is brewing up a batch at home, it may be difficult to figure out the exact consistency, but there are companies that sell premixed pest-fighting ingredients to take out the guesswork.

Since cannabis regulations vary by state, each state also dictates what marijuana and hemp growers can use on their crops to control pests and weeds. In California, marijuana growers are only allowed to use botanical, microbiological, or naturally sourced pesticide products.[63] The list of approved products largely includes botanical oils such as sesame, soybean, peppermint, and rosemary. After following these guidelines, each pot grower has to then pay $5,000 to $10,000 for a lab to test for purity and potency. Within the first two months of testing marijuana-infused products and marijuana buds in 2018, almost two thousand of approximately eleven thousand samples failed, but only four hundred of the eleven thousand products were flagged for unacceptable levels of pesticides.[64]

So if cannabis growers aren't legally allowed to use synthetic chemicals, how did pesticides get on those four hundred products? The same way those levels of glyphosate get into all-natural, organic General Mills products. If the cannabis is grown outside, it's just as susceptible to the fly-by-night coating of glyphosate as any other crop. Or, if we're talking about a cannabis-infused product such as chocolate, perhaps some of the other ingredients had trace amounts of pesticides in them, which caused the product to fail.

Interesting to note that *zero* is the precise amount of chemical pesticides legally allowed in marijuana and marijuana-infused products in California. Due to California's strict regulations on recreational marijuana, rest assured, dispensaries carry the equivalent of all-natural, non-GMO, organic products. So here we are: The FDA and EPA allow the breakfast cereals we feed our children to contain more synthetic chemicals than what California will allow in a recreational drug.

Just to expand on that hypocrisy, organic products are allowed to have what is known as a pesticide tolerance. That means the EPA sets a limit on the amount of pesticides that may remain in or on food, and these limits are referred to as "tolerances." Yet there is a zero-tolerance policy for synthetic pesticide residue on cannabis. However, thanks to these strict guidelines, regulators have been able to catch potentially fraudulent biopesticide products. At the expense of the cannabis grower, that is.

The Oregon Department of Agriculture (ODA) approved different biopesticides than California, but the state also has a zero-tolerance policy for pesticides on cannabis—these rules apply whether a grower is cultivating hemp or marijuana. The ODA website warns: "organic does not automatically make it approved for use on cannabis! Many organic products have tolerances established for use of food crops and would not be on ODA's guide list for pesticides and cannabis."[65]Then in 2018, when testing time came, regulators discovered Oregon marijuana was laced with banned chemicals. The cannabis growers themselves didn't add the chemicals to the crops, and after an investigation, it's now clear that the manufacturers did.

In 1996, the EPA allowed "up to 1,000 parts per million of contamination" in pesticides.[66] Organic biopesticides aren't excluded from this, so *six* of Oregon's approved biopesticide products actually wound up contaminating marijuana. These national distributed products didn't list the chemicals on their labels, including permethrin, bifenthrin, cypermethrin, cyfluthrin, chlorpyrifos, fenpropathrin, lambda-cyhalothrin, piperonyl butoxide and MGK-264. So how were marijuana growers supposed to know?

Unfortunately, a zero-tolerance policy is a zero-tolerance policy, and if the cannabis tested positive for these chemicals, it couldn't be sold legally in Oregon. The ODA was able to issue "stop sale" orders for the six biopesticide products, but the damage had already been done to the marijuana, and remember, those growers don't have crop insurance to turn to for recourse.

As if cannabis growers didn't have enough to be paranoid about, they now have to worry about companies lacing biopesticide with synthetic chemicals?

Luckily, growers have other forms of biopesticides to turn to that they can control, such as microbes and beneficial insects. As for myself, I have experience with using microbes to promote marijuana growth and to ward off pests. Granted, my family's (legal) outdoor marijuana farm in California wasn't a large operation, but we did successfully home-brew our own compost teas (otherwise known as microbial tea) to provide specific nutrients for the marijuana.

My husband incorporated neem seed meal in the microbial tea he sprayed on the marijuana plants. (Neem is an evergreen tree found in India that repels a wide variety of pests and their larva without being harmful to mammals, birds, earthworms, or beneficial insects such as honeybees or butterflies.) When the tea was absorbed by the marijuana, it made the plant less susceptible to pests. We found this to be a better method than spraying the plants with the neem oil (which is essentially a vegetable oil made from neem seeds).

We also grew sunflowers, peppers, grapes, green apples, and figs as companion plants to maintain pest control. During pollination, we used large industrial fans to keep airflow constant. Mosquitoes and other flying bugs were no match for the artificial breeze, which brushed them onto the blackberry bushes that were growing outside the garden (companion crop). Looking at the pictures we took, all the way up to harvest, the rows of marijuana bushes were weed-free—the weeds had sprouted up around the fig and apple trees instead.

Yes, marijuana is a needy crop compared to industrial hemp, but contrary to popular belief, even though hemp is resistant to many bugs, there are pesky little critters—such as the corn ear worm and grasshoppers[67]—that love munching on the plant. So what's the best solution to get rid of pests if hemp is supposed to be used to *repel* them?

CANADIAN HEMP

According to the Canadian Hemp Trade Alliance (CHTA):[68]

> *A number of insects have been found feeding in hemp. To date none have been found at economic levels that would require control. As acreages increase and production becomes more intensive it is expected common pests of other crops may become a problem. No insecticides are registered for use on hemp field crops.*

That's quite an endorsement for all-natural hemp. Considering Canada began growing hemp on an experimental basis in 1994, and started its commercial production of industrial hemp in 1998,[69] the country should have a thorough understanding of whether or not the crop needs glyphosate.

Canada's agriculture department goes even a step further by stating the crop is "environmentally-friendly because it doesn't need herbicides; in fact, most of Canada's hemp is certified organic."[70] Which means whether the hemp is ultimately used for food, clothing, plastics, or building materials, the nearly 140,000 acres of Canadian industrial hemp planted in 2017 was predominantly organic.[71]

And how do the Canadians ensure the hemp is certified organic? First of all, they prefer to grow it in the "clean soils" of Canada's prairies—farmland that hasn't been contaminated with chemicals from industry or agriculture.

Secondly, the Canadian Food Inspection Agency carefully regulates and monitors all Canadian commercial hemp strains to maintain THC levels and ensure genetic identity. In Canada, farmers are required to purchase "clean" hempseed from a registered seed establishment every year. In other words, they can't save seeds from last year's harvest and plant them for commercial use. While this may seem like an added expense for a farmer to purchase more seed every year, it actually cuts down the cost of testing. The plants don't need to be randomly tested to ensure the THC content is .3 percent or less because the seed is certified before it reaches the farmer.

Also contrary to popular belief, many farmers don't keep seeds from a previous harvest to grow in the next one. If a farmer is planting a GMO crop, they are planting seeds that are patented, and therefore they must purchase new seeds each year or face potential lawsuits from Monsanto.[72] Other farmers may choose not to plant seeds from a previous crop due to consistency concerns; the seeds may not be the same standard as certified seed, especially if a particular harvest generated a weak yield due to an unprecedented cold front or drought.

As far as Canada's traditional, nonorganic field farming is concerned, the country might wind up changing pesticide farming practices as a whole. In November 2018, Canada's CBC News reported that "troubling allegations" about Monsanto's products, particularly glyphosate, caused Health Canada to review studies claiming its safety.[73] Documents in Dewayne Johnson's lawsuit, now known as the "Monsanto Papers," contained emails between the corporation's executives and scientific experts that revealed Monsanto played a hand in writing the very academic papers Health Canada relied upon to determine if the pesticide was safe. A coalition of environmental groups including Equiterre, Ecojustice, and Canadian Physicians for the Environment called on Health Canada to reexamine the evidence in the hopes that new studies will be conducted. The agency is now reviewing hundreds of studies, originally presented by Monsanto as independent research to Health Canada's Pest Management Regulatory Agency, to determine if Monsanto did indeed influence them.

HEMP AS A NATURAL BIOPESTICIDE

One final reason why hemp doesn't require pesticides is because there are indications that the oils found in the plant can repel pests. A study published in April 2018 in *Environmental Science and Pollution Research* looked at hemp strain Futura 75. Researchers took essential oils from the leaves (the leaves are generally a waste product) and created an eco-friendly insecticide for three varieties of insects and determined that the oil "moderately inhibited" the insects.[74]

Obviously, an all-natural alternative to Roundup made entirely from hemp essential oils hasn't been invented yet, but a possible reason why essential hemp oil is successful at repelling insects and mosquitoes is that hemp is loaded with fatty acids, such as omega-3 and omega-6, and fatty acids have been proven to repel insects.

In India and Africa, locals burn the seed oil from *Jatropha curcas,* a traditional folk remedy plant used to ward off bugs. Interestingly, researchers have found the technique is extremely effective because the plant's oil is filled with fatty acids.[75] Suffice to say, this is perhaps another reason why companion farming works so well with hemp. If the natural oils on its stems and leaves repel most pests, then pests will stay away from the main crop.

So there you have it:

- Grow hemp to increase bee productivity.
- Grow hemp to fertilize the soil.
- Grow hemp as an all-natural solution to chemical weed killers.
- Grow hemp as an environmentally friendly way to ward off pests.
- Grow hemp as a way to increase and support organic farming.
- Grow hemp to disrupt our dependence on glyphosate, and simultaneously save tadpoles and bees (and ourselves) from dying.

The 2018 Farm Bill is the best indication to date that the time has come to disrupt modern farming techniques by growing hemp. We can make our food healthier and our agricultural practices more sustainable. The blueprint is already here—we just need to follow it.

NOTES

1. Christina Morales, "5 Unbelievable Facts About Roundup," Periscope Group, Oct 23, 2018, https://www.periscopegroup.com/roundup/facts-about-roundup.
2. Daniel Arkin, "Jury orders Monsanto to pay nearly $290M in Roundup trial," US News, *NBC News,* Aug 11, 2018, https://www.nbcnews.com/news/us-news/jury-orders-monsanto-pay-290m-roundup-trial-n899811.
3. Carey Gillam, "I Won a Historic Lawsuit, But May Not Live to Get the Money," Health, *Time.com,* http://time.com/5460793/dewayne-lee-johnson-monsanto-lawsuit/.
4. Daniel Arkin, "Jury orders Monsanto to pay nearly $290M in Roundup trial," US News, *NBC News,* Aug 11, 2018, https://www.nbcnews.com/news/us-news/jury-orders-monsanto-pay-290m-roundup-trial-n899811.
5. Beth Kaiserman, "General Mills To Remove 'Natural' Label From Granola Bars After Glyphosate Lawsuit," Food & Drink, *Forbes.com,* Aug 28, 2018, https://www.forbes.com/sites/bethkaiserman/2018/08/28/general-mills-to-remove-natural-label-from-granola-bars-after-glyphosate-lawsuit/#463ab03f73c9.
6. Ibid.

7. Jonathan Stempel, "General Mills changing Nature Valley labels after lawsuit's pesticide claim," Business news, *Reuters,* Aug 23, 2018, https://www.reuters.com/article/us-general-mills-nature-valley-settlemen/general-mills-changing-nature-valley-labels-after-lawsuits-pesticide-claim-idUSKCN1L82E5.
8. Jeff Mapes, "Marijuana and hemp growers duke it out in Oregon Legislature," Politics, *The Oregonian,* June 3, 2015, https://www.oregonlive.com/mapes/index.ssf/2015/06/after_marijuana_growers_compla.html.
9. Alex Tempkin, "Breakfast with a Dose of Roundup?" *EGW's Children's Health Initiative,* Aug 15, 2018, https://www.ewg.org/childrenshealth/glyphosateincereal/#.W6kvdJNKgdU.
10. "Pesticide National Synthesis Project, Estimated Annual Agricultural Pesticide Use," *U.S. Geological Survey, US Department of the Interior,* last modified Sept 11, 2008, https://water.usgs.gov/nawqa/pnsp/usage/maps/show_map.php?year=2015&map=GLYPHOSATE&hilo=L&disp=Glyphosate.
11. Ben Hewitt, "Why Farmers are using Glyphosate—And What It Might Mean For You," Features, *Ensia*, Dec 19, 2017, https://ensia.com/features/glyphosate-drying/.
12. Beth Kaiserman, "General Mills To Remove 'Natural' Label From Granola Bars After Glyphosate Lawsuit," Food & Drink, *Forbes.com,* Aug 28, 2018, https://www.forbes.com/sites/bethkaiserman/2018/08/28/general-mills-to-remove-natural-label-from-granola-bars-after-glyphosate-lawsuit/#463ab03f73c9.
13. Alex Tempkin, "Breakfast with a Dose of Roundup?" *EGW's Children's Health Initiative,* Aug 15, 2018, https://www.wg.org/childrenshealth/glyphosateincereal/#.W6kvdJNKgdU.
14. EPA, "EPA Releases Draft Risk Assessments for Glyphosate," *United States Environmental Protection Agency,* December 18, 2017, https://www.epa.gov/pesticides/epa-releases-draft-risk-assessments-glyphosate.
15. JMPR secretariat, "Frequently asked questions," *World Health Organization,* May 27, 2016, http://www.who.int/foodsafety/faq/en/.
16. International Agency for Research on Cancer, "IARC Monographs Volume 112: evaluation of five organophosphate insecticides and herbicides," *World Health Organization,* March 20, 2015, http://www.iarc.fr/en/media-centre/iarc-news/pdf/MonographVolume112.pdf.
17. Chemicals Considered or Listed Under Proposition 65, "Glyphosate," *OEHHA Science for a Health California,* March 28, 2017, https://oehha.ca.gov/proposition-65/chemicals/glyphosate.
18. "California's Proposed Limit vs. the Amount Allowed by EPA," Rethinking Cancer, *EWG.org,* 2016, https://www.ewg.org/research/california-proposes

-safe-level-roundup-more-100-times-lower-epa-limit/californias-proposed#.W6lbcZNKgdU.
19. "Monsanto Papers: Secret Documents," Baum Hedlund Aristei Goldman PC, *Baum Hedlund Law,* 2018, https://www.baumhedlundlaw.com/toxic-tort-law/monsanto-roundup-lawsuit/monsanto-secret-documents/.
20. Mesnage R, Defarge N, Spiroux de Vendômois J, Séralini GE. Major pesticides are more toxic to human cells than their declared active principles. *Biomed Res Int*. 2014;2014:179691, https://www.ncbi.nlm.nih.gov/pmc/articles/PMC3955666/.
21. "The History of Roundup," *Roundup,* accessed Jan 23, 2019, http://www.roundup.ca/en/rounduphistory.
22. Mary Shinn, "Geothermal domes nurture plants and learning in Pagosa Springs," Local News, *The Durango Herald*, Dec 28, 2017, https://durangoherald.com/articles/201364.
23. The Geothermal Greenhouse Partnership, "Phases," The Geothermal Greenhouse Project, 2015, http://pagosagreen.org/phases/.
24. Andrew Jenkins, "Food security: vertical farming sounds fantastic until you consider its energy use," *The Conversation,* Sept 10, 2018, https://theconversation.com/food-security-vertical-farming-sounds-fantastic-until-you-consider-its-energy-use-102657.
25. Ibid.
26. Abigail Klein Leichman, "Home robot grows cannabis . . . or other herbs," *Israel 21c,* Oct 10, 2018, https://www.israel21c.org/home-robot-grows-cannabis-or-other-herbs/.
27. Jelle Bruinsma, FAO, "The Resource Outlook to 2050: By How Much Do Land, Water and Crop Yields Need to Increase by 2050?," *Food Security and Nutrition Network,* June 2009, https://www.fsnnetwork.org/resource-outlook-2050-how-much-do-land-water-and-crop-yields-need-increase-2050.
28. National Statistics, "Agriculture in the United Kingdom 2017," *United Kingdom Government Statistics,* last modified, Sept 18, 2018, https://www.gov.uk/government/statistics/agriculture-in-the-united-kingdom-2017.
29. Peter Border, Ruth Barnes, "Security of UK Food Supply," Research Briefings, *Parliamentary Office of Science and Technology,* July 4, 2017 https://researchbriefings.parliament.uk/ResearchBriefing/Summary/POST-PN-0556#fullreport.
30. Jeanette Settembre, "This innovative $7,000 'indoor farm' may change how America eats forever," Moneyish, *Marketwatch,* Sept 13, 2018, https://moneyish.com/upgrade/this-innovative-7000-indoor-farm-may-change-how-america-eats-forever/.
31. Ibid.

32. EPA, "EPA Releases Draft Risk Assessments for Glyphosate," *United States Environmental Protection Agency,* December 18, 2017, https://www.epa.gov/pesticides/epa-releases-draft-risk-assessments-glyphosate.
33. Brian Bienkowski, "Active ingredient in Monsanto's Roundup hurts honey bee guts," *Environmental Health News,* Sept 25, 2018, https://www.ehn.org/monsanto-herbicide-roundup-hurts-bees-2607605097.html.
34. Renée Johnson and M. Lynne Corn, "Bee Health: Background and Issues for Congress," *Congressional Research Service*, Jan 20, 2015, https://fas.org/sgp/crs/misc/R43191.pdf.
35. Ash E. Samuelson, Richard J. Gill, Mark J. F. Brown, Ellouise Leadbeater, "Lower bumblebee colony reproductive success in agricultural compared with urban environments," 285, *Proceedings of the Royal Society B: Biological Sciences,* June 16, 2018, https://doi.org/10.1098/rspb.2018.0807.
36. Anamika Vaughan, "Portland Parks And Rec Continues To Use Roundup After Recent Cancer Controversy," *Williamette Week*, Sept 20, 2018, https://www.wweek.com/news/city/2018/09/20/portland-parks-and-rec-continues-to-use-roundup-after-recent-cancer-controversy/.
37. Rick A. Relyea, "Amphibians Are Not Ready for Roundup," Department of Biological Sciences, University of Pittsburgh, *Wildlife Ecotoxicology: Forensic Approaches Emerging Topics in Ecotoxicology 3*, pp. 267–300, https://www.biology.pitt.edu/sites/default/files/facilities-images/Relyea%20286.pdf.
38. Rick A. Relyea, "New effects of Roundup on amphibians: predators reduce herbicide mortality; herbicides induce antipredator morphology," Department of Biological Sciences, University of Pittsburgh, *Ecol Appl.*, March 2012, 22(2): 634-47.PMID: 22611860, https://www.ncbi.nlm.nih.gov/pubmed/22611860.
39. Rick A. Relyea, "Amphibians Are Not Ready for Roundup," Department of Biological Sciences, University of Pittsburgh, *Wildlife Ecotoxicology: Forensic Approaches Emerging Topics in Ecotoxicology 3*, DOI 10.1007/978-0-387-89432-4_9, pp. 267–300, https://www.biology.pitt.edu/sites/default/files/facilities-images/Relyea%20286.pdf.
40. Case 3:16-md-02741-VC, Document 246-2 , Filed April 20, 2017, https://usrtk.org/wp-content/uploads/2017/04/MDLLetNothingGomotion.pdf.
41. Baum Hedlund, "Monsanto Paid Internet Trolls to Counter Bad Publicity," Monsanto Roundup News, *Baum Hedlund Law,* May 4, 2017, https://www.baumhedlundlaw.com/monsanto-paid-internet-trolls/.
42. Bonnard L. Moseley, "Hardware Disease of Cattle," College of Veterinary Medicine, *University of Missouri Extension,* Oct 1993, https://mospace.umsystem.edu/xmlui/bitstream/handle/10355/3652/HardwareDiseaseOfCattle.pdf?sequence=1.

43. Princeton University. (2014, March 27). "A more potent greenhouse gas than carbon dioxide, methane emissions will leap as Earth warms." *ScienceDaily*. Retrieved January 24, 2019 from www.sciencedaily.com/releases/2014/03/140327111724.htm.
44. Sara Chodosh, "Cow farts are an even bigger problem than we thought," Animals, *Popular Science*, Oct 2, 2017, https://www.popsci.com/cow-farts-are-an-even-bigger-problem-than-we-thought.
45. Sam Lemonick, "Scientists Underestimated How Bad Cow Farts Are," Science, *Forbes.com,* Sept 29, 2017, https://www.forbes.com/sites/samlemonick/2017/09/29/scientists-underestimated-how-bad-cow-farts-are/#327cfa5978a9.
46. John M. McPartland, "Cannabis as repellent and pesticide," *Journal of the International Hemp Association* 4(2): 87-92, 1997, *Drug Library.net,* http://www.druglibrary.net/olsen/HEMP/IHA/jiha4210.html.
47. T. Randall Fortenbery, Thomas B. Mick, "Industrial Hemp: Opportunities and Challenges for Washington," School of Economic Sciences, *Washington State University,* Feb 2015, http://ses.wsu.edu/wp-content/uploads/2015/02/WP2014-10.pdf.
48. Derek Markham, "32 companion plants to grow with your peppers," Lawn & Garden, *Treehugger,* May 28, 2014, https://www.treehugger.com/lawn-garden/companion-plants-grow-your-peppers.html.
49. David Quick, "Urban gardening experts use companion planting to attract pollinators, deter pests, improve soil," Garden Bedfellows, *The Post and Courier,* April 14, 2018, https://www.postandcourier.com/features/urban-gardening-experts-use-companion-planting-to-attract-pollinators-deter/article_8b9d9c04-3cdd-11e8-af20-137299f02feb.html.
50. Koskey, Gilbert et al. "Potential of Native Rhizobia in Enhancing Nitrogen Fixation and Yields of Climbing Beans (*Phaseolus vulgaris* L.) in Contrasting Environments of Eastern Kenya" *Frontiers in plant science* vol. 8 443. 31 Mar. 2017, doi:10.3389/fpls.2017.00443 https://www.ncbi.nlm.nih.gov/pubmed/28408912#.
51. Arthur Hanks, "Hemp: an organic crop for farmers, an organic food for consumers," *The Hemp Report*, May 12, 2005, http://www.hempreport.com/2005/05/hemp-organic-crop-for-farmers-organic.html.
52. Justin Duncan, "Companion Planting & Botanical Pesticides: Concepts & Resources," *ATTRA Sustainable Agriculture,* last modified April 2016, https://attra.ncat.org/viewhtml/?id=72#3.
53. Bob Hammon, John Rizza, Doug Dean, "Current Impacts of Outdoor Growth of Cannabis in Colorado," Colorado State University Extension, July 2015, http://extension.colostate.edu/docs/pubs/crops/00308.pdf.
54. "North Dakota Department of Agriculture 2017 Industrial Hemp Pilot-Program," *North Dakota Department of Agriculture Plant Industries Division,*

March 2018, https://www.nd.gov/ndda/sites/default/files/2017%20NDDA%20 Industrial%20Hemp%20Pilot%20Project-Report%20final%20review.pdf.
55. Ibid.
56. Dr. Heather Darby, Abha Gupta, Erica Cummings, et. al, "2016 Industrial Hemp Weed Control Trial," Northwest Crops & Soils Program, *University of Vermont Extension,* Jan 2017, https://www.uvm.edu/sites/default/files /Northwest-Crops-and-Soils-Program/2016-ResearchReports/2016_Hemp _Weed_Control.pdf.
57. Minnesota Natural Resources Conservation Service, "Cover Crop," Conservation Practice Standard, *USDA,* July 2013, https://efotg.sc.egov.usda .gov/references/public/MN/340mn_Cover_Crop_Standard.pdf.
58. Folorunso O, Rolston D, Prichard P, Louie D, 1992. "Cover crops lower soil surface strength, may improve soil permeability," Calif Agr 46(6):26-2, Nov 1, 1992, http://calag.ucanr.edu/Archive/?article=ca.v046n06p26.
59. "Industrial Hemp Trial," *Rodale Institute,* accessed Jan 24, 2019, https: //rodaleinstitute.org/industrialhemp/.
60. Heidi Lindberg and Steven Arthurs, "What is a biopesticide?" *Michigan State University Extension,* June 22, 2017, http://msue.anr.msu.edu/news /what_is_a_biopesticide.
61. Janet Burns, "Cannabis Is Creating A Boom For Biological Pesticides," Consumer Tech, *Forbes.com,* Aug 19, 2018, https://www.forbes.com/sites /janetwburns/2018/08/19/cannabis-could-help-biopesticides-take-root-in -american-agriculture/#78be37f22c1d.
62. Kerrie and Kurt Badertscher, "Are You Ready for California's Pesticide Regulations?," How-to, *Cannabis Business Times*, March 2018, http://www .cannabisbusinesstimes.com/article/are-you-ready-for-californias-pesticide-regulations/.
63. Michael R. Blood, "California pot products seeing big safety testing failure rate," *The Associated Press,* Sept 11, 2018, https://www.denverpost.com/2018/09/11 /california-pot-product-testing/.
64. Pesticides Program, "Oregon Cannabis: Cannabis and Pesticides," Oregon Department of Agriculture, April 2018, https://www.oregon.gov/ODA /shared/Documents/Publications/PesticidesPARC/CannabisPesticides.pdf.
65. Mateusz Perkowski, "Cannabis testing reveals biopesticide contamination," *Capital Press,* June 13, 2018, http://www.capitalpress.com/Oregon/20180613 /cannabis-testing-reveals-biopesticide-contamination.
66. College of Agricultural Sciences, "Hemp Insect Factsheets," *Colorado State University*, accessed Jan 24, 2019, https://hempinsects.agsci.colostate.edu /hemp-insects-text/.

67. "Insects and Pests," *Canadian Hemp Trade Alliance*, last modified Jan 2019, http://www.hemptrade.ca/eguide/production/insects-and-pests.
68. "Industrial Hemp Production in Canada," Agriculture, *Alberta Agriculture and Forestry,* last modified Nov 13, 2015, https://www1.agric.gov.ab.ca/$department/deptdocs.nsf/all/econ9631.
69. "Canadian Hemp," *Agriculture and Agri-Food Canada*, accessed Jan 23, 2019 http://www.agr.gc.ca/resources/prod/Internet-Internet/MISB-DGSIM/ATS-SEA/PDF/4687-eng.pdf.
70. Mark Halsall, "Hemp area may stall in 2018," *Country Guide,* March 2, 2018, https://www.country-guide.ca/2018/03/02/are-canadas-hemp-acres-poised-for-a-downward-slide-in-2018/52710/.
71. Ibid.
72. "Why Does Monsanto Sue Farmers Who Save Seeds?," *Monsanto,* April 11, 2017, https://monsanto.com/company/media/statements/saving-seeds/.
73. Gil Shochat, "'Troubling allegations' prompt Health Canada review of studies used to approve popular weed-killer," *CBC News,* Nov 11, 2018, https://www.cbc.ca/news/technology/monsanto-roundup-health-canada-1.4896311.
74. Benelli, G., Pavela, R., Lupidi, G. et al., *Environ Sci Pollut Res* (April 2018) 25: 10515. pp 10515–10525, https://doi.org/10.1007/s11356-017-0635-5.
75. United States Department of Agriculture–Research, Education and Economics, "Insect-repelling compounds discovered in folk remedy plant, Jatropha," *ScienceDaily*, Nov 5, 2012, www.sciencedaily.com/releases/2012/11/121105140205.htm.

6

HEMP HEALTH AND NUTRITION

Hemp products were once stereotyped as something produced for hippies by hippies, right up there with patchouli oil. The classic striped hemp hoodie, once a staple among Phish and Grateful Dead fans, has now been mass-produced and embraced by mainstream surf companies such as Tommy Bahama, Billabong, Hurley, Vans, Abercrombie & Fitch, O'Neill, and Roxy. Hemp lotions, essential oils, shampoos, and conditioners have also been creeping into the health, beauty, and wellness aisles of major grocery-store chains.

Pop culture surrounding the nutritional benefits of hemp has also changed. Organic, non-GMO hemp protein smoothies and hemp energy bars, once considered to be fringe health products, are now easily accessible. We've accepted hemp products. Science has validated that hemp is a healthy and nutritious food, and there are benefits in adding it to our daily diets.

While locally sourced, grass-fed, and free-range meats without antibiotics are becoming more and more popular, hemp is another way to get protein into our diets and into the diets of the animals we consume. The Colorado Department of Agriculture is currently studying the benefits of adding industrial hemp to animal feed, which isn't a novel concept. In Europe, hempseed meal is currently used as a supplement for livestock. According to Hemp Foods Australia, the largest hemp food wholesaler, retailer, manufacturer, and exporter in the Southern Hemisphere: "hemp supplement improves digestion and increases life expectancy in livestock."[1] Since hemp contains a high content of omega-3 and omega-6 fatty acids, as well as all known amino acids, it's a one-stop solution to provide nutritional benefits to animals, including the ones that wind up on our dinner plates.

A Canadian study published in *Poultry Science* in 2012 determined that feeding hens either hempseed or hempseed oil doesn't affect the frequency of how many eggs they produce, but it does increase the amount of n-3 fatty acids (otherwise known as omega-3 fatty acids) in the eggs by 12 to 20 percent.[2] A similar German study determined that eggs from hens that were fed hempseed oil cakes had lower percentages of both saturated fatty acids and monounsaturated fatty acids (the bad fats) but an increased level of polyunsaturated fatty acids (the good fats), such as n-3 fatty acids.[3] Nutritional studies conducted on livestock have also shown:

- Adding hemp to the diet of pigs is beneficial when breeding livestock to increase the likelihood of newborn piglets' survival. When hempseed is fed to sows, it increases the fat, lactose, protein, and minerals in sow milk, and these added nutrients improve the survival rate of newborn piglets.[4]

- Japanese quail fed a dietary supplement of hempseed produced higher omega-3 content in their egg yolks. When the quails themselves were slaughtered for meat, there was a "decreased cooking loss," which means more edible meat per quail.[5]

- Sheep that are fed hempseed and hempseed cake increase their yields of milk, and hemp improves their milk quality by increasing protein and fatty acids.[6]

- When hemp oil is added to pig feed, it provides a source of n-3 fatty acids for the pig, which in turn increases the n-3 fatty acid content in pork (for human consumption) without modifying the growth or performance of the pigs.[7]

Overall, independent research shows that feeding hemp oil and hempseed to livestock makes their produce healthier and more nutrient-rich for humans. (Remember, hemp oil and hempseed do not contain THC or CBD.) But why stop there; why not add it to pet food?

In February 2018, pentobarbital, a drug typically used to euthanize animals, was found in at least twenty-seven varieties of Smucker's pet food, including Gravy Train and Kibbles 'N Bits products. While it is illegal for any food products to contain this ingredient, it still happens every few years, as the FDA has found the drug in dog food brands previously after conducting studies in 1998 and 2000.[8]

The FDA is currently conducting an investigation into Smucker's tainted pet foods to determine how pentobarbital got into the products, but really there isn't a mystery here. The protein source for the food contained pentobarbital because the meat came from euthanized animals. No company wants to admit to pet owners that they've been secretly feeding their furry friends dead furry friends, but veterinarians have known it for years. Pets have become more resistant to pentobarbital and require a larger dose when they're being euthanized due to the content they've been fed on a daily basis.

"The pet food industry is the wild west when it comes to pet food quality and ingredient quality," states Jackie Bowen, the director of consumer advocacy group Clean Label Project, which tests human and pet food for contaminants.[9]

The Clean Label Project has conducted 1,084 tests on pet food products from 80 manufacturers to find that pentobarbital isn't the only problem. There are actually multiple toxins hiding in the majority of dog and cat food products, including lead, arsenic, cadmium, pesticides, BPA, and carcinogenic contaminants.[10] Among the worst of the worst are fish products. Although the cleanest protein source overall is turkey-based, hemp protein (and plant proteins in general) could easily replace tainted, meat-based sources of protein.

Hemp is a superfood because of how many nutritional benefits it contains; due to its nutritional profile, it is often marketed as a substitute for fish. Ever since the Fukushima nuclear power plant accident, South Koreans have been modifying their fish-based diets. Locals are no longer relying on fish as their primary source of protein due to contamination concerns, and they are turning to hemp as a nutritional substitute.[11] Here are some of the benefits of incorporating hemp into daily meals.

HEMP PROTEIN

- At least 25 percent of the total calories found in hemp is from protein, which is a higher content than chia or flax seeds.
- Hemp protein is considered a "complete protein" because it offers every single amino acid the human body needs to survive.
- The protein contains twenty amino acids, including the nine essential amino acids our body can't produce on its own.
- The protein is also rich in fiber, which is known to help lower risk of heart disease, type 2 diabetes, and constipation.
- Fiber helps regulate blood glucose, which can fight against energy crashes and the craving for unhealthy sweets.
- One serving of hemp protein powder satisfies 50 percent of our daily recommended fiber serving.

- Hemp protein contains edestin—the most easily digestible protein.
- Hemp is also a great source of vegan protein (plant-based protein), which the body digests faster than animal protein.
- In general, protein also boosts metabolism, increases the body's fat-burning potential, and boosts the immune system.

OMEGA-3 FATTY ACIDS AND OMEGA-6 FATTY ACIDS

- Hempseed has an ideal balance of 3:1 omega-3 fatty acids and omega-6 fatty acids, which promotes cardiovascular health.
- Omega-3 has been suggested to prevent coronary heart disease.
- Omega-6 is shown to decrease risk for osteoporosis. Women over sixty-five years of age who took omega-6 supplements had less bone loss over three years and also increased their bone density.[12]
- Amino acids also improve immune function and can boost the immune system when it's already compromised.[13]

RICH IN GLA

- Hempseed is a prime source of Gamma-linolenic acid (GLA), a necessary building block for our bodies to control inflammation, body temperature, and muscle control.
- Because GLA helps control inflammation, a daily dose of hempseed oil is also known to help with joint pain and arthritis.[14]

Other nutrients found in hemp include:[15]

Vitamin E	Iron
Phosphorus	Copper
Potassium	Magnesium
Zinc	Thiamin

Hempseed is also called "heart healthy" because it can improve cardiovascular health, reduce cholesterol levels, and lower high blood pressure.[16] In just one ounce (two tablespoons) of hempseed, there are approximately 9.2 grams of protein and 2 grams of fiber.[17] Some people don't even get that much protein in one meal.

EATING RAW HEMP

The owners of the Milwaukee-based Selthofner Farm are involved in the state's hemp pilot program and they are looking to "offer an edible hemp-leaf salad mix."[18] I can't say that I've ever met anyone who has taken cannabis leaves off the plant and eaten them raw, but given the number of recipes out there for raw cannabis smoothies and other concoctions, this seems to be a developing health trend.

Unlike marijuana edibles, eating the plant raw *doesn't* result in any trippy experiences. This is because when cannabis is in its raw form, the cannabinoids have a carboxylic acid group attached to them. In the case of THC, this is designated as THC-A. For THC-A to become psychoactive, it must be decarboxylated to lose the A and become THC, and this only occurs when a high heat source has been added to the buds after they've been dried out. The plant must be smoked, cooked, or vaped (hence, the high heat source) for the euphoric process to work.[19]

Eating raw leaves is becoming increasingly popular because studies have shown the cannabinoid acids found in them can have the potential to produce health benefits including "anti-inflammatory activity."[20]

There's also a general consensus among nutritionists that eating dark green leafy vegetables such as kale and spinach have health benefits because they are good sources of vitamins A, C, E and K.[21] However, there aren't an abundance of scientific studies behind juicing raw dark green cannabis leaves to gain access to these vitamins.

Most nutritional information for ingesting raw hemp is found on a variety of pro-cannabis websites, such as CannabisInternational.org, and the majority of the information is anecdotal. For example, Dr. William Courtney's documentary on YouTube entitled *LEAF: The Health Benefits of Juicing Raw Cannabis* presents the results of patients who have used raw, nonpsychoactive cannabis to treat a number of conditions including lupus, rheumatoid arthritis, interstitial cystitis, hypoglycemia, anemia, chronic sinusitis, and chronic bacterial infections.[22] According to Dr. Courtney, the health benefits from the cannabis plant are drastically increased when it is juiced from the raw plant. He feels the carboxylic acid present in the raw form is a stronger medicinal dose than the one patients receive after the carboxylic acid is removed.

DOES THE CHLOROPHYLL IN HEMP HOLD NUTRITIONAL VALUE?

Chlorophyll content is another reason why people are ingesting the leafy greens on cannabis plants. While again there aren't a plethora of scientific studies to dive into on the topic of chlorophyll's nutritional benefits, proponents say it helps detoxify the body, and in doing so can do just about anything from boosting metabolism to fighting bad breath, herpes, and cancer.

One such research study that backs up the chlorophyll trend was published in 2014, and focused on weight loss over a three-month period. The Swedish researchers gave overweight women a smoothie with five grams of green-plant membranes (green-plant membranes are filled with chlorophyll as well as other nutrients) over a period of 90 days. The women experienced "significant weight reduction," a reduction in their total cholesterol levels "within 3 weeks of treatment," and a decreased urge for sweet and fatty foods.[23] The results were significant enough to conclude that "chlorophyll-containing parts of green plants . . . may thus be a new agent for control of appetite and body weight."

Although these studies do not extract chlorophyll from hemp plants, chlorophyll is chlorophyll whether it comes from seaweed, spinach, kale, or cannabis. Raw hemp protein powder is green, so it's presumably rich in chlorophyll, and organic health food markets are currently incorporating actual hemp greens as part of the leafy vegetable rotation.

For instance, in Eaton, New York, there are restaurants carrying "Organic Baby Hemp Greens" grown by local JD Farms. The hemp greens are mixed in with salads and pesto sauces or added to sandwiches instead of lettuce. There's even a photo on JD Farm's Facebook page of New York Assemblywoman Donna Lupardo holding a pot of raw hemp tea[24] to show her support of the movement to bring raw hemp to the masses, and this health trend is sure to expand, thanks to the 2018 Farm Bill.

WHERE DOES AMERICAN HEMP COME FROM?

States that don't participate in the industrial hemp farming pilot program still allow hemp products to be sold in stores, so most Americans can benefit from the plant's nutritional value. Edible products such as hemp protein powder and hulled hempseed can also be purchased online from numerous distributors including Amazon. However, most of these hemp products are imported from countries like Canada because, similarly to marijuana, hemp grown legally in a particular state usually remained within that state's borders (that is, until the 2018 Farm Bill passed). In most states that adopted the federal terms in the 2014 Farm Bill for their hemp pilot programs, farmers were prohibited from selling seeds to each other; the seeds were supposed to be imported and inspected before they could be planted.

In September 2018, a hemp farm in West Virginia challenged these rules. The owners of CAMO Hemp WV and Grassy Run Farms purchased hempseed in Kentucky and brought it over state lines to West Virginia to grow. Since West Virginia's hemp pilot program stated farmers can only obtain hempseed internationally, the companies were in violation of the rules, and the federal government is now suing them.

On the farmers' application to grow hemp (which was accepted by the West Virginia pilot program), they did state they'd purchase the seed internationally, and now they're being subjected to a fine of at least $250,000 in civil penalties. According to the lawsuit, the owners also "intend to harvest the hemp plants . . . and ship the top portions of those plants, which includes the seeds, to Pennsylvania"; this is also in violation of the policies put in place in West Virginia's pilot program because "the shipment of any part of a cannabis plant, including the seeds of a cannabis plant, across state lines is a violation of the CSA [Controlled Substances Act]."[25]

Seems a bit ridiculous that an American hemp farmer can't sell hemp to another American farmer, but this was the way the West Virginia hemp pilot program was structured: once a grower is approved by a state's hemp pilot program, the state's agricultural department works with the DEA to arrange for the importation of industrial hempseed. This is supposed to ensure the hemp is being purchased from a reputable, certified hemp distributor, but it also makes the seeds more expensive due to import taxes, and the product can be damaged in the transportation process. After the DEA approves the import permit and allows the seed into the United States, the state's agricultural department arranges for the seed to be shipped to its office so that it can then be transported to the farmer.

However, West Virginia governor Jim Justice signed a bill earlier in 2018 to allow hemp farmers to buy certified seeds within state lines, and the 2018 Farm Bill has granted interstate commerce to hemp, so it's possible that by the time the lawsuit against CAMO Hemp WV and Grassy Run Farms actually goes to court the charges will be outdated.

Regardless, this is a lot of regulation for a superfood, especially since the FDA requirement for multivitamins and supplements to be proven safe or effective leaves much to be desired.

HYPOCRISY OF THE MULTIVITAMIN INDUSTRY

For those who take a daily multivitamin or vitamin supplement to get the maximum amount of nutrients possible, studies have shown there isn't much (if any) health benefit, and there isn't a reason to continue to take them. In 2018, researchers at St.

Michael's Hospital and the University at Toronto noted that multivitamins, vitamin D, calcium and vitamin C supplements do not give any major advantage "in the prevention of cardiovascular disease, heart attack, stroke, or premature death."[26] And since 1999, the National Institute of Health (NIH) has spent $2.4 billion studying the effects of vitamins and supplements to report that neither has any real benefit.[27]

The NIH states that multivitamins "cannot take the place of eating a variety of foods that are important to a healthy diet," and evidence that multivitamins can provide health benefits for the general population "remains limited."[28]

The only exceptions seem to be for folks with a severe deficiency—which in most cases a change in diet can easily fix—and prenatal vitamins for pregnant women. There are conflicting studies here as well when it comes to exactly what prenatal vitamins accomplish. One study from Israel states taking prenatal vitamins (folic acid and multivitamin supplements) before and during pregnancy decreases the chances of having a child with autism,[29] and another study from the Norwegian Institute of Public Health found the same results,[30] but a collaborative study from the University of California, the University of Pittsburgh, the Aarhus University in Denmark, and the Odense University Hospital in Denmark found the exact opposite: folic acid and multivitamins taken before and during pregnancy have no bearing on reducing the risk of autism.[31] Another study conducted by researchers in England found that no multivitamin is needed—just folic acid and vitamin D—and the American Congress of Obstetricians and Gynecologists agree that a well-rounded diet will supply a pregnant woman all the vitamins and minerals necessary,[32] yet most doctors tell pregnant women to continue to take them.

Regardless of this research, many Americans take multivitamins. According to the NIH, more than one-third of all Americans take multivitamins,[33] and it's estimated that more than 50 percent of Americans 50 years of age and older take them,[34] even though there is "no standard or regulatory definition" for multivitamins—"such as what nutrients it must contain and at what levels."

Talk about the power of perception over the facts of science!

People *believe* that hemp is the same as marijuana and a harmful drug because the federal government has stated this. People *believe* vitamins and supplements will help them live longer, healthier lives because why else would they be on the shelves in every pharmacy?

Buying vitamins supports the growth, health, and well-being of the dietary supplements industry—an industry that pulled in an estimated $36.7 billion in 2014.[35] The annual sales for multivitamin supplements alone are over $11 billion.[36] While the FDA has established a "current Good Manufacturing Practice" for dietary supplement manufacturers, it's up to the manufacturers themselves to evaluate their products and test their purity, strength, and so on.[37] Unlike prescription drugs and over-the-counter medications, the FDA does not have to approve or even inspect dietary supplements, including vitamins. (This is because food, beverages, and dietary supplements do not require FDA approval.)

Imagine what could actually be in vitamin supplements when the entire industry has no third-party oversight. Think about what's been found in pet food—which also doesn't have to be approved by the FDA. The FDA states pet food is only required to be "pure and wholesome, safe to eat, produced under sanitary conditions, contain no harmful substances, and be truthfully labeled."[38]

Are all vitamin manufacturers going to be ethical?

In October 2018, a dietary supplement industry analysis was conducted by California public health scientists and published in *JAMA*. The report found that 776 dietary supplements are contaminated with unapproved ingredients, and in some cases even contain ingredients that are banned by the FDA.[39] This has been one of the largest and most comprehensive industry studies to date. The most common harmful ingredients included:

- Sildenafil (Viagra's active ingredient) found in sexual-enhancement supplements,
- Sibutramine (an appetite suppressant that's banned in many countries) found in weight-loss supplements,
- Synthetic steroids or steroid-like ingredients found in muscle-building supplements,
- And overall, 157 products (20.2 percent of the products tested) contained two or more unapproved ingredients.

While the FDA does test dietary supplements on an ongoing basis, the agency doesn't have the authority to test the efficacy or safety of any food product before it's sold to consumers. Only after the product is on shelves and being sold through infomercials can it be tested (when the FDA gets around to it).

The FDA does have a list of the tainted products it has tested on its website. Last checked, there were 923 products on the list, which includes the date tested, name of product, manufacturer, the harmful hidden ingredient, and what the supplement is used for.[40] The majority of these tainted dietary supplements fall into the sexual-enhancement, weight-loss, or muscle-building categories.

However, vitamins can also contain serious contaminants, usually in the form of bacteria. In 2017, the FDA put out a press release advising consumers and healthcare professionals not to use products made by PharmaTech LLC, including liquid vitamin D drops and liquid multivitamins marketed for infants and children, because they contained a bacteria (*B. cepacia*) that leads to serious respiratory infections that are resistant to common antibiotics.[41] The press release was to warn consumers because it's up to the manufacturer to take the product off the shelf, and many times the company isn't diligent in doing so. For example, in 2018, Kellogg's Honey Smacks breakfast cereal was recalled in thirty-three states due to salmonella (which was severe enough to land thirty people in the hospital), and the FDA again had to send out additional warnings to consumers because the product was still being sold in stores after the recall.[42]

More recent vitamin studies have shown is that if a person regularly consumes *more* vitamins than what is necessary, the vitamins can cause the body harm, even permanent damage. A 2018 study published in *The Journal of Nutrition* found that most non-prescription prenatal vitamins contained about 26 percent more iodine than expected, and the average adult multivitamin contained 40.5 percent more vitamin D than what was listed on the label[43]—which is concerning because vitamin D can damage the kidneys and cause disorientation and problems with heart rhythm if taken in excess.

According to the NIH, "vitamin D toxicity almost always occurs from overuse of supplements."[44] This is especially concerning because researchers found children's multivitamins contained more iodine, vitamin D, and vitamin E than listed. Another study published in the *Drug Testing and Analysis* journal found that eleven weight-loss supplements contained a synthetic amphetamine (speed) that has never been tested for safety in humans and can obviously put patients with heart conditions at risk.[45]

When companies like Monsanto state there is no harm from consuming products or handling products with glyphosate, let's make a simple comparison: too much of anything is not good, not even vitamins. If too many vitamins can cause kidney damage, then why wouldn't too much glyphosate cause problems as well? Is Monsanto seriously asking Americans to believe glyphosate is *safer* to consume than vitamins?

When it comes to educating the public about vitamins, researchers have deduced that the culture surrounding their benefits is the most difficult challenge:[46]

> *In the United States, unlike the case for drugs, human research is not required to prove that supplements are safe or effective. Only if the Food and Drug Administration (FDA) finds that supplements are unsafe, can they stop the*

> *distribution of the products. . . . Avoiding tasty, but unhealthy food, may be difficult, but taking a pill once a day is relatively easy. . . . Therefore, Americans who have been using MVMs [multivitamins] since the early 1940s, will most likely continue to use them in the foreseeable future, and the rest of the world will follow.*

That's right—those precious gummy multivitamins for kids that take up a whole aisle in the pharmacy don't necessarily *do anything.* Might as well give kids a lucky penny or four-leaf clover every day to ensure they grow strong and healthy. Plus, the sugar content in the gummy vitamins has been linked to cavities and tooth decay problems among children.[47]

Full disclosure: I'm just as guilty of buying into the vitamin hype as the average American, so please don't think I'm being preachy here. My mother was a nurse, my grandmother was a nurse, and they took vitamins, and made sure I did too. Of course I took prenatal vitamins when I was pregnant, and I currently have a big ol' bottle of Disney Princess multivitamins on my kitchen counter for my four-year-old daughter. (Well, I can't throw them away now. That would be *wasteful.* Wouldn't it?)

So what's the answer? Mom and Dad always said if you didn't get enough nutrients today, take your vitamin. Okay, then we'll stock up on these all-natural, organic products with heart-healthy oats and fiber instead to make sure our diets are more balanced. *Oh wait, these also have hidden ingredients such as glyphosate.*

The answer is simple: There is strong academic research proving hempseed products will give our bodies an exceptional amount of nutrients, especially if taken every day. If we just add a few tablespoons of hemp to our diets, we won't have to change the culture of multivitamins, we'll just be correcting it.

Hemp is a superfood, so make a hemp protein smoothie for breakfast, or add hempseed to your salad, or cook with hempseed oil or hemp butter if you want to add actual nutritional value to your daily meals.

Scientists theorize that reason we get more nutrients from actual fruits and vegetables but a pill with those exact ingredients doesn't work is because the food is doing something in our bodies that can't be replicated. Or, in other words, synthetic vitamins are no match for nature's nutrients. Taking that logic a step further and applying it to the pharmaceutical industry, perhaps that's also why plant-based CBD medications are better than pharmaceutical pills—pills that can have dangerous side effects, even if they're taken as prescribed.

NOTES

1. Trevor Reid, "State legislature approves study for hemp use in livestock feed," *The Tribune,* March 26, 2017, https://www.greeleytribune.com/news/local/state-legislature-approves-study-for-hemp-use-in-livestock-feed/.
2. N. Gakhar, E. Goldberg, M. Jing, R. Gibson, J. D. House; "Effect of feeding hemp seed and hemp seed oil on laying hen performance and egg yolk fatty acid content: Evidence of their safety and efficacy for laying hen diets," *Poultry Science*, Volume 91, Issue 3, March 1, 2012, pp 701–711, https://doi.org/10.3382/ps.2011-01825.
3. Halle, I. and Schöne, F., "Influence of rapeseed cake, linseed cake and hemp seed cake on laying performance of hens and fatty acid composition of egg yolk," *J. Verbr Lebensm*, 8: 185, May 22, 2013, https://doi.org/10.1007/s00003-013-0822-3.
4. Habeanu, M., Gheorghe, A., Surdu, I., "N-3 Purfa-Enriched Hemp Seed Diet Modifies Beneficially Sow Milk Composition and Piglets' Performances," *Scientific Papers Series Management, Economic Engineering in Agriculture and Rural Development,* Vol. 18, Issue 1, 2018, pp 181–190, http://managementjournal.usamv.ro/pdf/vol.18_1/Art23.pdf.
5. Konca, Y., Yuksel, T., "Effect of dietary supplementation of hemp seed (Cannabis sativa L.) on meat quality and egg fatty acid composition of Japanese quail (Coturnix coturnix japonica)," *Animal Physiology and Animal Nutrition,* Vol. 102, Issue 1, Feb 2018, https://onlinelibrary.wiley.com/doi/abs/10.1111/jpn.12670.
6. D Mierliță, "Effects of diets containing hemp seeds or hemp cake on fatty acid composition and oxidative stability of sheep milk," *South African Journal of Animal Science,* Vol. 48:3, 2018, https://www.ajol.info/index.php/sajas/article/view/172390.
7. Jacques Mourot, Mathieu Guillevic, "Effect of introducing hemp oil into feed on the nutritional quality of pig meat," *OCL,* Vo. 22:6, Oct 2, 2015, https://www.ocl-journal.org/fr/articles/ocl/full_html/2015/06/ocl150012/ocl150012.html.
8. Ed Cara, "How Does a Euthanasia Drug Keep Ending Up in Dog Food?" Animals, *Gizmodo,* Feb 21, 2018, https://gizmodo.com/how-does-a-euthanasia-drug-keep-ending-up-in-dog-food-1823186412.
9. Amy Martyn, "The long, bizarre history of euthanasia drugs in pet food," *Consumer Affairs,* Feb 16, 2018, https://www.consumeraffairs.com/news/the-long-bizarre-history-of-euthanasia-drugs-in-pet-food-021618.html.
10. Pet Food Project Summary, *Clean Label Project,* Sept 2017, https://www.cleanlabelproject.org/pet-food.

11. Robert Arnason, "Hemp Takes Off In S. Korea," *The Western Producer,* May 12, 2016, https://www.producer.com/2016/05/hemp-takes-off-in-s-korea/.
12. "Omega-6 fatty acids," Penn State Hershey, *Milton S. Hershey Medical Center,* August, 5, 2015, https://www.umms.org/ummc/health/medical/altmed/supplement/gammalinolenic-acid.
13. Daly J. M., Reynolds J., Signal R. K., et al, "Effect of dietary protein and amino acids on immune function," *Crit Car Med*, PMID: 2105184, Feb 18, 1990, pp 86-93, https://www.ncbi.nlm.nih.gov/pubmed/2105184.
14. Jeong M, Cho J., Shin J., et al, "Hempseed oil induces reactive oxygen species- and C/EBP homologous protein-mediated apoptosis in MH7A human rheumatoid arthritis fibroblast-like synovial cells," *Journal of Ethnopharmacology,* Vol 154: 3, July 3, 2014, pp 745-752, https://doi.org/10.1016/j.jep.2014.04.052.
15. "Hemp Protein," Examine.com, published on 22 July 2013, last updated on 14 June 2018, https://examine.com/supplements/hemp-protein/.
16. Rodriguez-Leyva, Delfin and Grant N Pierce, "The cardiac and haemostatic effects of dietary hempseed," *Nutrition & metabolism* vol. 7:32, April 21, 2010, https://www.ncbi.nlm.nih.gov/pmc/articles/PMC2868018/.
17. "Hemp Seeds Nutrition Facts & Calories," Nutritional Data, *Conde Nast,* 2018, https://nutritiondata.self.com/facts/custom/1352377/1.
18. Mark Leland, "FOX 11 Investigates pros and cons of legalizing marijuana," Fox 11 Investigates, *Fox 11 News,* July 16, 2018, https://fox11online.com/news/fox-11-investigates/fox11-investigates-pros-and-cons-of-legalizing-marijuana.
19. "Activation and Metabolism of Cannabinoids," Pharmacology, *Prof of Pot,* Aug 2, 2016, http://profofpot.com/activation-metabolism-cannabinoids/.
20. Burstein, Sumner H., "The cannabinoid acids, analogs and endogenous counterparts," *Bioorganic & medicinal chemistry* vol. 22,10, PMID: 24731541, May 15, 2015, pp 2830-43, https://www.ncbi.nlm.nih.gov/pmc/articles/PMC4351512/.
21. Lin Yan, "Dark Green Leafy Vegetables," News 2013, USDA Agricultural Research Service, last modified Aug 13, 2016, https://www.ars.usda.gov/plains-area/gfnd/gfhnrc/docs/news-2013/dark-green-leafy-vegetables/.
22. K. Astre, "Raw Cannabis Juice is the Next Superfood," Edibles, *Cannabis Now,* Sept 1, 2015, https://cannabisnow.com/raw-cannabis-juice-is-the-next-superfood/.
23. Montelius C., Erlandsson D., Egzona V., Stenblom E., "Body weight loss, reduced urge for palatable food and increased release of GLP-1 through daily supplementation with green-plant membranes for three months in overweight women," *Appetite,* Vol 81, Oct 1, 2014, pp 295-304, https://www.sciencedirect.com/science/article/pii/S0195666314003493.
24. JD Farms Facebook Post, Dec 20, 2018, https://www.facebook.com/jdfarmsusa/.

25. Jake Zuckerman, "US attorney files civil suit against WV hemp farm," *Charleston Gazette-Mail,* Sept 22, 2018, https://www.wvgazettemail.com/news/politics/us-attorney-files-civil-suit-against-wv-hemp-farm/article_9a6ff103-7f6c-5f80-a028-92877ae1502f.html.
26. St. Michael's Hospital, "Most popular vitamin and mineral supplements provide no health benefit, study finds," *ScienceDaily*, May 28, 2018, www.sciencedaily.com/releases/2018/05/180528171511.htm.
27. Liz Szabo, "Older Americans are 'Hooked' on Vitamins," *The New York Times,* April 3, 2018, https://www.nytimes.com/2018/04/03/well/older-americans-vitamins-dietary-supplements.html.
28. "Vitamins and Minerals," National Center for Complementary and Integrative Health, National Institute of Health, *US Department of Health and Human Services,* last modified Feb 9, 2018, https://nccih.nih.gov/health/vitamins.
29. Levine S. Z., Kodesh A, Viktorin A, et al., "Association of Maternal Use of Folic Acid and Multivitamin Supplements in the Periods Before and During Pregnancy With the Risk of Autism Spectrum Disorder in Offspring," *JAMA Psychiatry,* Vol. 75(2), Feb 1, 2018, pp. 176–184. https://www.ncbi.nlm.nih.gov/pubmed/29299606.
30. Surén P, Roth C, Bresnahan M, et al., "Association Between Maternal Use of Folic Acid Supplements and Risk of Autism Spectrum Disorders in Children," *JAMA,* Vol. 309(6), Feb 13, 2013, pp. 570–577, https://www.ncbi.nlm.nih.gov/pubmed/23403681.
31. Virk, J., Liew, Z., Olsen, J., Nohr, E. A., Catov, J. M., & Ritz, B., "Preconceptional and prenatal supplementary folic acid and multivitamin intake and autism spectrum disorders," *Autism*, Vol. 20(6), Aug 2016, pp 710–718, https://doi.org/10.1177/1362361315604076.
32. Agata Blaszczak-Boxe, "Prenatal Multivitamins Don't Help Much, Study Says," Health, *Live Science,* July 11, 2016, https://www.livescience.com/55363-prenatal-multivitamins-dont-help-much.html.
33. Office of Dietary Supplements, "Multivitamin/mineral supplements," National Institutes of Health, *US Department of Health & Human Services,* last modified July 8, 2015, https://ods.od.nih.gov/factsheets/MVMS-HealthProfessional/.
34. Kamangar, Farin and Ashkan Emadi, "Vitamin and mineral supplements: do we really need them?" *International journal of preventive medicine*, Vol. 3,3, March 2012, 221–226, https://www.ncbi.nlm.nih.gov/pmc/articles/PMC3309636/.
35. Office of Dietary Supplements, "Multivitamin/mineral supplements," National Institutes of Health, *US Department of Health & Human Services,* last modified July 8, 2015, https://ods.od.nih.gov/factsheets/MVMS-HealthProfessional/.

36. Kamangar, Farin and Ashkan Emadi, "Vitamin and mineral supplements: do we really need them?," *International journal of preventive medicine*, Vol. 3,3, March 2012, 221–226, https://www.ncbi.nlm.nih.gov/pmc/articles/PMC3309636/.
37. FDA, "Fortify Your Knowledge About Vitamins," *US Department of Health and Human Services*, last modified Sept 26, 2018, https://www.fda.gov/ForConsumers/ConsumerUpdates/ucm118079.htm.
38. FDA, "Information on Marketing a Pet Food Product," Animal & Veterinary, *US Department of Health and Human Services,* last modified Oct 19, 2017, https://www.fda.gov/animalveterinary/resourcesforyou/ucm047107.htm.
39. Tucker J, Fischer T, Upjohn L, Mazzera D, Kumar M., "Unapproved Pharmaceutical Ingredients Included in Dietary Supplements Associated With US Food and Drug Administration Warnings," *JAMA Netw Open,* Vol. 1(6), Oct 12, 2018, e183337, https://jamanetwork.com/journals/jamanetworkopen/fullarticle/2706496?widget=personalizedcontent&previousarticle=2706489.
40. FDA, "Tainted Products Marketed as Dietary Supplements_CDER," *US Department of Health and Human Services,* last modified Oct 16, 2018, https://www.accessdata.fda.gov/scripts/sda/sdNavigation.cfm?filter=&sortColumn=1d&sd=tainted_supplements_cder&displayAll=true.
41. FDA News Release, "FDA warns of potential contamination in multiple brands of drugs, dietary supplements," News & Events, *US Department of Health and Human Services,* Aug 11, 2017, https://www.fda.gov/newsevents/newsroom/pressannouncements/ucm571328.htm.
42. WFLA, "FDA: Kellogg's Honey Smacks cereal still being sold after recall," US & World, *NBC 4 WCMH-TV,* July 12, 2018, https://www.nbc4i.com/news/u-s-world/fda-kellogg-s-honey-smacks-cereal-still-being-sold-after-recall/1299305399.
43. Karen W Andrews, Pavel A Gusev, Malikah McNeal, et al., "Dietary Supplement Ingredient Database (DSID) and the Application of Analytically Based Estimates of Ingredient Amount to Intake Calculations," *The Journal of Nutrition*, Vol. 148:2, August 1, 2018, pp 1413S–1421S, https://doi.org/10.1093/jn/nxy092.
44. NIH Office of Dietary Supplements, "Vitamin D: Fact Sheet for Consumers," National Institutes of Health, *US Department of Health and Human Services*, last modified April 15, 2016 https://ods.od.nih.gov/factsheets/VitaminD-Consumer/.
45. Cohen P., Bloszies C., Yee C., Gerona R., "An amphetamine isomer whose efficacy and safety in humans has never been studied, β-methylphenylethylamine (BMPEA), is found in multiple dietary supplements," *Drug Testing and*

Analysis, Vol 8, Issue 3-4, March-April 2016, pp 328-333, https://onlinelibrary.wiley.com/doi/full/10.1002/dta.1793.

46. Kamangar, Farin and Ashkan Emadi, "Vitamin and mineral supplements: do we really need them?," *International journal of preventive medicine*, Vol. 3,3, March 2012, 221–226, https://www.ncbi.nlm.nih.gov/pmc/articles/PMC3309636/.
47. Harvey K., Li E., Stanton R., Dashper S., "Kids' vitamin gummies: unhealthy, poorly regulated and exploitative," Health + Medicine, *The Conversation*, May 28, 2017, https://theconversation.com/kids-vitamin-gummies-unhealthy-poorly-regulated-and-exploitative-76466.

7

HEMP-CBD: A SUPER MEDICATION

Hempseed is a superfood because it contains a rare combination of healthy nutrients, but can cannabidiol (CBD) be considered a super medication? Without a doubt. Yes. But it's taken the medical community as a whole a long time to come around to accepting this fact.

Prior to the 2018 Farm Bill, there were three classifications of CBD medication in the United States:

- CBD derived from CBD-rich marijuana (Schedule I narcotic)
- CBD derived from CBD-rich hemp (Schedule I narcotic)
- Federally legal CBD with THC content below .1 percent (Schedule 5 narcotic), which is only currently found in FDA-approved Epidiolex (a pharmaceutical drug that treats two rare and severe forms of epilepsy, Lennox-Gastaut Syndrome (LGS) and Dravet Syndrome)

Thanks to the 2018 Farm Bill, CBD derived from hemp is legal, as the federal government now defines it as part of the hemp plant, and all parts of *Cannabis sativa* L have been taken off the Controlled Substances Act. The plant and all its derivatives are now listed as agricultural commodities—*not a drug*.

When scientists conduct CBD-related research studies, they're either working with CBD in its isolated form or full-spectrum CBD.

When CBD is referred to as "full-spectrum" or "whole-plant CBD" it means that the CBD contains all other cannabinoids found in the cannabis plant, such as:

- CBN—cannabinol
- CBG—cannabigerol
- THCV—tetrayhdrocannabivarian
- THC—tetrahydrocannabinol (the psychoactive element)

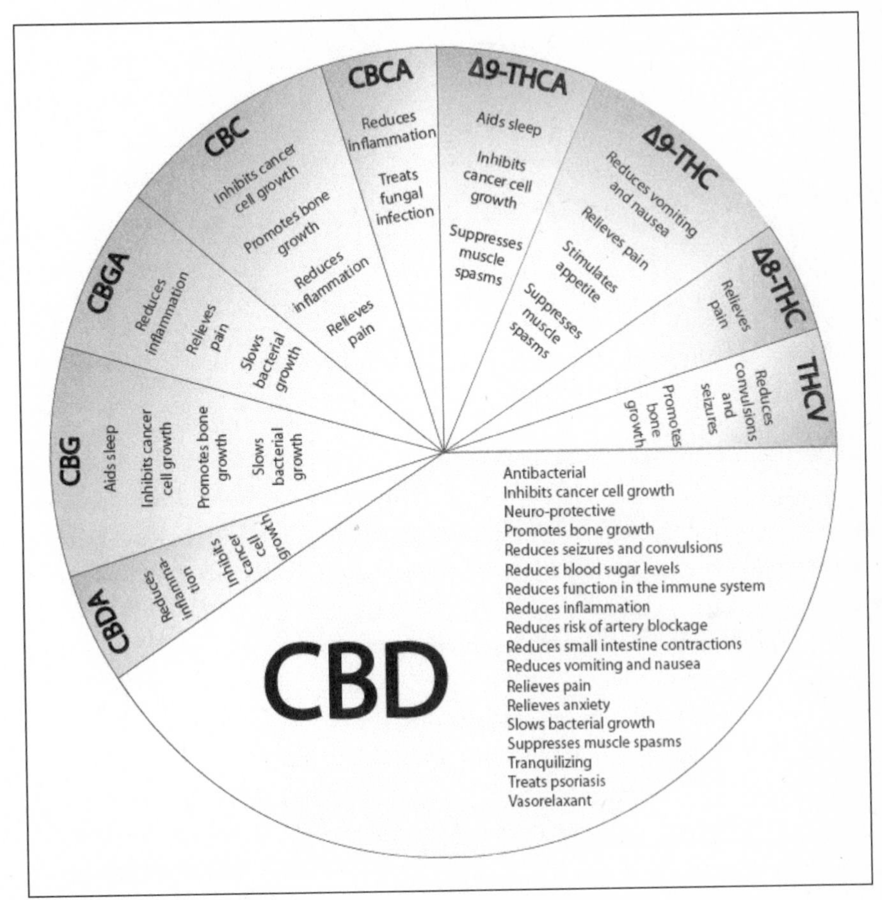

Understanding the different benefits of cannabinoids. *Source: medium.com, 2017 (https://medium.com/cbd-origin/cbd-isolate-vs-full-spectrum-cbd-b78a6eab319c)*

Researchers know that isolated CBD contains nearly all of the benefits of each cannabinoid combined. This led many to believe that CBD in its isolated form is more potent and concentrated when compared to full-spectrum CBD. However, in 2014, researchers at the Hebrew University of Jerusalem discovered that isolated CBD is not as effective as whole-plant CBD when administered at the same dose for relieving inflammation. Mice were given both the whole-plant CBD and the isolated CBD to find that the whole-plant CBD produced better anti-inflammatory results. When the dosage of the isolated CBD was increased, it provided the same results.[1]

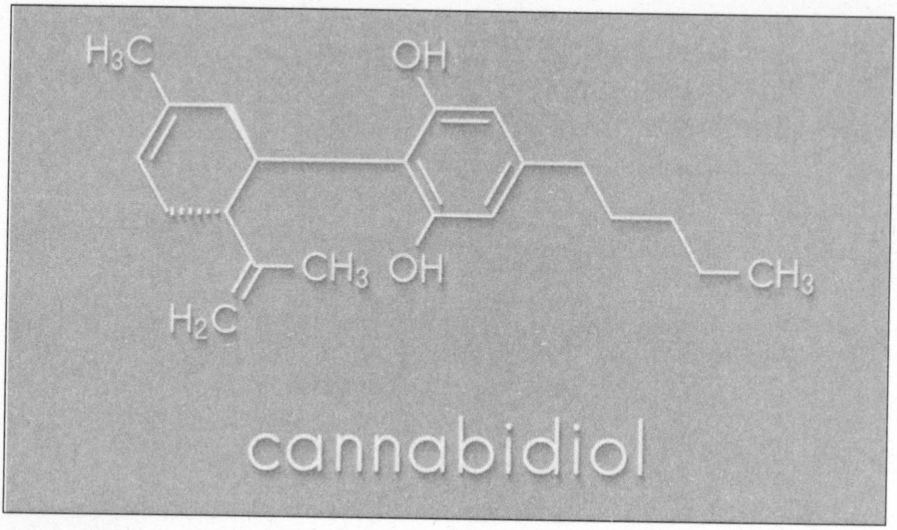

The structure of the cannabidiol compound. Its molecular formula is C21H30O2. *Source: (https://pubchem.ncbi.nlm.nih.gov/compound/cannabidiol#section=Top)*

The full-spectrum or whole-plant CBD extract is now known as the Entourage Effect. This phrase was coined by cannabis researchers to best describe the way cannabis compounds work in the human body.[2] Essentially, CBD by itself isn't as effective as when a combination of all chemical compounds are able to work together.

For CBD to be the most effective, it needs an entourage!

While research has shown both purified/isolated CBD and whole-plant CBD extract do the same thing, it takes lower dosages of whole-plant CBD to achieve results. A 2018 Brazilian study sums this up perfectly:[3]

> *CBD-rich extracts seem to present a better therapeutic profile than purified CBD, at least in this population of patients with refractory epilepsy. The roots of this difference is likely due to synergistic effects of CBD with other phyto-compounds (aka Entourage effect), but this remains to be confirmed in controlled clinical studies.*

This is also the basis of GW Pharma's success with Epidiolex and Sativex; their research determined that CBD needed the entourage effect of the other compounds, such as THC.[4]

Papers have been published in numerous scientific publications that indicate CBD can help treat fifty medical conditions including:[5]

Acne ADD/ADHD Addiction ALS Alzheimer's Disease Anorexia Anxiety Atherosclerosis Arthritis Asthma Autism Bipolar Affective Disorder Cancer Crohn's and Colitis Depression Endocrine Disorders Epilepsy and Seizures Fibromyalgia Glaucoma Heart Disease	Huntington's Disease Inflammation Irritable Bowel Syndrome Kidney Disease Liver Disease Metabolic Syndrome Migraines Mood Disorder Motion Sickness Multiple Sclerosis Nausea Neurodegeneration Neuropathic Pain Obesity OCD Parkinson's Disease Prion-MCD PTSD Rheumatism Schizophrenia	Sickle Cell Anemia Skin Conditions Sleep Disorders Spinal Cord Injury Stress Stroke/TBI Thyroid And many more

CBD has such a wide range of uses because it stimulates the endocannabinoid system (ECS), and in doing so, it helps to promote homeostasis in the body. While the circulatory system, digestive system, nervous system, and so on have a distinct role, the ECS is involved in nearly all bodily systems, particularly our cognitive functioning. From our mood, to our response to stress, to our sense of pain, to our metabolism, to our immune system . . . even our short-term and long-term memory are affected by the ECS.

When cannabinoids such as CBD bond with ECS receptors, they modulate the activity of the endocannabinoid system to balance deficiencies and/or adjust excessive activity. Studies have found that modulating the ECS holds "therapeutic promise" in a range of diseases and conditions: "from mood and anxiety disorders, movement disorders such as Parkinson's and Huntington's disease, neuropathic pain, multiple sclerosis and spinal cord injury, to cancer, atherosclerosis, myocardial infarction, stroke, hypertension, glaucoma, obesity/metabolic syndrome, and osteoporosis, to name just a few."[6] As scientists continue to explore the ECS, and as research into the possible uses for CBD (and other cannabis compounds) continues to grow, the list of cannabinoid benefits is likely to continue to grow as well, especially when current treatments don't cure or fully address the symptoms of debilitating conditions.

WHAT'S THE DIFFERENCE BETWEEN MARIJUANA-CBD AND HEMP-CBD?

Marijuana typically contains between 5 percent to 10 percent THC. CBD with this amount of THC is typically used for pain relief in patients with cancer and AIDS, and has been known to alleviate the negative side effects of chemotherapy (such as decreased appetite).

Marijuana-derived CBD is not an agricultural commodity. Since marijuana is still listed as a Schedule I narcotic under the Controlled Substances Act, CBD sourced from marijuana is still federally illegal.

CBD sourced from hemp has .3 percent or less THC—nowhere near enough to get a person high. The typical amount of CBD content in CBD-rich hemp plants is about 10 percent.[7] CBD hemp is typically associated with the ability to fight cancerous tumors, reducing nausea and inflammation, and combat depression, diabetes, and MS. This form of CBD is now federally legal, per the passing of the 2018 Farm Bill.

FEDERALLY LEGAL CBD

Prior to the passing of the 2018 Farm Bill, as of September 2018, there was only one form of federally legal CBD in the United States—and that's the CBD found in the prescription drug Epidiolex, which contains .1 percent THC.

The DEA classified the drug as a Schedule V controlled substance, which is the lowest level of drug classification on the Controlled Substances Act. Schedule V drugs have a proven medical use and a low potential for abuse; other drugs in this category include cough medications containing codeine.

Epidiolex is the first FDA-approved drug that contains CBD, but just because it has CBD in it, that doesn't mean the DEA was willing to reclassify CBD as a whole. This is the same process that occurred with Marinol (THC in a pill). When Marinol was rescheduled, many advocates assumed the DEA's next move was going to reclassify THC and marijuana as well, but that obviously never happened. The same holds true for Epidiolex. Here's the breakdown with federal law as it pertains to CBD:

- The DEA hasn't changed its stance on CBD or any raw extract from marijuana—it's all still a Schedule I narcotic.

- Just like when Marinol was approved by the FDA, an exception was made for Epiodiolex and only Epidiolex because the word EPIDIOLEX is listed in the Schedule V classification.

- The classification also states that *if* the FDA approves any other drugs with the same compound—CBD + .1 percent THC—then those drugs will also become Schedule V.

- This allows other pharmaceutical companies to create similar drugs for other uses aside from seizure disorders, since CBD + .1 percent THC can be used to treat other conditions.

- This also implies that when GW Pharmaceuticals' patent runs out on Epidiolex, other drug manufacturers can easily create generics and have them approved in the Schedule V classification (which is what happened with generic Marinol).

- As far as the DEA is concerned, the only legal substances derived from cannabis (aside from Epidiolex) are the ones that are produced from industrial hemp because they were legalized in the 2018 Farm Bill. This includes hemp-derived CBD.

- Up until the 2018 Farm Bill was signed by President Trump, CBD oil that was *not* in Epidiolex, whether derived from hemp or marijuana made no difference to the DEA—both were illegal because CBD oil contains cannabinoids, and cannabinoids are federally illegal. Therefore, as far as the DEA is concerned, it doesn't matter if CBD was extracted from hemp or marijuana or if the CBD is isolated so that there isn't any THC in it. The only federally legal, FDA-approved form of CBD is in Epidiolex.

When Epidiolex was approved by the FDA, the DEA put other stipulations on CBD products:[8]

- Epidiolex is to be sold through pharmacies and obtained with a doctor's prescription, meaning Epidiolex (and future drugs similar to it) are covered under health insurance.

- If other CBD products are approved by the FDA as medications, then the same rules will apply.

- This means marijuana dispensaries are not going to be able to sell Epidiolex or any other FDA-approved medication with CBD in it.

- Moving forward, a DEA permit is required to import or export CBD products.

- This is unfortunate because we all know the DEA isn't going to grant access to CBD products if a dispensary is trying to obtain them from Canada or Europe, given the fact that CBD is an illegal substance under the CSA.

- However, in September 2018, the DEA allowed capsules containing CBD and THC to be imported from Canada for clinical trials in a study on Essential Tremor (ET), a nervous system disorder. Since the NIDA's drug supply program at the University of Mississippi doesn't offer capsules (and this is where the researchers would've had to get them from to be FDA compliant), the DEA made an exception. Researchers spent "about two years going through regulatory hoops" to import the capsules from British Columbia-based Telray Inc.[9] The University of California's Center for Medical Cannabis Research is currently conducting the study.

STATE LAWS ON CBD CONFLICT WITH THE DEA

Granted, just because the DEA says CBD is illegal, that doesn't grant the agency the ability to ban CBD from the United States. If a state has already set up its own definition of CBD, or if a state decides to do so now, then state law can nullify federal law. We know this to be true because of medical marijuana legislation. Medical and recreational marijuana use is illegal under federal law, yet we have plenty of states that have defied this without repercussions, due to the concept of states' rights.

Congress also passed the Rohrabacher-Farr Amendment, which protects states' rights and cannabis businesses. According to the amendment (which must be renewed every year by Congress in order to remain in effect), the federal government—including the DOJ and DEA—can't go into a state where marijuana is legal and arrest everyone using and selling it. What this implies is that all CBD products derived from marijuana can still be sold within a state if that state's marijuana laws determine that source of CBD is legal. However, now that the 2018 Farm Bill has passed, hemp-CBD products will have to abide by FDA guidelines; a state like California that has already banned hemp CBD can still make the decision on whether it wants hemp products (and what kinds of hemp products) within its borders.

For instance, in July 2018, the California Department of Public Health ruled that hemp-derived CBD is not allowed in food or drinks for humans or pets in California.[10] (Food includes CBD gum, by the way.)

As Mitch McConnell explained: "My Hemp Farming Act as included in the Farm Bill will not only legalize domestic hemp, but it will also allow state departments of agriculture to be responsible for its oversight."[11]

THE EMERGING CBD INDUSTRY

Immediately following Donald Trump's signing of the 2018 Farm Bill, the FDA put out a press release indicating that it will continue to prohibit companies from adding CBD and THC to food, drinks, and supplements and from making any therapeutic claims about their products.[12] Then in late December 2018, FDA Commissioner Scott Gottlieb stated that although the FDA still prohibits the sale of food, supplements, and other products containing CBD across state lines, the agency is looking into drawing up new regulatory framework. As much as the FDA would prefer to continue to ban CBD products and ban CBD as an ingredient, the agency really can't ban hemp-CBD now that it's a legal substance.

According to Hemp Business Journal, CBD sales reached $820 million in the United States in 2017 and they are expected to grow to $2.1 billion by 2020.[12] Many CBD products (including CBD oil) are already marketed as dietary supplements, plus there's CBD lotions, bath salts, skin and hair-care products, beverages, and other food items.

In September 2018 (prior to the 2018 Farm Bill), Ohio's medical marijuana measure went into effect and it tackled the CBD issue in a unique way. The law made it illegal to sell any cannabis products—including CBD oil—unless the products were sold in a licensed medical marijuana dispensary, which means the products must list all ingredients and be tested to prove the ingredients are accurate.[13] Although some health food shops and smoke shops in Ohio have been selling CBD oil and CBD products for seven years or more,[14] as of September 2018, they were no longer able to sell them if they weren't registered with the state as a licensed medical marijuana dispensary.

I agree with testing the products and ensuring they are accurate. However, the problem Ohio's medical marijuana program faced was efficiency. The program was supposed to start on September 8, 2018, but none of the fifty-six approved dispensaries that spanned the state actually opened.[15] The businesses that had been selling CBD products were being told to stop, so customers had no way to purchase their CBD. Out of desperation, where could customers go to purchase CBD in the meantime? Most likely to the internet, where the FDA claims the authenticity of many CBD products remains uncertain.

Furthermore, if someone in Ohio had been using CBD prior to the medical marijuana measure, the only way to legally purchase it moving forward was to join the state's medical marijuana program and receive a medical marijuana patent ID card. There are obviously fees associated with this (in Ohio, the cost is $60),[16] and only certain medical conditions are covered under the state's medical marijuana program. I do see the value in organizing and regulating the medical marijuana program; however, restricting someone's access to a substance that doesn't have harmful side effects, especially now that the substance is being regulated for quality control, doesn't make much sense.

The main benefit and argument for organizing medical marijuana and CBD products lies in the power of the state-issued medical marijuana card. If a state-issued medical marijuana card is proof positive that a patient has legal access to marijuana as a medication, then that person should be able to use the medication in a state where marijuana isn't legal. And if you're a medical marijuana patient in Ohio, you should be able to purchase your medication in another state that has a medical marijuana program. This commonsense approach is called medical cannabis reciprocity. Several states including Michigan, Nevada, Rhode Island, and New Hampshire allow out-of-state visitors to purchase their medication with their out-of-state medical marijuana ID.[17]

Due to this distinction between marijuana-derived CBD and hemp-derived CBD, some businesses in Ohio are defying the medical marijuana law because they feel the Ohio Medical Marijuana Control Program's definition of CBD is inaccurate.

The current law states: "CBD extracted from hemp is not lawful. . . . Ohio law does not make a distinction between CBD extracted from hemp and CBD extracted from marijuana."[18]

This is because (1) Ohio does not participate in the industrial hemp pilot program, and (2) the state of Ohio's Board of Pharmacy concludes if there are cannabinoids in a product, then that product automatically falls under the medical marijuana program.

Meanwhile, the Ohio retailers selling CBD claim the products are derived from hemp, not marijuana, and hemp products now fall under the 2018 Farm Bill instead of Ohio's new medical marijuana laws.[19] This is where state and federal laws become confusing. Should industrial hemp products—everything from CBD-infused lotions to CBD oils—be subject to Ohio's marijuana legislation? Should they be subject to FDA approval instead? Should they be protected under the 2018 Farm Bill? In terms of hemp and marijuana legislation, states' rights have prevailed over federal law, so in the case of hemp-CBD, its legality is in the hands of each individual state, at least for the time being.

DOES THE ORIGIN OF CBD REALLY MATTER?

As thrilled as I am that hemp-CBD is officially legal due to the 2018 Farm Bill, it's a bit strange that our country has found a way to outlaw CBD and legalize CBD.

CBD is the same compound, whether it is in marijuana or hemp. THC is the same compound, whether it is in marijuana or hemp. The same goes for every other chemical compound—from chlorophyll to vitamin A, it's the same compound. Some plants have more vitamin A than others, just like some cannabis plants have more CBD than others, but vitamin A is vitamin A, regardless if it comes from broccoli or peas.

Of course, as far as federal law is concerned, the origin of CBD does matter from a legal perspective, but I really don't think it matters from a chemical compound perspective.

I haven't read every single CBD-related research study that's been published, but many do not specify if the CBD came from a hemp plant or a marijuana plant. They often list the breakdown of CBD and THC percentages from a particular strain when that's crucial to the research, but then again, when researchers rattle off a wide range of studies that show the promise of CBD's medicinal use, they don't differentiate if those medicinal uses are specific to CBD hemp or CBD marijuana. The reasoning behind this is because CBD contains the same properties, whether it is extracted from hemp or marijuana.

Yes, not all marijuana plants are bred to have the same THC/CBD content and not all hemp plants have the same CBD/THC concentrations. However, CBD and THC can be isolated from either plant and used in a lab at whatever concentration researchers prefer. For better or for worse, when the research community references CBD, they're referring to it in much the same way the DEA prefers to reference it: regardless of its origin, it's CBD.

Here's an example of what I'm talking about. Dr. Nora D. Volkow, director of the National Institute on Drug Abuse (a division of the NIH), defines CBD in the following way:[20]

> *CBD does not produce euphoria or intoxication. Cannabinoids have their effect mainly by interacting with specific receptors on cells in the brain and body: the CB1 receptor, found on neurons and glial cells in various parts of the brain, and the CB2 receptor, found mainly in the body's immune system. . . . There is also growing evidence that CBD acts on other brain signaling systems, and that these actions may be important contributors to its therapeutic effects. . . pre-clinical research (including both cell culture and animal models) has shown CBD to have a range of effects that may be therapeutically useful,*

> *including anti-seizure, antioxidant, neuroprotective, anti-inflammatory, analgesic, anti-tumor, anti-psychotic, and anti-anxiety properties.*

She doesn't differentiate here between marijuana-derived CBD and hemp-derived CBD because regardless of *where* the CBD comes from, it interacts in the human body the same exact way *and* she is stating that CBD itself—regardless of its origin—does not contain THC. She's talking about the compound here, and the compound is the same, regardless what plant it originated from.

Of course America's CBD laws have contradicted science for years. According to the DEA, Schedule I drugs—which CBD has been lumped into—are defined as having "no currently accepted medical use and a high potential for abuse."[21] This completely contradicts Volkow's medical assessment, and she goes as far as to state: "A review of 25 studies on the safety and efficacy of CBD did not identify significant side effects across a wide range of dosages, including acute and chronic dose regimens, using various modes of administration."[22]

So to recap:

1. Medical research that the federal government considers to be sound has shown there aren't any adverse side effects of CBD.
2. The federal government classifies CBD as a drug with a high potential for abuse, which means it's an addictive substance, but will now make an exception for hemp-CBD.
3. That makes no sense. When something has *no side effects* that means it is *not addictive,* so therefore, *all CBD* has medical value.

THE FIRST CBD PATENT IN THE UNITED STATES

In 1998, the U.S. Department of Health and Human Services (HHS) filed a patent on cannabinoids (which includes CBD). Patent 6,630,507 is titled "Cannabinoids as antioxidants and neuroprotectants" and it states that "cannabinoids have been found to have antioxidant properties," which can be useful in treating a variety of conditions including "oxidation associated diseases, such as ischemic, age-related, inflammatory and autoimmune diseases." The patent goes on to state cannabinoids are also useful in "limiting neurological damage following ischemic insults, such as stroke and trauma, or in the treatment of neurological diseases, such as Alzheimer's disease, Parkinson's disease and HIV dementia."[23] Again, the patent is referring to the compounds found in the cannabis plant—not in the marijuana plant, not in the hemp plant—in *all cannabis plants.* Therefore giving medical legitimacy to CBD found in *both* subspecies of cannabis, but again not differentiating between them.

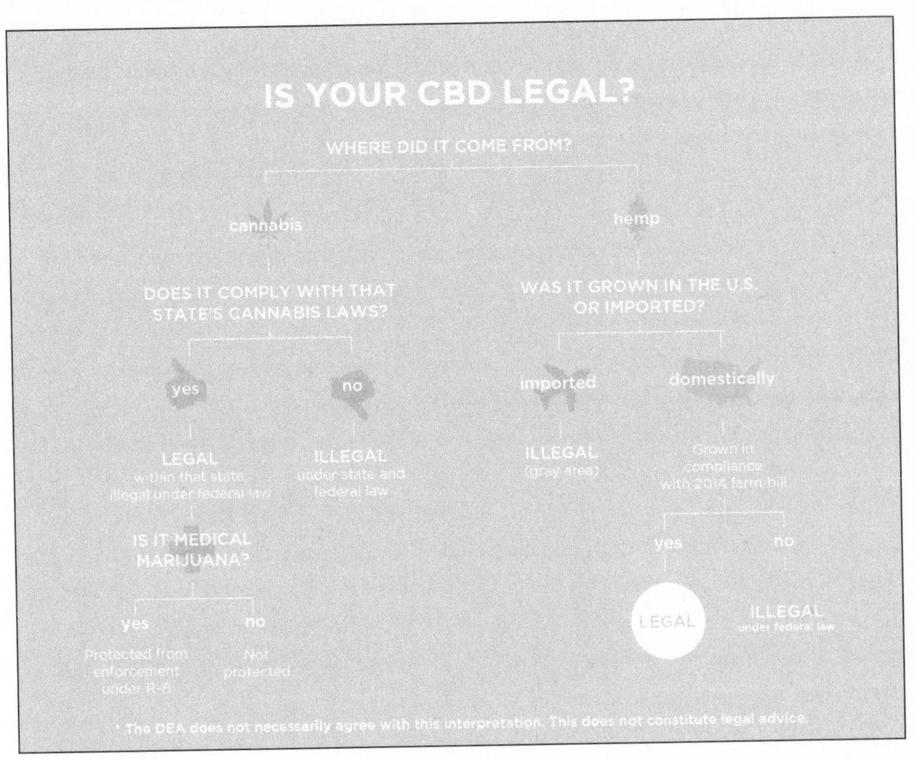

States that had legalized CBD prior to the 2018 Farm Bill. Note: "R-B" refers to the Rohrabacher-Blumenauer amendment, which prevents the DOJ from using its budget to interfere with state-legal medical cannabis. *Source: Elysse Feigenblatt/Leafly, 2018 (https://www.leafly.com/news/politics/is-cbd-oil-legal-now-with-epidiolex-approved-its-still-not-clear)*

This patent also speaks to the thought process behind eating raw cannabis as mentioned in chapter 6. If cannabinoids are in fact more potent when they're in their raw form (with the carboxylic acid attached to them), then those looking to use cannabis medicinally will be getting more benefits *without* any of the euphoric feelings. Therefore, the health benefits of juicing raw hemp or eating raw hemp protein could have its own entourage effect.

Just for clarification, the U.S. Department of Health and Human Services is listed as the *assignee* on the cannabinoid patent. This means the department has the right to license the patent to pharmaceutical companies, but they don't "own" the patent.

The Committee on Drug Addiction decided this was the best way to "avoid monopolizing drugs and purposefully increasing the price on essential pharmaceuticals,"[24] since typically a corporation must purchase the right to use intellectual property in a patent from the owners.

Since no one "owns" the rights to cannabinoids as an antioxidant or neuroprotectant, the theory here is that multiple companies will have the ability to request the use of the government's patent to develop new drugs instead of one company monopolizing all uses for CBD and all the other cannabinoids found in cannabis.

However, this doesn't stop a company such as GW Pharmaceuticals from patenting several uses for CBD after creating a product such as Epidiolex[25] (which, by the way, is of course patented). They also filed patents for CBD drugs to treat epilepsy, glioblastoma, schizophrenia, and spasticity due to multiple sclerosis.[26]

Although the 2018 Farm Bill legalizes hemp-CBD, it also leaves matters up to the Secretary of Agriculture to determine the path of the industrial hemp industry as a whole. As far as the FDA is concerned, the agency would prefer if CBD in general is subject to FDA approval (same as pharmaceutical drugs) rather than regulatory guidelines that apply to food, cosmetics, and dietary supplements.

THE 2015 SENATE DRUG CAUCUS HEARING ON CBD

On Wednesday, June 24, 2015, the Senate's Caucus on International Narcotics Control held a Drug Caucus Hearing to discuss the barriers to cannabidiol research. This was about a year after the FDA approved Epidiolex for research studies. Since the discussion did not distinguish between marijuana-CBD and hemp-CBD, for the rest of this chapter, if I refer to "CBD," I'm referring to the medical uses for CBD that is identical in both plants.

At the hearing, Dr. Nora Volkow presented her CBD facts in her testimony and detailed scientific studies that show CBD has:[27]

- Anti-seizure effects
- Neuroprotective properties to help those with multiple sclerosis (MS), Parkinson's disease, Alzheimer's disease, stroke, and neurodegeneration caused by alcohol abuse
- Analgesic effects (anti-inflammatory and pain reduction) for cancer pain, neuropathic pain (such as pain from MS), and rheumatoid arthritis.
- Antitumor effects—CBD either reduced cancer cells, increased cancer cell death, decreased tumor growth, or stopped cancer from spreading to other parts of the body
- Antipsychotic effects—some studies have shown the effects of schizophrenia and Parkinson's can be mediated by CBD
- Anti-anxiety effects—to lower social anxiety and PTSD management

- Efficacy for treating substance abuse—CBD reduces dependence on morphine and heroin in animal trials

She also referenced the need for more CBD research to develop:

- Treatment for substance abuse, such as opioids, methamphetamine, and alcohol
- Treatment for neuropathic pain due to spinal cord injury
- Schizophrenia and epilepsy treatments (epilepsy treatments such as Epidiolex, perhaps?)

The Senate also heard from Joseph Rannazzisi, the Deputy Assistant Administrator of Drug Diversion with the DEA. He thoroughly explained that the Controlled Substances Act:[28]

> *defined "marihuana" as all parts of the plant Cannabis sativa L., with certain exceptions for the parts of the plant that are not the source of cannabinoids. Among the parts of the cannabis plant included in the definition of marijuana are: the flowering tops, the leaves, viable seeds, and the resin extracted from any part of the plant, and every compound, manufacture, salt, derivative, mixture, or preparation of the plant, its seeds or resin. . . . CBD derived from the cannabis plant is controlled under Schedule I of the CSA because it is a naturally occurring constituent of marijuana. While there is ongoing research into potential medical uses of CBD, at this time CBD has no currently accepted medical use in the United States.*

Mr. Rannazzisi goes on to state that the DEA "supports research involving CBD and its potential capacity to treat multiple conditions," and the agency supports the FDA's decision to fast-track Epidiolex studies in the United States. Phenomenal answer. Except the DEA stalled for *seven years* on Dr. Sue Sisley's stateside study to see if smokable, plant-based cannabis containing different percentages of THC and CBD will help veterans with PTSD, so perhaps the DEA's definition of "support" should be evaluated. In his closing remarks, Rannazzisi states the DEA "will continue to make the review and approval of Schedule I researchers a top priority."

FDA's deputy director for regulatory programs at the Center for Drug Evaluation and Research spoke next, and it's obvious after reading his statement that the FDA would greatly prefer if CBD was only legal for medical use in pharmaceutical products. From the FDA's standpoint, the agency decides if a pharmaceutical product is safe for the public and has a three-phase drug trial for this purpose. If CBD is an ingredient in a cosmetic item or food item (including vitamins or dietary supplements), then it doesn't require the same robust FDA approval before it goes to market.

In his testimony, Dr. Douglas C. Throckmorton states the "FDA continues to believe that the drug approval process represents the best way to help ensure that safe and effective new medicines, including any such medicines derived from cannabidiol or other constituents of marijuana." He isn't enthused with all-natural, organic CBD products because "botanicals include herbal products made from leaves, roots, stems, seeds, pollen, or any other part of a plant," and they "pose challenges" including potential impurities consistency, plus it is "difficult to determine if the product is causing the change in a patient's condition, or the change is related to some other factor."[29]

Throckmorton is describing the issues he is currently facing with vitamins and dietary supplements—products that line the shelves of retail stores across America—which are rife with impurities and consistency concerns.

Granted, I get it is a "challenge" to determine an all-natural medication is consistent in potency and quality, but states like Oregon have already established those very means for cannabis, including tests that exceed the FDA's quality-control measures for food products, so if the FDA is concerned about how to regulate hemp-CBD, perhaps just adapt Oregon's regulations, which have already been effective in discovering synthetic pesticide chemicals in products marketed for organic farming.

Or perhaps look to Canada. I'm not saying any system is perfect, but they seem to have this cannabis thing under control.

Dr. Throckmorton continues his testimony by describing the results of FDA testing on CBD products:[30]

> *In late February 2015, FDA tested those products and, in some, did not detect any cannabidiol. It is important to note that these products are not approved by FDA for the diagnosis, cure, mitigation, treatment, or prevention of any disease. Often they do not even contain the ingredients found on the label. . . . These products and marketing can create false hope in a population especially vulnerable: those seeking relief from serious medical conditions for themselves or their loved ones, including their children. Moreover, it might divert patients from products with demonstrated safety and effectiveness.*

But, of course there wasn't any quality control. Of course there were snake-oil salesmen on the internet preying on people's desperation. This is what happens when there are no industry-wide standards or guidelines in place on a product. And for those products that already have FDA guidelines in place, we *still* have this issue. The meat of euthanized animals is in pet food—along with seriously toxic substances like lead and cadmium—and even breakfast cereals have had to be recalled due to harmful bacteria.

Pediatric neurologist Dr. John "Brad" Ingram from the University of Mississippi Medical Center was the next to give testimony at the 2015 Senate Drug Caucus Hearing, and he backed up Dr. Throckman's concerns about CBD by stating "popular culture, media reporting, and encouraging anecdotal reports suggest that CBD works well to treat epilepsy" but there isn't enough significant data in science to prove this.[31] He felt that if cannabis were changed from Schedule I to a lesser classification, it could be more readily accessible for researchers, and more research could be done to determine the true benefits of CBD.

While I agree that there haven't been a profound number of studies in the United States on CBD, there are plenty of studies from other countries to prove its value and significance as a medication. Quite frankly, we do know it treats epilepsy. There's no question of that, since CBD is a main ingredient in FDA-approved Epidiolex, which treats rare forms of epilepsy, but let's take Israel, for example. As the United States' ally in the Middle East, we give Israel $3.7 billion per year (or roughly $10.1 million per day) in economic and military aid.[32] Israel, Canada, and the Netherlands are the only three nations in the world with government-sponsored medical cannabis programs.

The Israeli Ministry of Health includes a Medical Cannabis Unit, and eight private companies are permitted by the government to produce and distribute plant-based medical marijuana to 25,000 licensed medical marijuana patients.[33]

In Israel, medical cannabis can be prescribed for many different conditions including cancer, chronic pain, PTSD, arthritis, pediatric epilepsy, Parkinson's disease, and Crohn's disease. Patients receive their medicine—which includes pre-rolled marijuana joints, baked goods, balms, liquid drops, and cannabis buds—from state-run hospitals and private clinics. They can also have it delivered to their homes.[34]

The cannabis isn't prescribed in pill form or in a nasal spray or any of the typical Big Pharma applications. All of the medication is considered high-quality medical grade because the plants are grown in a carefully monitored, controlled, strict environment to ensure the quality and consistency of the product. Tikun Olam Ltd., one of the first and largest medical cannabis supplier in Israel, claims to treat over 1,000 children suffering from cancer, epilepsy, and autism.[35] The company was established in 2006, and is vertically integrated—meaning they cultivate the plants, conduct the research, and administer the products to patients. So again, there isn't necessarily a need to worry about inconsistencies with botanical medications or the ingredients of CBD oil if our government stepped up and provided the regulatory platform such as Israel's—especially since that platform is proving to be effective.

In October 2018, Bazelet—one of Israel's largest medical cannabis companies—signed a manufacturing and licensing agreement with Israeli-based Alvit LCS Pharma, a medical cannabis pharmaceutical product developer.[36] The goal here is to manufacture, market, and sell Israeli medical cannabis by converting it to pharmaceutical products such as oral sprays and extended-release tablets so that they can be distributed in Israel and the EU, and thus have a longer shelf life than a brownie. Since other countries probably won't agree to import pot plants from Israel, this is the easiest way for Bazelet to expand abroad. Although Bazelet has only been in business since 2014, the company claims to already meets the needs of 9,000 patients per month and is scaling up to reach 100,000 patients.[37] Their current products include extended-release cannabis capsules, an asthma inhaler, prerolled cannabis cigarettes, ground cannabis, buds, and cannabis oil. Interestingly, the extended-release capsules are patented by Cannabics, an American company based in Israel, which is researching cannabinoid-based therapies for cancer.[38]

Since our federal government has been funding cannabis research in Israel since the 1960s, why not reap the benefits here in the United States?

According to its website, the National Institutes of Health is "the largest public funder of biomedical research in the world, investing more than $32 billion a year."[39] From 1964 to 2010, one particular Israeli-based cannabis researcher named Dr. Raphael Mechoulam received $100,000 per year in grant money from the NIH.[40] Dr. Mechoulam was the first scientist to isolate and synthesize THC, he was the first to decode the structure of CBD, and he was the first to test and develop the medicinal properties of THC. He also determined how cannabis interacts with the human body—through what is now known as the endocannabinoid. He also discovered that the human brain produces its own cannabinoids, which stimulate the body's receptor system.[41]

In 2006, the NIH published a paper called "The Endocannabinoid System as an Emerging Target of Pharmacotherapy."[42] The paper compiles much of Dr. Mechoulam's research (including how cannabis's cannabinoids interact with the endocannabinoid system), plus other leading studies that show how cannabinoid compounds including CBD have the ability to alleviate illnesses such as schizophrenia, diabetes, cancer, multiple sclerosis and spinal cord injuries, obesity, pain and inflammation, nervous system disorders such as strokes, Parkinson's disease, Alzheimer's disease, and epilepsy. Since the NIH funded some of these studies under research grants, doesn't that make the value of CBD more than "anecdotal"?

Or is it so easily dismissed because the research wasn't conducted in the United States via FDA-approved studies? Even though our tax dollars funded Mechoulam's research, British GW Pharmaceuticals (the makers of Epidiolex) sought him out and acquired the rights to a number of cannabinoid patents he invented—and he's invented over twenty-one patents, including several synthetic cannabinoids.[43] When it comes to CBD, some of the more recent research conducted by Israeli scientists includes:

- Marijuana strains with high levels of CBD and low levels of THC resulted in "reduction in seizure frequency"[44] among children with epilepsy.

- At the University of Guelph, rats were injected with cocaine until addicted, and then they were injected with CBD. After the CBD injections, they no longer showed a preference for cocaine,[45] thereby showing CBD in large dosages may actually be a key ingredient in curing addiction.

- Cardiologists at Hebrew University discovered that mice that were given CBD treatment at the time of a heart attack had reduced the amount of dead tissue (infarct size) by 66 percent.[46]

- Another test conducted on mice showed that CBD can block the development of diabetes. The mice were given a course of CBD injections at 14 weeks of life, then tested at 24 weeks, and only 30 percent of the mice still had diabetes.[47]

The list of CBD-related studies in Israel (and around the world) goes on and on, and stateside studies are under way, all echoing similar results. Plus, there are studies that focus specifically on hemp-CBD.

Dr. Stephanie McGrath, a neurologist at Colorado State University's James L. Voss Veterinary Teaching Hospital, found that full-spectrum hemp-CBD extract can help epileptic dogs suffering from seizures.[48] In her research, she used CBD oil made by a Colorado company called Applied Basic Science (their products are already available to the public). In her initial study of thirty dogs, an epileptic dog named Ferguson had two to three seizures a day, but after taking the CBD oil for about two weeks, the seizures were down to two to three per week.[49] She's currently conducting a similar trial with sixty dogs to determine the proper dosage for pets.

Researchers at Sullivan University College of Pharmacy in Kentucky are focusing their cancer studies on KY-hemp, a hemp strain produced in Kentucky. KY-hemp significantly slowed the growth of ovarian cancer cells and had the potential to stop or significantly slow down cancer spreading to other parts of the body.

According to the study, KY-hemp's anti-cancer effects proved to be "comparable to or even better than the current ovarian cancer drug Cisplatin."[50] Which is great news, considering Cisplatin is essentially toxic and has many side effects including dehydration, loss in ability to taste food, and temporary hair loss.[51]

When Dr. John "Brad" Ingram stated we need more CBD research, what I acknowledge in his defense is that the conclusions of many of these studies recommend *more* studies take place to not only replicate the results, but to replicate them among larger populations with more variables. The subjects of CBD studies start with mice, rats, human tissue samples; they don't start with humans. We're still tinkering with the percentages too—what percentage of CBD to THC works best for each condition? We know CBD *works,* but we don't know the exact magic ratio that will work on *most* people that share a common ailment, and even if that ratio is proven effective for *most people*, that doesn't mean it will cure all people, so we have to figure out how to scale the percentages correctly for larger dosages.

However, that's also true of any pharmaceutical product. While there are no adverse side effects of CBD, there are side effects of Epidiolex, including drug interactions, weight loss, rash, insomnia, and suicidal behaviors.[52]

An FDA-approved drug for children two years of age and older can produce suicidal behaviors, but CBD in its pure form has no adverse side effects.

Dr. Thomas Minahan, an emergency-room physician based in California, also gave testimony at the 2015 Senate Caucus on International Narcotics Control Drug Caucus Hearing. He agreed with Dr. John "Brad" Ingram about the need to reclassify cannabis as a whole to make it easier for researchers in the United States to conduct more studies.

To back up his points, Minahan gave largely anecdotal information regarding his own personal experiences with CBD. His daughter was diagnosed with seizures at fourteen months old, and he reluctantly turned to CBD oil as his last option, only to find out it worked better as a medication than any of the other drugs she tried previously. Here's an excerpt from Dr. Minahan's testimony:[53]

> *At work I consistently see the harmful effects of drug abuse, the results of which typically fills up my trauma room on Saturday nights. Today I address you not as a doctor, but as a dad. . . . My twelve-year-old daughter Mallory*

had her first seizure when she was fourteen months old. We immediately saw a pediatric neurologist and had all the necessary testing performed. Over the next ten years Mallory's seizures persisted, despite trying about a dozen medications. In desperation we tried many "alternative" therapies, all with no success. During one dark period we even entertained surgical resection of part of her brain.

In August of 2013, Mallory was experiencing thirty to forty seizures a month and as many as up to a dozen in a night. It was typical for us to put a diaper on our functional eleven-year-old daughter when going to bed. And every morning, we would go into her room and dreadfully put our hand on her chest to ensure she was still breathing.

We are fortunate that my daughter has had the ability to see the world's best physicians for the treatment of her epilepsy. But even with their expertise, no prescribed medications kept her seizures under control and allowed her to have any quality of life. Some medications had such toxic side effects that my wife and I often wondered if it was better for her to just try and live with constant seizures.

Our day of reckoning came when our world-renowned epilepsy specialist suggested we "try" a drug called Felbatol. This medication has the potential "black box" side effect of aplastic anemia, which essentially shuts down one's bone marrow production of blood. The incidence of death if someone is affected by this is 20–30 percent. As a physician faced with giving his daughter this potentially deadly medicine, I just couldn't bring myself to risk my daughter's life. It was around this time that a family member suggested we try an oil extracted from the marijuana plant that can be ingested orally. . . .

Now, you have to know that I am very conservative. My personal lifelong beliefs and my professional experience up to this point made me the biggest skeptic you could ever imagine. However, out of desperation, I started to investigate it. . . . Again and again I found reports that many kids had showed improvement after taking the oil, and that there was not a high risk of death with using marijuana. Importantly, this particular type of marijuana oil had a very low THC content, so there was no "high" associated with taking it. And it wasn't smoked. It was ingested.

Reluctantly, cautiously, and knowing that we really didn't have too many options left, we started my daughter on it. She took the oil every eight hours. To our surprise, almost immediately my daughter demonstrated amazing results.

Today, a year and a half after starting Mallory on cannabis oil, she has had a decrease in her seizure frequency by 90%. She's more alert than ever before, and she's back in school for the first time in three years. . . .

We truly feel that thanks to cannabis oil, we have our daughter back. I'm often asked what my ultimate dream is in regards to this subject. My answer is simple—rescheduling this drug to a lower category would allow greater research and development opportunities, which could ultimately result in kids like Mallory having safe and regulated access to this life-changing medicine.

In closing, I want to thank you for this opportunity to share our story. When I received the invitation to appear before this committee, I immediately thought how neat it would be to bring Mallory with me, introduce her to you and have her experience this once-in-a-lifetime opportunity. But then I realized that this would not be possible without breaking the law. You see, Mallory and many kids like her cannot legally travel or cross state lines because their medication is against the law on a federal level.

It took an educated leap of faith for me to get where I am today on this subject, and I urge you to keep an open mind as you look into this matter because our kids' lives are truly in your hands.

Yes, this testimony is largely if not entirely anecdotal, but it is coming from a doctor, and it does back up research studies around the world that state CBD can alleviate severe epilepsy and seizure disorders. There's no doubt that cannabis is disrupting our perception of modern medicine and disrupting the multibillion-dollar pharmaceutical industry as a whole, and that can be threatening to a lot of people.

We take pills, damn it. That's what we do. They have side effects and we live with them or we take another pill to alleviate them. That's. How. It's. Done.

Plant-based medication? That can't possibly be more effective. I don't care what research studies have shown. It came from the ground. The ground is dirty. Pills come from a lab. A sterile work environment that ensures every pill has the same exact dose. Need I say more?

The closing statements at the Senate Caucus on CBD were from Kevin A. Sabet, PhD. He's the president of Smart Approaches to Marijuana (SAM), the director of the University of Florida Drug Policy Institute, the author of *Reefer Sanity: Seven Great Myths about Marijuana*, and he served as adviser to three US presidents (Clinton, Bush, and Obama) at the White House Office of National Drug Control Policy.[54] He started off by saying that he "vigorously" opposes marijuana legalization, so perhaps placing his testimony as the closing remarks was strategic. I do agree with Sabet's summarization:[55]

Right now the current situation can be summed up this way: Most CBD manufacturers get away with selling whatever they say is CBD; researchers and other groups who want to follow the FDA/DEA rules are being stifled by bureaucracy; parents are left confused and frustrated; FDA-approved CBD products could very well be held up through a lengthy DEA scheduling process; and state elected officials with absolutely no background in these issues are hastily putting laws together in the absence of robust federal action. This must change.

He goes from here to outline a six-point plan for how he'd like to see CBD be "legalized" so that research can be expanded and CBD can be accessed for the seriously ill. While the 2018 Farm Bill has gone above and beyond these measures for hemp-CBD, his perceptions are worth noting:

1. Allow multiple licenses to grow marijuana for research purposes, aside from the farm at the University of Mississippi.

Though the University of Mississippi is now growing CBD-rich strains of marijuana (or what they consider CBD-rich strains), it remains the only facility where researchers can obtain cannabis for FDA-approved studies. Sabet reasonably suggests that other NIDA-approved sites should become available so that more strains of marijuana can be grown for different studies. Since states are already doing this themselves through their own state-run medical marijuana programs, there's clearly a need, so why not expand the program? Remember, an FDA-approved study is the starting point. There are typically three phases of testing before the FDA can approve a drug for consumers. This means that if KY-hemp is found to be better than any other means at killing ovarian cancer cells, it still will not be approved by the FDA as a prescription drug treatment option unless the strain comes straight from the University of Mississippi for FDA-approved drug trials. However, the 2018 Farm Bill *has already* changed this. Since hemp-CBD is now a federally legal substance, there is no reason to get it from the University of Mississippi, and since the university doesn't follow any particular guidelines for growing cannabis aside from simply planting it in the ground, watering it, and harvesting it once it is fully grown, then any research facility nationwide should be able to grow hemp-CBD for FDA-approved studies (or at least, this should be the logical route for the FDA to take). While this doesn't solve the problem for marijuana research, it's a step in the right direction.

2. Waive (or lessen) DEA registration requirements for handling CBD.

Sabet claims the main reason researchers want marijuana rescheduled is because they have difficulty working with the DEA due to "long delays between getting FDA approval for handling CBD and checking the boxes to fulfill DEA registration requirements." Since Epidiolex was fast-tracked by the FDA, there's no reason other research can't be fast-tracked, given there are rules already in place to allow this to occur. He adds that if the FDA approves a CBD drug, then that drug should be automatically moved out of a Schedule I classification, without the DEA reviewing the FDA's decision and adding more paperwork. This will allow a quicker turnaround for the drug to be prescribed by doctors and available at pharmacies. Which is great, but where does that leave marijuana dispensaries if the goal is to take medical marijuana and turn it into a substance that only doctors can prescribe to pharmacies?

3. Eliminate the Public Health Service (PHS) review for marijuana research applications.

In 1999, the Department of Health and Human Services (HHS) intended to streamline the review process for marijuana studies to ensure there would always be a sufficient amount of research-grade marijuana available. However, most researchers found the process delayed their research because these protocols added extra steps for no reason. The phrase "too many cooks in the kitchen" applies here, especially since marijuana-related research is the only category of research subjected to this process. Since the FDA and DEA already have enough parameters in place, Sabet suggests it's time to eliminate the PHS review process for marijuana research applications.

4. Establish compassionate CBD research programs for the seriously ill.

Sabet explains that in the 1980s, the National Cancer Institute worked with the DEA to allow over 20,000 patients access to Marinol (the first pill to contain pharmaceutical-grade THC). The Group C program gave Marinol to specific chemotherapy patients over a period of four years. This study concluded once Marinol was approved, and it played a major role in proving the value of the drug treatment. Sabet states that the DOJ/DEA could collaborate with the National Institute for Neurological Diseases and Stroke on a similar research project for CBD.

However, Sabet didn't mention the Investigational New Drug Program (Compassionate Access IND), which surprised me. This program is detailed on the FDA's website and is specifically for "investigational" drugs, or substances that aren't legal. The FDA-approved program allows a sponsor to determine "if a substance is reasonably safe for initial use in humans, and if the compound exhibits pharmacological activity that justifies commercial development."[56] Clinical trials in the IND are exempt from legal requirements, and the program can be used for "experimental drugs showing promise in clinical testing for serious or immediately life-threatening conditions."[57] In other words, the FDA will allow patients to be treated with a drug while the FDA is reviewing the drug and determining if it is going to be approved.

As of 2017, the FDA receives approximately 1,500 requests per year for the IND program and approves nearly 99 percent of the requests.[58] The program began in 1978 as a response to a medical need for marijuana. Technically, it was the direct result of the *Randall v. United States* lawsuit against the FDA, DEA, NIDA, DOJ, and the Department of Health, Education, and Welfare. Robert Randall had glaucoma, and was successful in arguing that he had a medical need for cannabis to treat his condition, so the government had to find a way to supply him the medication without making it legal for everyone else. The Compassionate Access IND was founded.

At the height of the program, fourteen people were legally supplied government-grown marijuana from the University of Mississippi. Even though the FDA stopped accepting new patients in 1992, there are still four people alive from the initial fourteen, and they are *still* receiving free marijuana cigarettes from the government for their conditions—which range from glaucoma to painful bone tumors. The four surviving patients receive a steel tin containing three hundred joints, each containing about 3 percent THC, on a monthly basis.[59]

One reason why Sabet might not have mentioned this program is because the medical-marijuana patients diligently completed biannual progress reports with their doctors (which was a stipulation for participation in the program), but the reports were never used to assess the impact of long-term marijuana use—even though some of the patients have now been receiving medical-grade cannabis for *forty years.*

In 2001, the four remaining patients agreed to undergo a series of exams led by Dr. Ethan Russo to try to determine the long-term effects of cannabis. The result was a research paper titled the "Chronic Cannabis Use Study," often referred to as the "Missoula Chronic Use Study."[60] Not to go into Dr. Russo's entire medical background, but he is a board-certified neurologist, psychopharmacology researcher, and currently the medical director of PHYTECS, "a biotechnology company researching and developing innovative approaches targeting the human endocannabinoid system."[61] Soon after completing the "Chronic Cannabis Use Study," he served as senior medical adviser and study physician to GW Pharmaceuticals from 2003 to 2014. He was also part of the clinical trials of Sativex and Epidiolex, so suffice it to say he was more than qualified to conduct this research.

What Russo discovered in 2001 was that the four surviving patients received poor-quality cannabis from the University of Mississippi—marijuana that was sometimes fourteen years old, containing seeds and stems. Regardless of the inferiority of the medication, the four patients were "able to reduce or eliminate other prescription medicines"; "no malignant deterioration" was observed; "no consistent or attributable neuropsychological or neurological deterioration" was observed; and "no endocrine, hematological or immunological sequelae" were observed.[62] To put it bluntly, there was no evidence of harm from extensive cannabis smoking (aside from minor bronchitis).

According to Russo, three of the four patients did try Marinol, but "found it inadequate or a poor substitute for cannabis in symptomatic relief of their clinical syndromes." Which makes sense to me because of what we now know about CBD.

First of all, Marinol is a pharmaceutical-grade THC, and the NIDA was shipping them cannabis cigarettes with varying amounts of THC, all *lower* than a 10mg Marinol pill. The joints contained 1.8 percent, 2.8 percent, 3 percent, or 3.4 percent THC.[63] The THC content in Marinol may have been too strong, plus the pill doesn't contain any CBD, and all of the patients were receiving regular doses of CBD every time they medicated by smoking a joint.

Unfortunately, due to when the study was conducted, Dr. Russo didn't focus on the CBD content of the marijuana cigarettes. THC was all the rage in the medical community at the time, so the report made no mention of CBD at all. Clearly we can assume the patients were receiving the medical benefits of CBD along with the THC, since the marijuana joints came directly from a marijuana plant. Since CBD isn't in Marinol, the drug was (and still is) a poor substitute for many conditions because it lacks the entourage effect of all the cannabinoids.

5. **Begin federal-state partnerships to allow a pure CBD product to be dispensed/explored by board-certified neurologists and/or epileptologists to appropriate patients under a research program.**

Since the Controlled Substances Act authorizes the DOJ/DEA to carry out educational and research programs, Sabet argues that if CBD that meets FDA quality standards is available, patients can choose to participate in research studies and therefore gain access to CBD. Again, all the FDA would really have to do is reopen the Compassionate Access IND for medical marijuana use (or specifically CBD), then actually follow through and *do something* with the research collected from the patients to show the true medical benefits. However, the 2018 Farm Bill can potentially make this a reality for hemp-CBD.

In May 2018, President Trump signed into law the "Right to Try" bill, which allows terminally ill patients to have access to experimental drugs that have not yet been approved by the FDA. When this bill was first discussed in the media, many jumped to conclusions and assumed the legislation would allow CBD and other cannabis medications, since surely they're considered "experimental" to many physicians.

However, in order for a drug to qualify for "Right to Try," it must already be in the FDA's wheelhouse, meaning the bill specifically pertains to drugs that have cleared the initial phase of FDA review (known as an FDA-approved Phase 1 clinical trial), but have not been fully cleared of the entire review process.[64] Therefore, the only way cannabis could potentially be eligible for "Right to Try" is if it is already in an FDA drug trial.

The federal government specifically states that the only way a terminally ill patient has the right to try cannabis medication is if it has already completed an FDA-approved Phase 1 clinical trial; and even then, once the drug enters Phase 2 or 3, it must be defined as a treatment for "an underlying terminal condition," *and only then,* under those circumstances, "it may qualify." This means Dr. Sue Sisley's FDA-approved cannabis/PTSD trial won't qualify under "Right to Try," since PTSD isn't a terminal condition. And overall, the answer is *maybe* because stating a drug "may" qualify is as good as saying *maybe.*

Maybe, someday, we'll allow terminally ill patients access to cannabis medication. *Maybe, someday,* we'll make their final days a little easier.

6. Shut down rogue "medical marijuana" companies that do not play by the rules.

Essentially, once the FDA decides what a pure-prescription-quality CBD product should contain, the FDA and DEA can enforce legal action on any companies that are selling inferior or false products.

While I agree with the majority of Sabet's points, I don't see CBD as a medication that only benefits the seriously ill:

- A 2011 study found people with generalized Social Anxiety Disorder (SAD) who received CBD prior to speaking in front of a large audience had significantly less anxiety than those who were given a placebo.[65]

- There are also several studies indicating that CBD can be a treatment for addictive behaviors—such as opioid, cocaine, tobacco, alcohol, and psycho-stimulant addiction—in animals and in humans. For instance, a 2015 study from researchers in Montreal, Canada concludes that "CBD is an exogenous cannabinoid that acts on several neurotransmission systems involved in addiction."[66]

- This backs up preliminary findings from a 2013 study conducted by the University College London's Clinical Psychopharmacology Unit. Researchers determined the endocannabinoid system plays a role in nicotine addiction. They gave twelve out of twenty-four tobacco smokers an inhaler of CBD (the other twelve received a placebo). In one week, those who used the CBD inhaler whenever they felt the urge to smoke "significantly reduced the number of cigarettes smoked by 40%" while the control group showed no difference in the number of cigarettes smoked.[67]

- In a 2018 study published in *Neuropsychopharmacology,* researchers determined that CBD could potentially prevent relapses in drug use. Rats that were addicted to cocaine or alcohol and showed signs of relapsing (meaning they would self-administer the drug if it was left in front of them and exhibited characteristics of a relapsing addict including dependency, anxiety, and impulsivity) received transdermal CBD (transdermal = they absorbed it through their skin) once every twenty-four hours for seven days. After seven days of treatment, the rats no longer displayed their addict-like characteristics *and* they did not relapse for five months.[68]

- Yasmin Hurd, a neurobiologist who studies addiction, has found that cannabidiol has decreased cravings and anxiety for opioid addiction in animals. As of February 2018, she's in Phase 2 of clinical trials in New York to test CBD's ability to reduce cravings in people addicted to heroin.[69]

According to the American Society of Addiction Medicine, "drug overdose is the leading cause of accidental death in the US, with 52,404 lethal drug overdoses in 2015. Opioid addiction is driving this epidemic, with 20,101 overdose deaths related to prescription pain relievers, and 12,990 overdose deaths related to heroin in 2015."[70] As of March 2018, the National Institute on Drug Abuse claims that "every day, more than 115 people in the United States die after overdosing on opioids."[71] While clearly those addicted to heroin and other substances have a serious illness—and those with SAD have a true debilitating condition—I doubt this is what Sabet was referring to when he said exceptions should be made for CBD access.

Regardless, as an expert on cannabis, Mr. Sabet is under the impression that if CBD and only CBD were legalized, there'd be complete and total chaos:[72]

Descheduling CBD . . . would simply encourage a "free-for-all" of concoctions and mixtures claiming to be "high-CBD" but with absolutely no regulation or oversight. This would result in hazardous conditions for parents and patients.

In the countries that have taken the time to safely and efficiently roll out a national cannabis program, there are regulations to stop this from occurring. In the states like Oregon that have already legalized cannabis, I've already detailed just a small fraction of the strict regulations for consumers, growers, and sellers, as well as examples of the consequences of defying the laws. So what Mr. Sabet is describing is a world where a product would be legal, but *no* guidelines, parameters, or restrictions would be placed on it. That statement doesn't even hold true for dietary supplements because there are FDA best practices and standards for the industry. So how could Mr. Sabet conclude that the CBD industry will go completely unchecked when this substance was previously defined as a Schedule I narcotic?

The cannabis industry that exists in the United States today is filled with regulations, so legalizing CBD won't ever mean that it's completely free of all restrictions, and the FDA has already alluded to the fact that the agency will continue to be strict on the CBD industry as a whole, despite the fact that hemp-CBD is now legal. So one thing is definitely certain: restrictions and regulations on top of restrictions and regulations will be coming down on hemp-CBD now that the 2018 Farm Bill has passed.

All of that aside, my main concern regarding the trustworthiness of domestically grown hemp-CBD and hemp foods wasn't even mentioned in the 2015 Senate Drug Caucus Hearing. These edible products are only as safe as the ground from which they came, making the health of American soil the biggest threat when consuming CBD.

NOTES

1. Gallily R., Yekhtin Z., Hanus L., "Overcoming the Bell-Shaped Dose-Response of Cannabidiol by Using Cannabis Extract Enriched in Cannabidiol," *Pharmacology & Pharmacy,* February 7, 2015, pp 75-85, http://file.scirp.org/pdf/PP_2015021016351567.pdf.
2. Piomelli, D., & Russo, E. B., "The *Cannabis sativa* Versus *Cannabis indica* Debate: An Interview with Ethan Russo, MD," *Cannabis and cannabinoid research*, Vol.1:1, pp. 44-46, https://www.ncbi.nlm.nih.gov/pmc/articles/PMC5576603/.
3. Pamplona F. A., da Silva L. R., Coan A. C., "Potential Clinical Benefits of CBD-Rich Cannabis Extracts Over Purified CBD in Treatment-Resistant Epilepsy: Observational Data Meta-analysis," *Front Neurol,* Vol 9:759, Sept 12, 2018, https://www.ncbi.nlm.nih.gov/m/pubmed/30258398/?i=6&from=hemp%20cbd.
4. GW Pharma CBD studies, accessed Jan 24, 2019, https://www.ncbi.nlm.nih.gov/m/pubmed/?term=GW+Pharma+Ltd.

5. "Research Shows CBD Benefits 50+ Conditions," *CBD Origin,* Aug 15, 2017, https://cbdorigin.com/cbd-benefits-many-conditions/ .
6. Pacher, Pál et al., "The endocannabinoid system as an emerging target of pharmacotherapy," *Pharmacological reviews,* vol. 58,3, Sept. 2006, 389-462, https://www.ncbi.nlm.nih.gov/pmc/articles/PMC2241751/.
7. Jeff Rice, "Hemp can be lucrative, but there are drawbacks," Local News, *Journal-Advocate*, Feb 27, 2018, http://www.journal-advocate.com/sterling-local_news/ci_31699721/hemp-can-be-lucrative-but-there-are-drawbacks.
8. DEA, "Schedules of Controlled Substances: Placement in Schedule V of Certain FDA-Approved Drugs Containing Cannabidiol; Corresponding Change to Permit Requirements," *Federal Register,* Sept 28, 2018,https://www.federalregister.gov/documents/2018/09/28/2018-21121/schedules-of-controlled-substances-placement-in-schedule-v-of-certain-fda-approved-drugs-containing.
9. Gene Johnson and Julie Watson, "Canadian marijuana imports OK'd by US for California study," *Associated Press,* Sept 18, 2018, https://www.apnews.com/837a1c64ad744ca2ae4a97131080fd54.
10. California Department of Public Health, "FAQ—Industrial Hemp and Cannabidiol (CBD) in Food Products," Food and Drug Branch, July 6, 2018, https://www.cdph.ca.gov/Programs/CEH/DFDCS/CDPH%20Document%20Library/FDB/FoodSafetyProgram/HEMP/Web%20template%20for%20FSS%20Rounded%20-%20Final.pdf.
11. "McConnell Signs Farm Bill Conference Report with Kentucky Hemp Pen," Press Releases, *Majority Leader Mitch McConnell US Senator for Kentucky,* Dec 11, 2018, https://www.mcconnell.senate.gov/public/index.cfm/pressreleases?ID=0F98C0E4-2E6B-4EED-AB14-60F754DA10E7.
12. FDA, "Statement from FDA Commissioner Scott Gottlieb, M.D., on signing of the Agriculture Improvement Act and the agency's regulation of products containing cannabis and cannabis-derived compounds," Press Announcements, *US Department of Health and Human Services,* Dec 20, 2018, https://www.fda.gov/NewsEvents/Newsroom/PressAnnouncements/ucm628988.htm.
13. Randy Tucker and Anne Saker, "CBD oil: Crackdown on Ohio sales hasn't stopped some retailers from defying state regulators," *Cincinnati Enquirer,* Sept 10, 2018, https://www.cincinnati.com/story/money/2018/09/10/some-retailers-defying-ban-cbd-oil-sales/1204989002/.
14. "CBD oil, sold for years locally, now illegal in Ohio," *WHIO 7,* Oct 1, 2018, https://www.whio.com/news/local/cbd-oil-sold-for-years-locally-now-illegal-ohio/gH1ab0bhv1xVi9LaiLaAMJ/.
15. Ibid.

16. "Frequently Asked Questions," Ohio Marijuana Card, *Ohio Medical Alliance LLC,* 2018, https://www.ohiomarijuanacard.com/frequently-asked-questions.
17. Abby Hutmacher, "Medical Cannabis Reciprocity: Who Accepts Out-of-State MMJ Cards?" *PotGuide.com,* Aug 10, 2018, https://potguide.com/pot-guide-marijuana-news/article/medical-cannabis-reciprocity-who-accepts-out-of-state-mmj-cards/.
18. "Ohio Medical Marijuana Control Program CBD Oil FAQ," *State of Ohio Board of Pharmacy*, accessed Jan 25, 2019, https://www.pharmacy.ohio.gov/Documents/Pubs/Special/MedicalMarijuanaControlProgram/CBD%20Oil%20FAQ.pdf.
19. Randy Tucker and Anne Saker, "CBD oil: Crackdown on Ohio sales hasn't stopped some retailers from defying state regulators," *Cincinnati Enquirer,* Sept 10, 2018, https://www.cincinnati.com/story/money/2018/09/10/some-retailers-defying-ban-cbd-oil-sales/1204989002/.
20. Nora D. Volkow, "Cannabidiol: Barriers to Research and Potential Medical Benefits," *National Institute on Drug Abuse,* June 24, 2015, https://www.drugabuse.gov/about-nida/legislative-activities/testimony-to-congress/2015/biology-potential-therapeutic-effects-cannabidiol.
21. DEA, "Drug Scheduling," United States Drug Enforcement Administration, *US Department Of Justice*, accessed Jan 24, 2019, https://www.dea.gov/drug-scheduling.
22. Nora D. Volkow, "Cannabidiol: Barriers to Research and Potential Medical Benefits," *National Institute on Drug Abuse,* June 24, 2015, https://www.drugabuse.gov/about-nida/legislative-activities/testimony-to-congress/2015/biology-potential-therapeutic-effects-cannabidiol
23. US Patent, "Cannabinoids as antioxidants and neuroprotectants," accessed Jan 24, 2019, https://patents.google.com/patent/US6630507B1/en.
24. "US Government Has Patent for Saliva," *Think Hempy Thoughts,* April 17, 2018, https://www.thinkhempythoughts.com/patent-6630507./
25. "GW Pharmaceuticals Announces Receipt of Notices of Allowance by the United States Patent and Trademark Office (USPTO) for Five New Epidiolex® (cannabidiol) Patents," Press releases, *GW Pharmaceuticals,* March 13, 2018, http://ir.gwpharm.com/news-releases/news-release-details/gw-pharmaceuticals-announces-receipt-notices-allowance-united. https://www.gwpharm.com/about/news/gw-pharmaceuticals-plc-announces-us-patent-allowance-use-cannabidivarin-cbdv-treating.
26. "Patents Assigned to GW Pharma Limited," *JUSTIA Patents,* accessed Jan 24, 2018, https://patents.justia.com/assignee/gw-pharma-limited.

27. Nora D. Volkow, "Cannabidiol: Barriers to Research and Potential Medical Benefits," *National Institute on Drug Abuse,* June 24, 2015, https://www.drugabuse.gov/about-nida/legislative-activities/testimony-to-congress/2015/biology-potential-therapeutic-effects-cannabidiol.
28. Joseph T. Rannazzisi statement before the Caucus on International Narcotics Control US Senate, June 24, 2015, https://www.drugcaucus.senate.gov/sites/default/files/DEA%20Rannazzisi%20CBD%20Testimony%20%2824June15%29.pdf.
29. Douglas C. Throckmorton statement before the Caucus on International Narcotics Control US Senate, June 24, 2015, https://www.drugcaucus.senate.gov/sites/default/files/6-24-15%20Throckmorton%20hearing.final_.doc.pdf.
30. Ibid.
31. Dr. John Bradford Ingram testimony before the Caucus on International Narcotics Control US Senate, June 24, 2015,https://www.drugcaucus.senate.gov/sites/default/files/Ingram%20Testimony%20to%20Senate%20Drug%20Caucus.pdf.
32. Jeremy M. Sharp, "US Foreign Aid to Israel," *Congressional Research Service,* April 10, 2018, https://fas.org/sgp/crs/mideast/RL33222.pdf.
33. Yardena Schwartz, "The Outsourcing of American Marijuana Research," *Newsweek Magazine,* Dec 17, 2015, https://www.newsweek.com/2015/12/25/outsourcing-american-marijuana-research-406184.html.
34. Yardena Schwartz, "The Outsourcing of American Marijuana Research," *Newsweek Magazine,* Dec 17, 2015, https://www.newsweek.com/2015/12/25/outsourcing-american-marijuana-research-406184.html.
35. "Pediatric Patients," *Tikun Olam*, accessed Jan 23, 2019, https://www.tikunolam.com/article.php?id=1042.
36. Alvit LCS Pharma, "Alvit LCS Pharma: Manufacturing and Licensing Agreement With Israel's Largest Medical Cannabis Company Bazelet," PR newswire, Oct 24, 2018,https://www.prnewswire.com/news-releases/alvit-lcs-pharma-manufacturing-and-licensing-agreement-with-israel-s-largest-medical-cannabis-company-bazelet-874668389.html.
37. "The Bazelet Group Companies," *Bazelet Group,* accessed Jan 24, 2019, https://www.bazelet-n.com/.
38. "Circulating Tumor Cells," Platform, *Cannabics Pharmaceuticals,* 2019, https://cannabics.com/platform/#CirculatingTumorCells.
39. "About Grants," National Institutes of Health, *US Department of Health & Human Services*, last modified March 17, 2017, https://grants.nih.gov/grants/about_grants.htm.

40. Yardena Schwartz, "The Outsourcing of American Marijuana Research," *Newsweek Magazine,* Dec 17, 2015, https://www.newsweek.com/2015/12/25/outsourcing-american-marijuana-research-406184.html.
41. Yardena Schwartz, "The Holy Land of Medical Marijuana," Best Countries, *US News,* April 11, 2017, https://www.usnews.com/news/best-countries/articles/2017-04-11/israel-is-a-global-leader-in-marijuana-research.
42. Pacher P., Batkai, S., Kunos G., "The Endocannabinoid System as an Emerging Target of Pharmacotherapy," The National Institutes of Health, *Pharmacol Rev.*, Sept 2006, Vol 58:3, pp 389-462, https://www.ncbi.nlm.nih.gov/pmc/articles/PMC2241751/pdf/nihms38123.pdf.
43. "GW to develop new Cannabinoid Opportunities with Professor Raphael Mechoulam," *GW Pharmaceuticals,* accessed Jan 24, 2019, https://www.gwpharm.com/about/news/gw-develop-new-cannabinoid-opportunities-professor-raphael-mechoulam.
44. "Epilepsy," *Tikun Olam*, accessed Jan 24, 2019, https://www.tikunolam.com/article.php?id=1053.
45. Fred Gardner, "Mechoulam on Cannabidiol," An Overview at the IACM meeting in Cologne, *O'Shaughnessy's,* 2008, https://www.projectcbd.org/sites/projectcbd/files/downloads/mechoulam-iacm-07.pdf.
46. Ibid.
47. Ibid.
48. Mary Guiden, "Preliminary data from CBD clinical trials 'promising,'" Colorado State University, July 19, 2018, https://cvmbs.source.colostate.edu/preliminary-data-from-cbd-clinical-trials-promising/.
49. "Colorado Researchers Studying CBD Oil in Dogs," CBS 4, May 18, 2018, https://denver.cbslocal.com/2018/05/18/colorado-cbd-oil-dogs/.
50. Experimental Biology, "Hemp shows potential for treating ovarian cancer: Researchers demonstrate hemp's ability to slow cancer growth and uncover mechanism for its cancer-fighting ability," *ScienceDaily*, April 23, 2018. www.sciencedaily.com/releases/2018/04/180423155046.htm.
51. Ibid.
52. FDA, "Highlights of Prescribing Information," Epidiolex, June 2018, https://www.accessdata.fda.gov/drugsatfda_docs/label/2018/210365lbl.pdf.
53. Dr. Thomas Minahan testimony before the Caucus on International Narcotics Control US Senate, June 24, 2015, https://www.drugcaucus.senate.gov/sites/default/files/Minahan%20Testimony.pdf.
54. "Biography," *Kevin Sabet*, 2011, http://kevinsabet.com/biography.
55. Kevin Sabet testimony before the Caucus on International Narcotics Control US Senate, June 24, 2015, https://www.drugcaucus.senate.gov/sites/default

/files/Testimony%20CBD%20Drug%20Caucus%20-%20Sabet%20-%20FINAL.pdf.

56. FDA, "Investigational New Drug (IND) Application," Drugs, US Department of Health and Human Services, Oct 5, 2017, https://www.fda.gov/drugs/developmentapprovalprocess/howdrugsaredevelopedandapproved/approvalapplications/investigationalnewdrugindapplication/default.htm.
57. Ibid.
58. FDA "Expanded Access INDs and Protocols 2009–2017," Expanded Access (Compassionate Use), US Department of Health and Human Services, last modified Feb 21, 2018, https://www.fda.gov/newsevents/publichealthfocus/expandedaccesscompassionateuse/ucm443572.htm.
59. Associated Press, "4 Americans get medical pot from the feds," *CBS News,* Sept 28, 2011, https://www.cbsnews.com/news/4-americans-get-medical-pot-from-the-feds/.
60. Russo E., Mathre M. L., Byrne A., et al., "Chronic Cannabis Use in the Compassionate Investigational New Drug Program: An Examination of Benefits and Adverse Effects of Legal Clinical Cannabis," Journal of Cannabis Therapeutics, Vol. 2(1), 2002,http://www.cannabis-med.org/jcant/russo_chronic_use.pdf.
61. Piomelli D, Russo E. B., "The Cannabis sativa Versus Cannabis indica Debate: An Interview with Ethan Russo, MD," Cannabis Cannabinoid Res., Vol. 1(1), Jan 1, 2016, pp 44-46, https://www.ncbi.nlm.nih.gov/pmc/articles/PMC5576603/.
62. Russo E., Mathre M. L., Byrne A., et al., "Chronic Cannabis Use in the Compassionate Investigational New Drug Program: An Examination of Benefits and Adverse Effects of Legal Clinical Cannabis," *Journal of Cannabis Therapeutics,* Vol. 2(1), 2002,http://www.cannabis-med.org/jcant/russo_chronic_use.pdf.
63. Ibid.
64. Goldwater Institute, "Federal Right to Try: Questions and Answers," *Right to Try,* accessed Jan 23, 2019, http://righttotry.org/rtt-faq/.
65. Bergamaschi M. M., Queiroz R. H., Chagas M. H., et al., "Cannabidiol reduces the anxiety induced by simulated public speaking in treatment-naïve social phobia patients," *Neuropsychopharmacology,* Vol 36:6, May 2011, pp. 1219-26, https://www.ncbi.nlm.nih.gov/pubmed/21307846.
66. Prud'homme M, Cata R, Jutras-Aswad D., "Cannabidiol as an Intervention for Addictive Behaviors: A Systematic Review of the Evidence," *Subst Abuse,* Vol.9, May 21, 2015, pp 33-8, https://www.ncbi.nlm.nih.gov/pmc/articles/PMC4444130/.

67. Morgan C., Das R., Joye A., et al., "Cannabidiol reduces cigarette consumption in tobacco smokers: Preliminary findings," *Addictive Behaviors,* Vol. 38, Issue 9, Sept 2013, pp 2433-36, https://www.sciencedirect.com/science/article/abs/pii/S030646031300083X.
68. Gonzalez-Cuevas G., Martin-Fardon R., Kerr T., et al., "Unique treatment potential of cannabidiol for the prevention of relapse to drug use: preclinical proof of principle," *Neuropsychopharmacology,* March 22, 2018, pp 2036-2045, https://www.nature.com/articles/s41386-018-0050-8.
69. Megan Thielking, "This scientist is testing a marijuana ingredient as a way to prevent relapse. It's a daunting task," In the Lab, *Stat News,* Feb 28, 2018, https://www.statnews.com/2018/02/28/marijuana-cannabidiol-opioids-addiction/.
70. "Opioid Addiction 2016 Facts & Figures," American Society of Addiction Medicine, accessed Jan 24, 2019, https://www.asam.org/docs/default-source/advocacy/opioid-addiction-disease-facts-figures.pdf.
71. NIH, "Opioid Overdose Crisis," *National Institute on Drug Abuse*, last modified Jan 2019, https://www.drugabuse.gov/drugs-abuse/opioids/opioid-overdose-crisis#one.
72. Kevin Sabet testimony before the Caucus on International Narcotics Control US Senate, June 24, 2015, https://www.drugcaucus.senate.gov/sites/default/files/Testimony%20CBD%20Drug%20Caucus%20-%20Sabet%20-%20FINAL.pdf.

8

HEMP CURES POISONED LAND

The only time hemp can be potentially dangerous to mankind is if it is grown for consumption and the farmer isn't aware of what the plant is absorbing. A 2005 study conducted in the Niger Delta region of Nigeria found that hemp exposed to arsenic, cadmium, chromium, iron, nickel, lead, mercury, and manganese stored those toxins in the leaves, and a smaller percentage of the toxins (mostly manganese) did enter the seeds. Nigerians who were growing hemp in this particular region were chewing on the leaves and creating cigarettes from the plant. They were also smoking it to provide relief from pains, aches, stresses, and so on.

Researchers conducted the study to determine if hemp consumption had any correlation with the "continuous and hostile crisis in the region" that has existed for decades between various religious groups and ethnicities.[1] What they concluded was that the hemp was indeed laced with extremely dangerous heavy metals that exceed health limits set by the World Health Organization. Hemp was being grown on soil that was highly polluted by the oil industry, which was why the plants stored so much of these particular toxic materials. Locals had no idea that they were poisoning themselves with heavy metals every time they ingested hemp.

The study didn't go so far as to make a direct correlation between the violence in the region and the consumption of heavy metals because there are a long list of historical and cultural reasons that factor into the Delta State situation, but if we take lead as an example and look at the impact of exposure to just one heavy metal, it's easy to deduce that mass consumption of heavy metals could be exacerbating the hostilities.

In general, farmers have to be cautious of where they grow their hemp if the intent is to eat it or to extract CBD. Since CBD extraction methods typically use as much of the hemp plant as possible, it's definitely not a good idea to produce CBD from hemp if the soil could be contaminated in any way. This is why when experts expressed concern for CBD-based products during the 2015 Senate Drug Caucus Hearing, I didn't believe they were completely unwarranted, but they also certainly didn't voice this concern, which was troubling, considering that they're *experts* and all.

Come to think of it, this might be the one difference between hemp-CBD and marijuana-CBD: The hemp plant absorbs toxins from the ground as it grows, whereas the marijuana plant does not.

PHYTOREMEDIATION

Just as hemp absorbs nitrogen and other nutrients from the soil, it also absorbs heavy metals and toxins through a process called phytoremedation. There are other wild weeds that have the same ability, including Indian grass (*Sorghastrum nutans*), kenaf and canola,[2] as well as black nightshade (*Solanum nigrum*), mustard plants such as *Rorippa globosa*,[3] and even sunflowers. Since weeds can grow in a variety of climates, without maintenance, they're ideal for cleaning up industrial waste sites. The root systems extract lead, copper, nickel, cadmium, zinc, arsenic, mercury, selenium, and other contaminants from soil and/or groundwater to lessen the toxicity. Hemp's deep root system—which can grow as long as eight feet belowground—can extract these toxins from farther down in the soil than many other plants, making it the focus of many scientific studies on phytoremediation, predominantly in Europe.

In 1998, Ukraine's Institute of Bast Crops planted hemp near the Chernobyl site exclusively for the purpose of removing contaminants, and proved that hemp can extract large amounts of radiation and heavy metals from the soil.[4] The term "phytoremediation" was actually coined by Dr. Ilya Raskin, a plant biologist at Rutgers University, whose research has now been featured in over 200 scientific publications.[5] He was part of the team that tested various plants' ability to absorb cesium and strontium (radioactive metals), including Indian mustard plants and sunflowers.

"For the specific contaminants that we tested, hemp demonstrated very good phytoremediation properties," stated Vyacheslav Dushenkov, a team member involved in the research.[6]

While this might've been breaking news at the time, hemp was never used to clean up the site. Perhaps it was too dangerous for humans to enter the area to plant the crop, remove the contaminated hemp, and replant more every 180 days. (Of course, this would've been the perfect opportunity to invent robotic technology to plant and rotate hemp until the soil was cleaned, but I digress.) Since plutonium has a 24,000-year half-life, the land won't be safe for human habitation for another 24,000 years.[7] Regardless, the site isn't completely uninhabited. As of October 2018, a solar energy plant now sits on the site to provide renewable energy to Ukraine. The country intends on expanding the site, which produced enough energy to power two thousand apartments when it first powered up.[8]

Even though the phytoremediation research gathered at Chernobyl wasn't actually implemented, the results were remarkable enough for researchers to continue to research hemp's ability to extract various heavy metals as a low-cost solution for toxic cleanup—here are some of the more recent examples:

- A study published in 2017 in *Waste and Biomass Valorization* noted that hemp can absorb and remove the toxic waste in sanitary landfills (waste known as landfill leachate), a major source of pollutants worldwide. For the study, hemp was grown hydroponically in "raw leachate" liquid collected from a sanitary landfill in Czech Republic. Depending on the amount of leachate added to the plant (25 percent, 50 percent, 75 percent, 90 percent, or 100 percent), hemp absorbed between 6.48 to 75.78 percent of the toxins.[9]

- Eating foods containing high levels of cadmium can cause major health problems, including respiratory diseases, anemia, renal failure, and deformations in the joints and bones. When most plants absorb cadmium, the metal negatively affects plant growth—specific to how chloroplast enters the leaf cells and the way the plant receives water (and how much). However, hemp can absorb high concentrations of cadmium and "it does not affect the growth of the plant,"[10] so it can clean the soil and grow to its full potential, thus extracting as much cadmium as possible throughout its growth cycle.

- Similarly, a 2017 study conducted at Colorado State University planted hemp in soil contaminated with selenium to discover "not a single plant died, and only a few, exposed to the highest doses, showed signs of stress."[11]

- A study from 2003 published in *Industrial Crops and Products* studied how fiber crops can absorb heavy-metal deposits from industrially polluted areas in Bulgaria. The researchers exposed flax, hemp, and cotton plants to soil contaminated with heavy metals to determine:[12]

  Flax and hemp are . . . suitable for growing in industrially polluted regions—they remove considerable quantities of heavy metals from the soil with their root system and can be used as potential crops for cleaning the soil from heavy metals.

- In Silesia, a region of Poland with large deposits of minerals, hemp has been "deliberately cultivated in wastelands contaminated with cadmium and copper" for generations to mitigate the severity of contamination by these heavy metals.[13]

- Italian researchers from the University of Padova and the University of Pisa published an article in *Plant Science* in 2011 called "Heavy Metal Hyper-Accumulating Plants: How and Why Do They Do It? And What Makes Them So Interesting?" They theorize that the hyper-accumulation of toxins found in hemp's leaves can actually work as a defense mechanism to ward off its enemies.[14] Perhaps that's another reason why pests are repelled by hemp.

FIGHT POLLUTION WITH HEMP

In Italy, hemp production is increasing each year by leaps and bounds for fiber, food, and environmental cleanup. Italian hemp fiber is typically processed into shoes, handbags, clothes, even bricks for construction. Italians also grind the seeds into high-protein, high-fiber, gluten-free flour. However, hemp grown for food and hemp grown for fiber processing aren't necessarily happening on the same farm.

In Taranto, Italy, farmers were once known for producing cheeses and meats. Over time, contaminants from a nearby steel plant made the soil toxic. In 2012, researchers discovered that the soil within a twelve-mile radius of the steel plant was so contaminated that the cows weren't safe for human consumption. Taranto dairy farmers were told by the Italian government that they could no longer produce meat or any dairy products. Farmers now grow hemp to decontaminate the land, and then they sell the hemp for processing into non-edible goods.

The steel plant responsible for the contamination is one of the largest in Europe, and at one time produced almost one-third of Italy's total steel. As the factory expanded substantially over the years, some farmers found themselves less than a mile from the now 15-million-square-meter facility.[15] A reporter from *Slate* described Taranto's air as having "a heavy metallic scent" due to the toxic chemicals the factory has been spewing into the air since the 1960s.[16]

From 2005 to 2012, eleven thousand local residents died due to exposure to the toxins in the air, and a study found high levels of lead and dioxins in the urine and blood of those who lived near the steel facility.[17] So when President Trump says he wants to bring back the steel industry, keep that visual in mind.

The problem with heavy metals and dioxins produced by major industry is that they do not degrade over time. The particles can break apart and become smaller, but they aren't biodegradable, so they continue to build up in the soil, air, and water. Heavy metals such as cadmium, lead, and thallium can bond to soil particles that can then be absorbed by the roots of edible plants. Eating food that contains heavy metals can cause negative effects on the entire body. The nervous, renal, gastrointestinal, reproductive, cardiovascular, skeletal, and muscular systems can be affected.[18]

Taking this concern a step further, when plastics degrade, they break apart into smaller pieces until they reach the size of microplastics, or plastic debris that is less than five millimeters in length (about the size of a sesame seed or a speck of glitter). Plastic does eventually decompose, but one plastic bottle can take approximately 450 years (the same amount of time as a disposable diaper).[19] As plastic breaks apart into smaller and smaller pieces, each small piece is absorbing chemicals and toxins.

Think about the way plastic Tupperware looks after tomato sauce (or gravy, as many Italians say) has been stored inside of it. Yes, tomato sauce stains just about everything, but *plastic* also absorbs just about everything, from tomato sauce to pollutants. A 2013 article titled "Classify Plastic Waste as Hazardous" published in *Nature: International Journal of Science* stated that at least 78 percent of pollutants listed by the EPA and 61 percent listed by the European Union can be found in plastic.[20] When plastics become microplastics that are accidentally eaten, the pollutants attached to the plastic enter internal tissues and are stored there. Scientists have discovered that these toxins are also accumulating because microplastics are being passed up the food chain, starting with mosquitoes. The stomachs of mosquitoes contain undigested microplastics that they've ingested either from birth, as larva on contaminated water, or by ingesting contaminated water. Since mosquitoes are eaten by other animals and mammals, the microplastics in the mosquitoes are passing from one species to the next, up the food chain.[21]

Microplastics have been found in our oceans, in the Great Lakes,[22] and even in human poop. Scientists think we're ingesting microplastics when we eat food wrapped in plastic, when we drink from plastic bottles (even plastic cups), or when we eat seafood—tuna, lobster, shrimp, and mussels in particular ingest the greatest percentages of microplastics.[23] Some have gone so far as to theorize we're actually breathing in microplastic dust, which is really concerning because these plastic particles are so small they can enter our bloodstream, lymphatic system, and liver.[24] Researchers from the Medical University of Vienna and the Environment Agency Austria found some people had up to *nine* different kinds of plastics in their poop, and theorized the main source was from drinking water from plastic bottles.[25]

As scary as it might be to think that plastic is *inside* our bodies, the medical community still doesn't know exactly what the long-term effects are. What we do know is that exposure to certain man-made chemicals and heavy metals can irreversibly affect human health. However, when hemp is planted to absorb toxins, phytoremediation can put a huge dent in the amount of poisonous land and water mankind has created. If hemp is lessening the amount of pollutants in the environment, then it can potentially lessen the amount of pollutants these microplastics are absorbing, and therefore decrease the amount of toxins we unknowingly consume.

AMERICA'S LEAD EPIDEMIC

Since my daughter started attending school, I've noticed teachers are encouraging students to bring their own water. Every day, I pack up her stainless steel water bottle. Every day, I explain to her that she can drink from it whenever she wants; I don't think she even knows where the school's water fountain is (I've never seen one).

Granted, her school wasn't built with lead pipes, but I'd still prefer her to avoid the habit of drinking from any public source of water. Water fountains are a game of Russian roulette. We don't know what the plumbing conditions are, we don't know what the pipes are made out of, we don't know what the pipes that bring the water to the building are made out of, we don't know how often the water is tested (if it is tested regularly), and we definitely don't know what is swimming around in the water at any given time. School is where kids are supposed to go to learn. They aren't supposed to come home with irreversible brain damage due to drinking from water fountains, but that's a definite possibility for many children living in America.

In 2017, the EPA stated lead poisoning is the number one environmental health threat that American children face.[26] Here are just a few recent signs that United States has a lead epidemic:

- An estimated 37 million housing units in the United States contain lead-based paint.[27]

- After six hundred children in Syracuse were poisoned in 2017 by lead paint in homes, the Central New York Community Foundation donated $2 million to renovate homes, build new housing, conduct lead inspections, and train contractors over a four-year period.[28]

- In 2018, the state of Indiana tested 915 schools and found that 61 percent of them had elevated lead levels.[29]

- In the beginning of the 2018 school year, the Detroit public school system actually shut off its water before its 47,000 students started school due to lead and copper contamination.[30]

- In September 2018, the metal embellishments on Boy Scout uniforms were recalled due to lead-based paint.[31]

- Even the children of military families are being poisoned. Housing offered to US military families has been found to be contaminated with lead-based paint or lead-ridden soil, including houses in Fort Benning (Georgia), Fort Knox (Kentucky), and the United States Military Academy at West Point (New York).[32]

- A July 2018 report by the US Government Accountability Office noted only 43 percent of US school districts tested for lead in 2016 or 2017. Out of that 43 percent, there was elevated lead in the drinking water in 37 percent of the school districts.[33]

- The last time all school water fountains in Howard County, Maryland, were tested for lead was 1989. Between September 2018 and October 2018, nine schools were retested to find that each of them tested positive for lead. Lead particles were in the range of 22 to 288 parts per billion, which is well above the EPA's federal standard of 15 parts per billion.[34]

Unfortunately, when a school finds lead in its water supply, school officials don't always provide complete transparency on the approach to solve the problem. Los Angeles Unified is the second-largest school district in the nation. It has spent more than $30 million on lead testing and water system repairs in the last ten years, but the schools still have dangerous amounts of lead in the water. Their solution in the battle against lead? To flush all toilets and run the tap water for thirty seconds at the start of each day to remove contaminants in the water. If a water source still has too many lead particles, it's shut off. Keep in mind: no pediatrician actually believes this "flushing" system actually gets rid of lead for good because the original source of the lead is still there, waiting to recontaminate the water.

Even though the school district claimed the water was safe, none of the water sources with lead in them became lead-free for any amount of time after the water supply system was "flushed." Los Angeles Unified received $20 million in 2015 to end the flushing program and either repair the water fixtures or permanently shut down water sources that contained lead leaves above fifteen parts per billion. However, some schools were still practicing this flushing technique in 2018,[35] so, suffice it to say, Los Angeles, instead of literally flushing millions of dollars down the drain, encourages students to come to school with their own water. Or provide water bottles for them if you're too incompetent to fix the problem.

AMERICAN SOIL: LACED WITH LEAD

Lead pipes aren't the only concern. Schools should also test the soil from soccer fields and playgrounds, because the majority of US soil contains particles from leaded gasoline (which wasn't completely banned in the United States until 1996).

Professor Howard Mielke from the Tulane University School of Medicine studies the adverse effects lead has on the brain. The lead particles in soil are so small that he refers to them as dust, and the dust particles can recirculate from the soil into the atmosphere, particularly during hot, dry weather.

"The dust is extremely fine particles—it's mere nanoparticles—and that becomes very potent," Mielke explained in a *ThinkProgress* article. "It goes right through the cell wall into the bloodstream."[36]

Sadly, even household dust can contain nanoparticles of lead from paint chips or contaminated soil. Kids can track contaminated soil from school back home on their shoes, cleats, or sports uniforms.[37] But don't be so quick to blame the school's soccer field. Studies have found alarming amounts of lead in backyards across America. Greenpoint, Brooklyn is just one of many examples.

In the spring of 2017, a graduate student from Columbia's Department of Earth and Environmental Sciences tested fifty-two private backyards in the neighborhoods of Greenpoint. Out of the 264 soil samples collected, approximately 92 percent had lead content exceeding what the EPA designates as safe. The analysis explained why children who lived in Greenpoint were much more likely to have elevated levels of lead in their blood than children living in any other New York City neighborhood.[38]

The advice given to Greenpoint residents by Cornell University and the New York State Department of Health? Don't go out onto the lawn. Don't let your kids go near the dirt. Don't try to replace the soil by digging it up because that will increase the likelihood of nanoparticles entering your bloodstream, getting on your clothes, entering your home, and leading to further contamination. Use an elevated garden with clean soil from a landscaping store if you want to plant vegetables. Or add landscaping fabric and add at least six inches of clean soil on top of the contaminated soil so that it won't continue to mix with new soil. Although the majority of vegetables don't absorb a high quantity of lead, it is still important to wash them thoroughly as nanoparticles can still stick to them. In other words, we can't help you get rid of the lead in your backyard. You have to live with it and hope it doesn't poison your family, which unfortunately it's already doing.

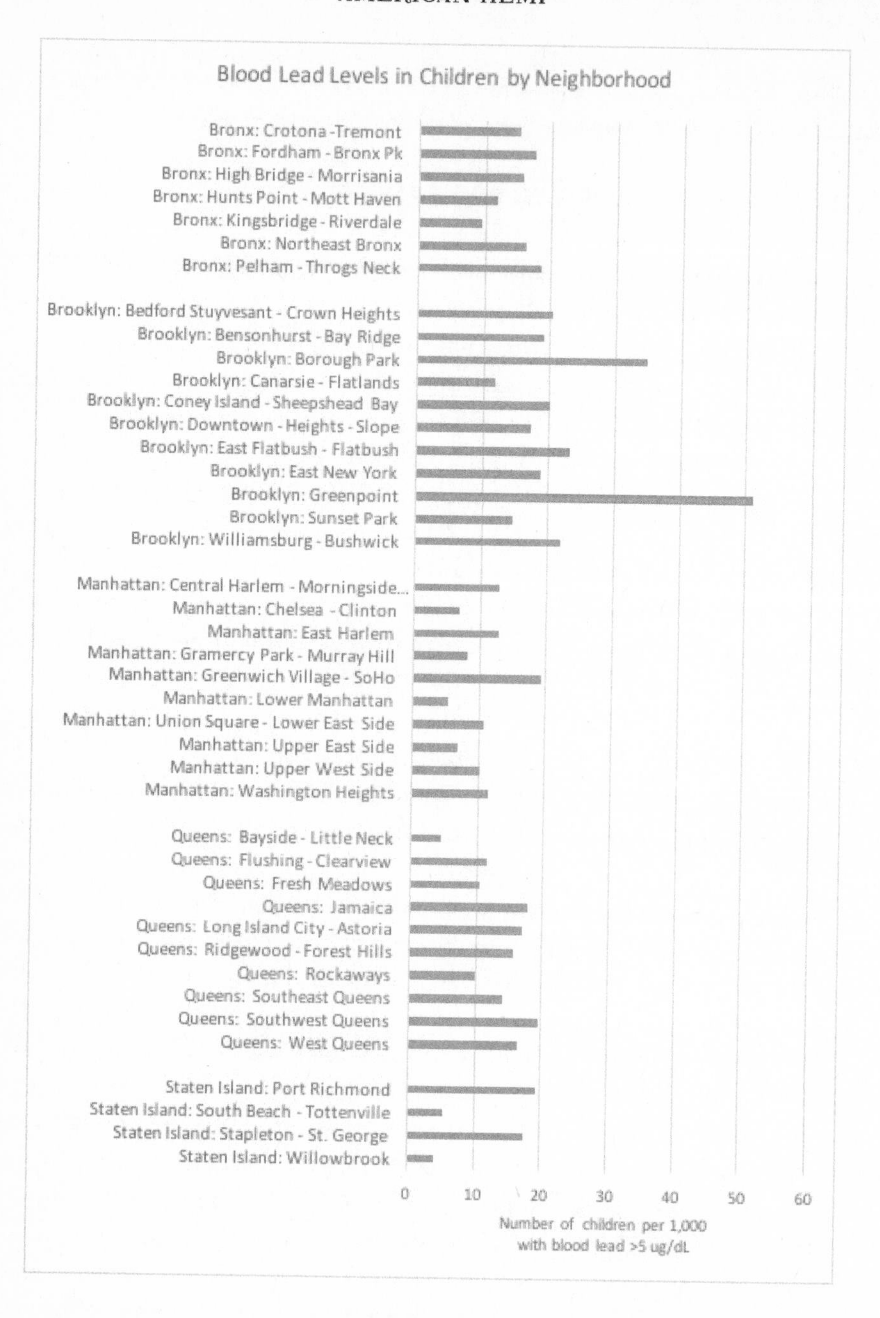

Greenpoint children are more likely to have elevated levels of lead in their blood than children in other NYC neighborhoods. This chart shows the number of children per one thousand expected to have blood lead levels exceeding five micrograms per deciliter, the level at which the CDC recommends taking action. *Source: NYC Dept of Health, 2015*

AMERICAN WATER: LACED WITH LEAD

On April 25, 2014, in an effort to save $5 million a year for two years, Flint City Council changed its water source from Lake Huron and the Detroit River to the Flint River. In October 2014, the General Motors Plant in Flint switched its water supply to a neighboring township because GM was concerned that the now highly chlorinated city water could corrode engine parts.[39] (Flint lost $400,000 in revenue when GM switched its water supply.) From there, the situation soon became one of the worst tragedies inflicted on a city by its elected officials.

Sure enough, Flint's lead and galvanized steel water service lines quickly corroded—the high levels of chlorine put into the water to make it safe to drink soon deteriorated the old pipes. By February 2015, the drinking water was contaminated with lead. Every single water line in the city was exposed to lead levels that were *twice* as high as what the EPA classifies as hazardous waste.

In March 2017, a federal judge approved a $97 million lawsuit settlement in which Michigan must pay to replace the eighteen thousand lead and galvanized steel water service lines in Flint by 2020. The total cost of the project is estimated at more than $100 million.[40]

Even though the project isn't fully completed, on April 6, 2018, Michigan governor Rick Snyder said the free bottled water program to the residents of Flint would end. He declared the Flint water crisis was over and the problems with the drinking water are now fixed after four and a half years. However, that didn't mean residents believed the water was safe.

In October 2018, the City of Flint announced that Elon Musk's foundation donated $480,350 to provide schools with a water filtration system for water fountains. The water filters will disinfect water tainted by lead and bacteria, and they're scheduled to be installed by the end of January 2019.[42]

Sadly, the people of Flint were not the first and are not the last to be poisoned with heavy metals in the United States. In 2017, more than three thousand children living in New Jersey were found to have high blood lead levels,[42] and that number spiked significantly after a water crisis in Newark, a city of 285,000 people. City officials are in the process of replacing 15,000 lead service lines, but the damage has already been done. These lead-based pipes connect the city's water main (two reservoirs) to the plumbing systems in houses, schools, and other buildings.

The problem in Newark was similar to Flint: the water in the reservoirs and the water leaving the treatment plants were safe, but once the water entered the distribution system and met the lead lines, the water became contaminated. The *New York Times* reported in October 2018 that some houses had water with lead levels of 42.2 parts per billion, which is nearly three times the federal action threshold.[43]

As early as 2016, a quarter of the more than 14,000 children under six years of age were found to have measurable levels of lead in their blood. This is where the problem started. When lead was discovered in the schools' drinking water, city officials claimed the water was safe to drink anyway. Most residents weren't even fully aware of the extent of the problem until city officials began handing out water filters. Unfortunately, using water filters to purify the main tap used for drinking water doesn't solve the full extent of the problem. Unless a filtration system is placed on the main line of the house or a filter is placed on every single faucet, people could still be taking showers, brushing their teeth, and washing their clothes with contaminated water.

Even though the lead pipes are being removed from Flint and Newark, debris from those pipes can certainly find its way into the ground and contaminate the soil. There's no way to guarantee every single particle of lead will be removed from the city; however, if Americans are able to grow hemp like the farmers in Italy are currently doing, we could one day declare victory over the lead crisis in Flint, Michigan, and Newark, New Jersey.

EFFECTS OF LEAD EXPOSURE

Researchers from the Department of Environmental Medicine and Public Health at the Icahn School of Medicine at Mount Sinai discovered the earliest evidence of lead exposure in the teeth of 250,000-year-old Neanderthals.[44] While further research into this discovery is still underway, there is enough current research to indicate that many socioeconomic issues in America are rooted in exposure to this toxin:

- A 2008 study conducted by the Cincinnati Children's Environmental Health Center tracked 250 children who were born in Cincinnati between 1979 and 1984 who had been exposed to lead during childhood, and whose mothers were also exposed to high concentrations of lead due to lead-contaminated housing.[45] The study followed the children into adulthood and found that there is a correlation between violent crime and lead exposure.[46] The risk of being arrested for a violent crime increased by nearly 50 percent for every blood level increase of five micrograms per deciliter, starting at six years of age. When the participants reached eighteen years of age or older, the male participants were arrested an average of five times, and the women were arrested an average of one time.[47]

- *ThinkProgress* conducted a study in Santa Ana, California, to determine if certain zip codes with the highest rates of juvenile offenders also have a high percentage of lead in the soil. According to the Orange County Probation Department's records from 2009 to 2015, the Santa Ana zip codes with the highest booking rates among minors were also the same zip codes with the highest percentages of lead in the soil,[48] thus proving there is a lead-to-prison pipeline, and it starts at a young age.

- The National Bureau of Economic Research looked at school suspensions and juvenile detention among 120,000 children born from 1990 to 2004 in Rhode Island.[49] The study compared children who lived closer to high-traffic areas and those who lived in less busy neighborhood, but all children who participated in the study went to the same school and their families had similar household incomes. Researchers determined if children are exposed to lead in preschool, then they are more likely to be suspended from school, and "children who have been suspended [from school] are ten times more likely to be involved in criminal activity." The students who lived closer to high-traffic areas were more likely to be suspended from school and incarcerated as juveniles than those who lived farther away.[50]

- Since leaded gasoline was largely being phased out between 1985 and 1988, researchers were also able to suggest that the decline in crime from 1990 to 2000 was due to the decreased exposure to lead, as cars were no longer pumping out the toxic chemical tied to an increase in violent tendencies. So those nationwide statistics about violent crime being the lowest it's ever been? When you *remove* a toxin from the entire nation that *causes violent tendencies* those statistics are gonna go down, man, way down.

- To back this up further, a 2016 study conducted by researchers at Harvard and Berkeley determined that cities *with* lead-based water service pipes have a *higher rate* of homicides than other cities, dating back to 1921.[51]

- Studies have also found that lead lowers IQ. Not only do children have lower IQs than their peers, but IQ points can actually decline over time as the child enters adulthood, even if the child is no longer exposed to lead, due to the long-term effects of the toxin.[52]

Although lead toxicity "can affect every organ system,"[53] the most immediate health concern in children is the neurological impact because lead disrupts the area in the brain responsible for rational thinking. Lead exposure is linked to various types of brain damage including a loss of self-control, loss of memory/forgetfulness, and the loss of the ability to deduce the consequences of one's actions or feel guilt or remorse for one's actions.

We know that children with a high blood lead concentration develop behavior problems including impulsivity, aggression, and an elevated risk of ADHD. They also have what's known as "conduct disorder"—including increased criminal behaviors—and their lead toxicity can be passed on genetically.[54] The CDC outlines these injuries, which (depending on the severity of the case) can take root in mild, seemingly average behavioral problems or severe neurological issues, including:[55]

- Problems with thinking (cognition)
- Difficulties with organizing actions, decisions, and behaviors (executive functions)
- Abnormal social behavior (including aggression)
- Difficulties in coordinating fine movements, such as picking up small objects (fine motor control)
- Decreased IQ, lower academic achievement, and reductions in specific cognitive measures
- Developmental delays and learning disorders; decreased reaction time
- Loss of memory
- Poor attention span
- Muscular tremor
- Dullness, irritability, depression/mood changes, lethargy, weakness
- Dizziness, fatigue, forgetfulness, impaired concentration, increased nervousness, headaches

- Malaise
- Paresthesia

Perhaps with the passage of the 2018 Farm Bill, the Secretary of Agriculture could work with the EPA and environmental groups to start a nationwide program to remove lead from our everyday lives. A 2004 Italian study published in the *Advances of Horticultural Science* looked specifically at hemp's ability to extract cadmium, lead, and thallium. Researchers noted that phytoremediation is "an extremely interesting and promising technique" when compared to other soil-purification techniques (chemical, thermal, and mechanical) that involve expensive, time-consuming, and labor-intensive means to extract toxins. They preferred phytoremediation because:[56]

> *i) it does not interfere with the efficiency of the ecosystem, ii) it is considerably cheaper. . . . iii) it concentrates the pollutant in the plant tissues, considerably reducing the volume of the polluted biomass and thus enabling the pollutants to be extracted using simple equipment.*

Researchers also noted that using hemp to improve the quality of soil had economic advantages. Hemp has a short life cycle of 180 days, so contaminants can be removed at a relatively fast rate, and it grows easily in a variety of climates.

THE (SOMEWHAT) SECRET HISTORY OF LEAD

Just as William Randolph Hearst and corporations like DuPont had a hand in outlawing cannabis, the United States has had a long, ugly history between politicians and the lead industry that kept lead alive and well, regardless of the harmful effects.

In a history of lead plumbing titled "The Lead Industry and Lead Water Pipes: 'A Modest Campaign'" ("modest" being thinly veiled sarcasm), published in the *American Journal of Public Health* in September 2008, Richard Rabin, an expert in occupational and environmental health at the Massachusetts Department of Labor's Occupational Lead Poisoning Registry, states the following:[57]

> *Lead pipes for carrying drinking water were well recognized as a cause of lead poisoning by the late 1800s in the United States. By the 1920s, many cities and towns were prohibiting or restricting their use. To combat this trend, the lead industry carried out a prolonged and effective campaign to promote the use of lead pipes. Led by the Lead Industries Association (LIA), representatives were*

> *sent to speak with plumbers' organizations, local water authorities, architects, and federal officials. The LIA also published numerous articles and books that extolled the advantages of lead over other materials and gave practical advice on the installation and repair of lead pipes. The LIA's activities over several decades therefore contributed to the present-day public health and economic cost of lead water pipes.*

Whether or not the LIA was aware, there's been evidence dating back to Roman times about lead's harmful side effects. The Romans used lead to make their water pipes, and lead miners often went mad or wound up dead.

"Water conducted through earthen pipes is more wholesome than that through lead," wrote Vitruvius, a Roman civil engineer from over two thousand years ago. "This may be verified by observing the workers in lead, who are of a pallid color."[58]

In 1983, a Canadian research scientist named Jerome Nriagu examined the diets of thirty Roman emperors from 30 BC to 200 AD to conclude that nineteen "had a predilection" for food and wine that was tainted with lead, and they all probably suffered from lead poisoning as a result.[59]

When Nriagu tested the ancient recipes of the Romans, he discovered the syrups they used to sweeten their food contained lead concentrations of 240 to 1000 milligrams per liter. Not only was the entire city of Rome drinking from public water fountains contaminated with a lead piping system, but the Roman people cooked in lead pots or lead-lined copper kettles.

According to Nriagu, "One teaspoon (5 ml) of such syrup would have been more than enough to cause chronic lead poisoning." He went so far as to surmise that the real reason why Rome fell is because its elected officials, its emperor, and all of its residents were poisoning themselves with obscene amounts of lead on a daily basis.

At the time, many researchers were quick to debunk Nriagu's theory, but studies from 2017 have shown that the Romans' entire aqueduct system was in fact affected by lead. In a paper compiled by the *Proceedings of the National Academy of Sciences,* researchers discovered that the Romans started using lead pipes two hundred years earlier than initially thought *and* the concentration of lead deposits found at the harbors where the water pipes expelled wastewater was compelling enough for researchers to determine the lead water pipes were probably "the primary source of lead pollution"[60] for the entire city.

While scientists are still determining the full effects of that much exposure to lead and how it factored into issues of historical significance at the time, there's already an unfortunate similarity between ancient Rome and Flint, Michigan.

In 2017, researchers took a look at fertility rates among women living in Flint who became pregnant between November 2013 through March 2015 (during the beginning stages of the water crisis). The overall fertility decreased by 12 percent and fetal death rates increased by 58 percent. When birth rates in Flint were compared with other cities in Michigan that did not have lead in the water system, it became clear that lead was the culprit in "reducing the number of expected births in the city" and decreasing the "overall health of those babies born."[61] The prevalence of miscarriages, stillbirths, and health complications in newborns were also present in Roman times, but to a much more alarming degree. In 18 BC, Augustus passed a law known as *lex Iulia de maritandis ordinibus* which penalized the unmarried and the childless in the hopes of raising the birth rate.[62]

However, America's Big Lead industry was certainly aware of the dangers of lead-based pipes, and lead-based paints, as early as the 1600s. Dating back to 1678, workers who made *lead white*, a pigment for paint, suffered from "dizziness in the head, with continuous great pain in the brows, blindness, stupidity."[63] Yet by 1900, more than 70 percent of American cities with populations greater than 30,000 used lead water lines.[64] Why lead? According to Richard Rabin, even though lead was more expensive than iron (the former material of choice), lead pipes lasted longer; they're also more malleable, so the pipes could be bent around existing structures. By the 1920s, lead was in telephones, iceboxes, vacuums, irons, washing machines, dolls, painted toys, beanbags, baseballs, fishing lures, gasoline, and paint.

Dr. Alice Hamilton, the first woman appointed to the faculty of Harvard University, was a leading expert in the field of occupational health. She studied the effects of industrial metals and chemical compounds on the human body, including lead. In 1911, she was a special investigator for the United States Bureau of Labor.[65]

According to Dr. Hamilton: "Where there is lead, some case of lead poisoning sooner or later develops, even under the strictest supervision."[66]

EVERY BREATH YOU TAKE: LEAD IS IN THE AIR

As Jamie Lincoln Kitman, New York bureau chief for *Automobile* magazine and member of the Society of Automotive Historians, wrote in an article for *The Nation* (which won an investigative reporting award): "You can choose whether to smoke, but you can't pick the air you breathe."[67]

Americans—and quite frankly all nations—have come a long way in reducing exposure to lead simply because we stopped putting it in everyday products, but the problem the United States now faces is what to do about the estimated 7 million tons of lead that originated from the leaded gasoline used to fuel motor vehicles in the twentieth century—all of which still remains in our soil, air, and water—not to mention it's still accumulating in plants, animals, and humans. According to Kitman's article:[68]

> *A 1983 report by Britain's Royal Commission on Environmental Pollution concluded that lead was dispersed so widely by man in the twentieth century that "it is doubtful whether any part of the earth's surface or any form of life remains uncontaminated by anthropogenic [man-made] lead."*

"Lead is not only bad for the planet and all its life forms, it is actually bad for cars and always was," stated Kitman. "For more than four decades, all scientific research regarding the health implications of leaded gasoline was underwritten and controlled by the original lead cabal—DuPont, GM and Standard Oil."

A 1985 EPA study estimated that as many as five thousand Americans died annually from lead-related heart disease prior to US lead phaseout, which was completed on January 1, 1996, when the Clean Air Act banned gasoline that contained even small amounts of lead. However, the act still allowed fuel containing lead to be sold for off-road uses "including aircraft, racing cars, farm equipment, and marine engines."[69] The act stated up to .05 grams of lead per gallon is permitted in everyday "unleaded gasoline" for motor vehicles, which is included in the EPA's definition of unleaded gasoline:[70]

> *Gasoline which is produced without the use of any lead additive and which contains not more than 0.05 gram of lead per gallon and not more than 0.005 gram of phosphorus per gallon. . . . After December 31, 1995, no person shall sell, offer for sale, supply, offer for supply, dispense, transport, or introduce into commerce for use as fuel in any motor vehicle (as defined in Section 216(2) of the Clean Air Act, 42 U.S.C. 7550(2)), any gasoline which is produced with the use of lead additives or which contains more than 0.05 gram of lead per gallon.*

Although lead can't be intentionally added to gasoline, the EPA allowed this percentage in unleaded gasoline to account for cross-contamination. While leaded gasoline is nearly completely phased out of every aspect of society, companies weren't barred from using the same pipes to carry and pump leaded gasoline and unleaded gasoline, so it is conceivable that *there is still lead* in today's unleaded gasoline, since the heavy metal doesn't degrade.

According to Kitman's research, since the 1920s, independent scientists who debunked the leaded-gas industry's studies were "threatened and defamed by the lead interests and their hired hands."[71] Even a *Washington Post* article from April 1987 titled "The End of Leaded Gasoline" didn't mention any of the known health risks associated with lead, but instead stated that the EPA's new laws for gas will bring "a sometimes-bewildering basket of new choices at the gas pump," all of which would be more expensive than the then-current price of leaded gas.[72]

So while scientists have spent decades studying lead and blowing the whistle on the dangers of this poisonous neurotoxin, *and have finally* convinced authorities to ban it, they're now faced with another problem: what to do with all the lead now that it's been accumulating?

WHAT IS THE PRACTICAL SOLUTION TO LEAD?

In 1991, Dr. Ilya Raskin received a grant from the EPA to explore how phytoremediation can purify soil and water in Superfund sites; he also cofounded Phytotech, Inc., a company focused on developing more phytoremediative plants, including sunflowers bred to have greater genetic abilities to detoxify contaminants. Despite Raskin's academic success in proving the value of phytoremediation, investors eventually lost interest in his company and promising opportunities in government funded research also diminished. To this day, phytoremediation is still a popular field of study among academic institutions, but there aren't many examples of it being used in commercial applications, such as how Italy's farmers are using hemp to clean the soil.

According to an article in *Plant Science*, there were over 600 academic papers published on phytoremediation in 2015, but only forty included references to field trials.[73] A July 2017 article published in the Proceedings of the National Academy of Sciences of the United States of America (PNAS) references a rare commercial application for phytoremediation that took place in a Washington, DC, neighborhood in early 2000. The Army Corps of Engineers discovered unsafe levels of arsenic in the soil on 177 Spring Valley properties (the arsenic deposits were remnants of World War I weapons testing that was done in the area). In the yards with higher arsenic levels, the Corps dug up the dirt, hauled it away, and replaced it with clean soil. For yards with lower arsenic levels, homeowners were given the option to plant brake ferns (known as Pteris) on their property to naturally draw arsenic from the soil. Only twenty-two residents chose this option. On sixteen of the properties, the soil was cleaned within a single growing season (which lasted about five months). The remainder of the properties required repeat plantings due to higher levels of contamination, but all homeowners had safe levels of arsenic within five years.

"We were able to save a lot of money in restoration cost because we didn't have to come back in and restore the landscape," explained Michael Blaylock, president and CEO of Edenspace, the company that trademarked Edenfern (the name for the brake fern).[74]

While the process to restore the soil via ferns cost less money and was less disruptive to the yards, it took five years in some cases. Let's face it, most people would want arsenic removed immediately if it was found in their yards, regardless of the cost, which is why most Spring Valley residents didn't opt for the more economical, ecological approach. Unlike lead poisoning, arsenic poisoning is treatable, but it can cause several forms of cancer and even multi-system organ failure, and the longer it lingers in the soil, the greater the possibility of health risks. Who has the patience to wait for ferns to remove the contaminant? Who wants to worry about health risks for five months or five years when the soil can be picked up and moved immediately?

This is why most cleanup processes involve the costly process of removing the soil and starting fresh, but what most people don't realize is that this isn't necessarily faster nor does it ensure that the contaminant is completely gone, especially when the contamination (such as lead) is found in more than one neighborhood. In Norway, researchers devised a nationwide soil cleanup method for the polluted soil in daycare centers and playgrounds. Sites had concentrations of heavy metals including lead, cadmium, copper, mercury, and nickel, as well as man-made toxins (PCBs). The solution was to remove half a meter of the polluted soil, deliver it to a landfill, then place a geotextile cloth over the soil before filling the area with half a meter of clean soil.[75] In areas with greater concentrations of toxins, landscaping techniques, such as large hills of clean dirt, were also added. While this didn't actually clean the original soil, it was effective in keeping the schoolyards relatively free from contamination.

In 2007, Norway's environmental protection agency was granted a budget of $40 million euros to identify two thousand of the most polluted daycare centers in the ten largest Norwegian cities and use this soil removal process to clean the soil. The goal was to clean the two thousand sites by 2010, and the country is now invested in using this technique to clean up parks and other public spaces, including industrial sites, such as shipyards and loading docks. City officials are to monitor soil conditions to ensure heavy rainfall or other weather conditions don't push toxins from the original soil up into the newly laid landscapes. Plus, the new soil has to be replenished from time to time.

However, Norway's national soil remediation process took years (and is still ongoing). The first research project initiated in Trondheim in 1996. From that city, cleanup was extended to Oslo and then to Bergen (where 45 percent of the eighty-seven childcare centers were contaminated with lead and PCBs). It took a year to identify the two thousand polluted sites, and then two years to complete the soil-remediation process at all of the sites. While daycares and schoolyards waited years to be cleaned, children remained at risk. What's the harm in using phytoremediation in the meantime? Sure, it only takes a few days to remove the toxic soil and redo the landscaping, but if the school is on a two-year waiting list, and if plants can remove the toxins months or even a year before the cleanup crew arrives, that is clearly a solution that benefits everyone.

Even if a national phytoremediation project is able to reduce the amount of toxins in the soil, we have another problem: the drinking water delivered to many of our houses passes through lead pipes. According to a 2004 report by the EPA, the American Water Works Association conducted a national survey to estimate the cost of replacing lead water lines. In Washington, DC, alone, "water authority appropriated $300 million to replace 23,000 lead service lines, plus some portion of 27,000 lines of unknown material."[76] So the question becomes, when will the federal government come up with a solution to remove the lead in our water-delivery systems? What will it take for every American to gain equal access to clean water?

In this case, perhaps the United States could follow Norway's example. Instead of removing and replacing all lead pipes (which could expose more people to the heavy metal), we could leave them where they are and take them out of commission. As Manhattan's subway technology evolved, the city didn't continue to update existing lines. The transit authority buried deteriorating or obsolete tunnels by building new tunnels over the old ones. So maybe we should leave the pipes alone, engineer an entirely new water-delivery system—one where we can access the water before it enters reservoirs and treatment centers laced with lead service lines. Not entirely a perfect solution, but better than continuing to drink contaminated water.

What would a new water-delivery system look like? Changing the way we collect and process rainwater is certainly a question for engineers and environmentalists to weigh in on, but there are osmosis systems currently in place that guarantee pure water. An FDA-approved company called Richard's Rainwater has been bottling rainwater since 2002. Rain is captured before it ever touches the ground, so it doesn't come into contact with any natural or man-made contaminants. The company then triple-filters it, using a proprietary process that includes ultraviolet light and reverse osmosis. The purification process "has zero waste, speeds up the hydrologic cycle and decreases the demand for water from wells, lakes, rivers and streams."[77] While I realize this is one company and a small-scale solution when compared to how many water treatment centers we have nationwide, it's a method that currently works, and a method that could be modified and expanded, just as Norway slowly modified and expanded their soil remediation.

Getting back to hemp's role in all of this, there haven't been too many large-scale phytoremediation projects in the United States, but multiple hemp pilot programs are in the process of studying it:

- In 2017, the 22nd Century Group, Inc, a biotechnology company focused on genetic engineering and plant breeding, partnered with the University of Virginia to "pinpoint the best variety of industrial hemp" to work toward healing Virginia's contaminated coal-mining lands and convert them into farmland.[78] The University of Virginia is also conducting several other ongoing studies on phytoremediation.[79]

- The Pennsylvania Hemp Industry Council was approved to evaluate use of hemp for erosion control and phytoremediation of heavy metals in March 2017.[80] Researchers at Lehigh University and Jefferson University are currently working on the studies at the Lambert Center for the Study of Medicinal Cannabis and Hemp.[81]

- The Hemp Farmacy, based in Wilmington, North Carolina, gives soils class about phytoremediation and using hemp to clean polluted soils as part of the state's hemp pilot program.[82]

- In June 2012, Colorado passed HB12-1099: Phytoremediation Hemp Remediation Pilot Program to study how hemp can rejuvenate contaminated soil and water.[83] The College of Natural Sciences at Colorado State University has a department dedicated to phytoremediation, and has conducted studies on hemp's ability to absorb selenium.[84]

A 2015 study supported by Higher Education Commission (HEC) of Pakistan showed that hemp has specific stress-tolerance genes. These genes work as antioxidant enzymes to protect it against the damage caused by heavy metals.[85] This is why the plant isn't easily stressed by the effects of industrial contaminants, including lead, zinc, copper, cobalt, nickel, and chromium. The study (as well as many others) noted that once toxins such as lead enter hemp's root system, they travel up the shoots and accumulate in the leaves. This hyper-accumulation of heavy metals is unusual among plants, and researchers are currently looking at hemp's genetic structure to determine how to breed super-hyper-accumulative varieties[86] that can then be used in phytoremediation cleanup efforts.

There's no doubt that it's going to take a multifaceted approach to reverse the damage caused by industrial pollution, but don't rely on government agencies such as the EPA to present proactive solutions. Corporate interests, specifically those of the chemical and oil industries, take precedence over the health and well-being of

Americans. Historically speaking, these companies aren't going to clean up the toxins caused by their industries. Instead, they'll continue to deny their toxicity, and ghostwrite scientific studies and present them to regulators as independent research. If we want cleaner air, soil, and water, we're going to have to take matters into our own hands. Gaining access to hemp and other plants with phytoremediative abilities can at the very least accomplish something, especially if the EPA doesn't pressure corporations to do anything.

NOTES

1. Eboh L.O., Boye T., "Analysis of Heavy Metal Content in Canabis Leaf and Seed Cultivated in Southern Part of Nigeria," *Pakistan Journal of Nutrition,* Vol 4(5), pp 349-351, 2005, http://citeseerx.ist.psu.edu/viewdoc/download?doi=10.1.1.630.3759&rep=rep1&type=pdf.
2. USDA, "Kenaf and Canola–Selenium Slurpers," *AgResearch Magazine,* June 2000, https://agresearchmag.ars.usda.gov/2000/jun/kenaf.
3. Girdhar, Madhuri et al., "Comparative assessment for hyperaccumulatory and phytoremediation capability of three wild weeds," *3 Biotech*, vol. 4,6, Jan 19, 2014, pp. 579-589, https://www.ncbi.nlm.nih.gov/pmc/articles/PMC4235884/#CR1.
4. Ahmad, Rafiq & Tehsin, Zara & Tanveer Malik, et al., (2015). "Phytoremediation Potential of Hemp (Cannabis sativa L.): Identification and Characterization of Heavy Metals Responsive Genes," *CLEAN—Soil Air Water*, Vol. 44 (2), Aug 2015, https://www.researchgate.net/publication/281651509_Phytoremediation_Potential_of_Hemp_Cannabis_sativa_L_Identification_and_Characterization_of_Heavy_Metals_Responsive_Genes.
5. "Dr. Ilya Raskin," Faculty, Rutgers, accessed Jan 28, 2019, https://plantbiology.rutgers.edu/faculty/raskin/Ilya_Raskin.html.
6. Andrew Leonard, "Can hemp clean up the Earth?" Politics Features, *Rolling Stone,* June 11, 2018, https://www.rollingstone.com/politics/politics-features/can-hemp-clean-up-the-earth-629589/.
7. Deorge Dvorsky, "Ukraine found the perfect use for the radioactive land of Chernobyl," *Gizmodo,* Oct 5, 2018, https://gizmodo.com/ukraine-found-the-perfect-use-for-the-radioactive-land-1829559033
8. Pavel Polityuk, "Three decades after nuclear disaster, Chernobyl goes solar," *Cnbc.com,* Oct 5, 2018, https://www.cnbc.com/2018/10/05/reuters-america-three-decades-after-nuclear-disaster-chernobyl-goes-solar.html.
9. Vaverková, M.D., Zloch, J., Adamcová, D. et al., *Waste Biomass Valor,* Vol.10 (2), pp 369–376, Feb 2019, https://link.springer.com/article/10.1007%2Fs12649-017-0058-z .

10. Girdhar, Madhuri et al. "Comparative assessment for hyperaccumulatory and phytoremediation capability of three wild weeds." *3 Biotech*, Vol.(4) 6, pp 579-589, Jan 29, 2014, https://www.ncbi.nlm.nih.gov/pmc/articles/PMC4235884/.
11. Andrew Leonard, "Can hemp clean up the Earth?" Politics Features, *Rolling Stone,* June 11, 2018, https://www.rollingstone.com/politics/politics-features/can-hemp-clean-up-the-earth-629589/.
12. Angelova V., Ivanova R., Delibaltova V., Ivanov K., "Bio-accumulation and distribution of heavy metals in fibre crops (flax, cotton and hemp)," *Industrial Crops and Products,* Vol 19 (3), May 2004, pp 197-205, http://www.sciencedirect.com/science/article/pii/S0926669003001110.
13. Jon Michael McPartland, Robert Connell, and David Paul Watson, *Hemp: Diseases and Pests: Management and Biological Control,* CABI Publishing 2000, New York, p. 165.
14. Rascio N., Navari-Izzo F., "Heavy metal hyperaccumulating plants: how and why do they do it? And what makes them so interesting?" *Plant Science,* Vol 180 (2), Feb 2011, Pp 169–181, https://www.sciencedirect.com/science/article/pii/S0168945210002402.
15. Seth Doane, "Farmers in Italy fight soil contamination with cannabis," *CBS News,* March 12, 2017, https://www.cbsnews.com/news/cannabis-plant-soil-decontamination-italy-vincenzo-fornaro/.
16. Sara Manisera, "Hemp and Change," Roads & Kingdoms, *Slate,* July 18, 2016, https://slate.com/news-and-politics/2016/07/taranto-italy-is-decontaminating-its-land-by-cultivating-hemp.html.
17. Ibid.
18. Di Candito M., Ranalli P., Dal Re L., "Heavy metal tolerance and uptake of Cd, Pb and Ti by hemp," *Advances in Horticultural Science,* Vol. 18 (3), 2004, pp 138-144. https://www.jstor.org/stable/42882328?seq=1#page_scan_tab_contents.
19. "Coastal Cleanups," Coastal Program, *New Hampshire Department of Environmental Services,* 2017, https://www.des.nh.gov/organization/divisions/water/wmb/coastal/trash/index.htm.
20. M Rochman, Chelsea & Browne, Mark Anthony & Halpern, et al., "Policy: Classify plastic waste as hazardous," *Nature,* Vol 494, Feb 2013, pp. 169-71, https://www.researchgate.net/publication/235620600_Policy_Classify_plastic_waste_as_hazardous.
21. Katherine J. Wu, "Mosquitoes are passing microplastics up the food chain," Smart News, *Smithsonian.com,* Sept 21, 2018, https://www.smithsonianmag.com/smart-news/mosquitoes-are-passing-microplastics-up-food-chain-180970373/.
22. "What are microplastics?" National Ocean Service, accessed Jan 28, 2019, https://oceanservice.noaa.gov/facts/microplastics.html.

23. Colin Fernandez, "Proof humans ARE eating plastic: Experts find nine different types of microplastic in every sample taken from human guts with water and drinks bottles blamed as the source," *The Daily Mail,* Oct 22, 2018, https://www.dailymail.co.uk/sciencetech/article-6303337/Experts-nine-different-types-microplastic-stool-samples-water-bottles-blamed.html.
24. Dennis Thompson, "Is there plastic in your poop? New study on microplastics in stool raises concerns," *CBS,* Oct 23, 2018, https://www.cbsnews.com/news/plastic-in-poop-study-microplastics-in-stool-raises-concerns/.
25. Colin Fernandez, "Proof humans ARE eating plastic: Experts find nine different types of microplastic in every sample taken from human guts with water and drinks bottles blamed as the source," *The Daily Mail,* Oct 22, 2018, https://www.dailymail.co.uk/sciencetech/article-6303337/Experts-nine-different-types-microplastic-stool-samples-water-bottles-blamed.html
26. EPA, "All News Releases By Date," Newsroom, last modified Dec 16, 2016, https://archive.epa.gov/epapages/newsroom_archive/newsreleases/b11c70fbd709f488852570d900462ae9.html.
27. Council on Environmental Health, "Prevention of Childhood Lead Toxicity," *American Academy of Pediatrics,* Vol. 138, Issue 1, July 2016, http://pediatrics.aappublications.org/content/138/1/e20161493.
28. Michelle Breidenbach, "Community foundation to spend $2 million to fight lead paint poisoning in Syracuse," Central NY Health, *Syracuse.com,* Oct 24, 2018, https://www.syracuse.com/health/index.ssf/2018/10/community_foundation_to_spend_2_million_to_fight_lead_paint_poisoning_in_syracus.html.
29. Kris Maher, "Schools across the US find elevated lead levels in drinking water," US, *Wall Street Journal,* Sept 5, 2018, https://www.wsj.com/articles/schools-across-the-u-s-find-elevated-lead-levels-in-drinking-water-1536153522.
30. Ibid.
31. Sara Magalio, "Potentially toxic lead content in Boy Scouts uniform accessory triggers recall," Breaking News Reporter, *Dallas News,* Sept 27, 2018, https://www.dallasnews.com/news/news/2018/09/27/potentially-toxic-lead-content-boy-scouts-uniform-accessory-triggers-recall.
32. Joshua Schneyer and Andrea Januta, "Children poisoned by lead on US army bases as hazards go ignored," *Reuters,* Aug 16, 2018, https://www.reuters.com/investigates/special-report/usa-military-housing/.
33. Kris Maher, "Schools across the US find elevated lead levels in drinking water," US, *Wall Street Journal,* Sept 5, 2018, https://www.wsj.com/articles/schools-across-the-u-s-find-elevated-lead-levels-in-drinking-water-1536153522.

34. Jess Nocera, "Lead found in water at nine Howard schools," *Howard County Times,* Oct 31, 2018, https://www.baltimoresun.com/news/maryland/howard/ph-ho-cf-lead-water-1023-story.htm.l
35. Nico Savidge and Daniel J. Willis, "Lead problems in water linger at Los Angeles schools, despite years of testing and repairs," Edsource Special Report, *EdSource,* Sept 25, 2018, https://edsource.org/2018/lead-problems-in-water-linger-at-los-angeles-schools-despite-years-of-testing-and-repairs/602870.
36. Yvette Cabrera, "Urban children are playing in toxic dirt," *Think Progress,* July 12, 2017, https://thinkprogress.org/urban-children-are-playing-in-toxic-dirt-41961957ff23/.
37. "Lead poisoning," Diseases and Conditions, *Mayo Clinic,* accessed Jan 28, 2019, https://www.mayoclinic.org/diseases-conditions/lead-poisoning/symptoms-causes/syc-20354717.
38. Sara Fecht, "High levels of lead contaminate many backyards in Brooklyn neighborhood," Health, State of the Planet, Earth Institute, *Columbia University*, Oct 9, 2017, https://blogs.ei.columbia.edu/2017/10/09/many-backyards-in-brooklyn-neighborhood-are-contaminated-with-high-levels-of-lead/.
39. Don Hopey, "Flint water crisis timeline," *Pittsburgh Post-Gazette,* Oct 20, 2018, http://www.post-gazette.com/news/health/2018/10/21/Flint-water-crisis-timeline-contamination-lawsuits-lead-exposure-children/stories/201810170150.
40. Don Hopey, "1,600 days on bottled water: Flint still swamped by water woes," *Pittsburgh Post-Gazette,* Oct 20, 2018, http://www.post-gazette.com/news/health/2018/10/20/Flint-water-lead-crisis-michigan-neurotoxin-legionnaires-miscarriages/stories/201810150196.
41. Mark Matousek, "Elon Musk's $480,000 donation to Flint public schools will help provide students clean drinking water starting next year," *Business Insider,* Oct. 9, 2018, https://www.businessinsider.com/elon-musk-donation-to-flint-will-provide-clean-water-for-schools-2018-10.
42. Tom Johnson, "Add $10M to the budget to fight lead poisoning in NJ's kids, advocates urge," Energy & Environment, *NJ Spotlight,* Feb 2, 2016, https://www.njspotlight.com/stories/16/02/01/add-10m-to-budget-to-fight-lead-poisoning-in-nj-s-kids-advocates-urge/
43. Liz Leyden, "In Echoes of Flint, Mich., Water Crisis Now Hits Newark," *New York Times,* Oct 30, 2018, https://www.nytimes.com/2018/10/30/nyregion/newark-lead-water-pipes.html.
44. The Mount Sinai Hospital, "Researchers discover earliest recorded lead exposure in 250,000 year old neanderthal teeth," Archaeology & Fossils, *Phys.org,* Oct 31, 2018, https://phys.org/news/2018-10-earliest-exposure-year-old-neanderthal-teeth.html.

45. Wright J., Dietrich K., Ris M., et al., "Association of prenatal and childhood blood lead concentrations with criminal arrests in early adulthood," *PLOS Medicine,* May 27, 2008, https://journals.plos.org/plosmedicine/article?id=10.1371/journal.pmed.0050101.
46. Ibid.
47. Yvette Cabrera, "The lead-to-prison pipeline," *Think Progress,* Jul 14, 2017, https://thinkprogress.org/the-lead-to-prison-pipeline-2707d218b6a3/.
48. Ibid.
49. Anna Aizer and Janet Currie, "Lead and Juvenile Delinquency: New Evidence from Linked Birth, School and Juvenile Detention Records," *National Bureau of Economic Research,* May 2017, https://www.nber.org/papers/w23392.pdf.
50. Jennifer L. Doleac, "New evidence that lead exposure increases crime," Up Front, *Brookings Institute,* June 1, 2017, https://www.brookings.edu/blog/up-front/2017/06/01/new-evidence-that-lead-exposure-increases-crime/.
51. Feigenbaum J, Muller C., "Lead exposure and violent crime in the early twentieth century," *Explorations in Economic History,* Vol 62, Oct 2016, pp. 51–86, https://www.sciencedirect.com/science/article/abs/pii/S0014498316300109.
52. Kendra Pierre-Louis, "The devastating effects of childhood lead exposure could last a lifetime," Health, *Popular Science,* March 29, 2017, https://www.popsci.com/lead-effects-can-last-lifetime#page-4
53. CDC, "Lead Toxicity: What are possible health effects from lead exposure?" *Agency for Toxic Substances & Disease Registry*, June 12, 2017, https://www.atsdr.cdc.gov/csem/csem.asp?csem=34&po=10.
54. Council on Environmental Health, "Prevention of Childhood Lead Toxicity," *American Academy of Pediatrics,* Vol. 138, Issue 1, July 2016, http://pediatrics.aappublications.org/content/138/1/e20161493.
55. CDC, "Lead Toxicity: What are possible health effects from lead exposure?" *Agency for Toxic Substances & Disease Registry*, June 12, 2017, https://www.atsdr.cdc.gov/csem/csem.asp?csem=34&po=10.
56. Di Candito M., Ranalli P., Dal Re L., "Heavy metal tolerance and uptake of Cd, Pb and Ti by hemp," *Advances in Horticultural Science,* Vol. 18 (3), 2004, pp 138–144. https://www.jstor.org/stable/42882328?seq=1#page_scan_tab_contents.
57. Rabin, Richard. "The Lead Industry and Lead Water Pipes: 'A Modest Campaign'" *American journal of public health*, Vol. 98(9) , Sept 2008, pp. 1584–92, https://www.ncbi.nlm.nih.gov/pmc/articles/PMC2509614/
58. Tim Harford, "Why did we use leaded petrol for so long," *BBC World Service,* Aug 28, 2017, https://www.bbc.com/news/business-40593353.

59. Lenny Bernstein, "Lead poisoning and the fall of Rome," Feb 17, 2016, https://www.washingtonpost.com/news/to-your-health/wp/2016/02/17/lead-poisoning-and-the-fall-of-rome/?utm_term=.f656268058d5.
60. Delile H.,Keenan-Jones D., Blichert-Toft J., et al., "Rome's urban history inferred from Pb-contaminated waters trapped in its ancient harbor basins," *Proceedings of the National Academy of Sciences*, Aug 28, 2017, http://www.pnas.org/content/early/2017/08/22/1706334114
61. Daniel S. Grossman, David J. G. Slusky, "The effect of an increase in lead in the water system on fertility and birth outcomes: the case of Flint, Michigan," Aug 7, 2017, http://www2.ku.edu/~kuwpaper/2017Papers/201703.pdf.
62. A.M. Devine, "The low birth-rate in ancient Rome: a possible contributing factor," University of New England, pp 313–317, http://www.rhm.uni-koeln.de/128/Devine.pdf.
63. Tim Harford, "Why did we use leaded petrol for so long?" *BCC World Service,* Aug 28, 2017, https://www.bbc.com/news/business-40593353.
64. Rabin, Richard. "The Lead Industry and Lead Water Pipes: 'A Modest Campaign'" *American journal of public health*, Vol. 98(9) , Sept 2008, pp. 1584–92, https://www.ncbi.nlm.nih.gov/pmc/articles/PMC2509614/.
65. "Dr. Alice Hamilton," Biography, *US National Library of Medicine,* last modified June 3, 2015, https://cfmedicine.nlm.nih.gov/physicians/biography_137.html
66. Tim Harford, "Why did we use leaded petrol for so long?" *BCC World Service,* Aug 28, 2017, https://www.bbc.com/news/business-40593353.
67. Jamie Lincoln Kitman, "The Secret History of Lead," Environment, *The Nation,* March 2, 2000, https://www.thenation.com/article/secret-history-lead/#1.
68. Ibid.
69. EPA, "EPA Takes Final Step in Phaseout of Leaded Gasoline," Jan 29, 1996, https://archive.epa.gov/epa/aboutepa/epa-takes-final-step-phaseout-leaded-gasoline.html.
70. EPA, "Part 80-Regulation of Fuels and Fuel Additives," July 1996, https://www.gpo.gov/fdsys/pkg/CFR-2013-title40-vol17/pdf/CFR-2013-title40-vol17-part80.pdf
71. Jamie Lincoln Kitman, "The Secret History of Lead," Environment, *The Nation,* March 2, 2000, https://www.thenation.com/article/secret-history-lead/#1.
72. Mark Potts and Michael Specter, "The end of leaded gasoline," *The Washington Post,* April 13, 1987, https://www.washingtonpost.com/archive/business/1987/04/13/the-end-of-leaded-gasoline/2eb8ef4b-a158-45f2-990a-fc672695a78e/?utm_term=.74665c86fb03

73. Gerhardt K.E, Gerwing P.D., Greenberg B.M. (2017) "Opinion: Taking phytoremediation from proven technology to accepted practice." *Plant Sci* 256:170–185.
74. Carolyn Beans, "Core Concept: Phytoremediation advances in the lab but lags in the field," *PNAS,* Vol. 114 (29), pp. 7475-7477, July 18, 2017, https://www.pnas.org/content/114/29/7475.
75. Norwegian Institute for Public Health, "Soil pollution in day-care centers and playgrounds in Oslo, Norway; National action plan for mapping and remediation," accessed Jan 28, 2019, http://www.ngu.no/upload/Arrangement/Internasjonale%20dager%202008%20-%20foredrag/Pollution_Ottesen.pdf
76. Rabin, Richard. "The Lead Industry and Lead Water Pipes: 'A Modest Campaign'" *American journal of public health*, Vol. 98(9) , Sept 2008, pp. 1584–92, https://www.ncbi.nlm.nih.gov/pmc/articles/PMC2509614/.
77. Claudia Alarcon, "Austin's Water Crisis Reveals Deeper Problems—And a Potential Solution," *Forbes.com,* Oct 25, 2018, https://www.forbes.com/sites/claudiaalarcon/2018/10/25/austins-water-crisis-reveals-deeper-problems-and-a-potential-solution/#2e18721f1d60.
78. Dylan Brown, "Hemp could clean up—if people stop mistaking it for pot," *E&E News*, Oct 18, 2017, https://www.eenews.net/stories/1060063977/print.
79. Sandra J. Adams, "Annual Report on the Status and Progress of the industrial Hemp Research Program," *Virginia Department of Agriculture and Consumer Services,* Nov 1, 2017, https://rga.lis.virginia.gov/Published/2017/RD394/PDF.
80. "Pennsylvania Greenlights 16 Industrial Hemp Research Projects," *Lancaster Farming,* March 25, 2017, https://www.lancasterfarming.com/farming/field_crops/pennsylvania-greenlights-industrial-hemp-research-projects/article_73eb806c-8b04-5b91-8f48-6226c23d1754.html.
81. Andrew Wagaman, "Lehigh University wants to be at forefront of hemp, pot's straight-laced sibling," *The Morning Call,* Dec 30, 2017, https://www.mcall.com/business/mc-biz-pennsylvania-hemp-lehigh-university-opportunity-20171228-story.html.
82. Meghan Corbett, "Hemp Health," *Wilma,* July 2017, http://www.wilmaon-theweb.com/July-2017/Hemp-Health/.
83. "HB12-1099—Phytoremediation Hemp Remediation Pilot Program," *Colorado General Assembly Tracking,* last modified June 4, 2012, http://www.legispeak.com/bill/2012/hb12-1099.
84. Andrew Leonard, "Can Hemp Clean Up the Earth?" Politics Features, *Rolling Stone,* June 11, 2018, https://www.rollingstone.com/politics/politics-features/can-hemp-clean-up-the-earth-629589/

85. Ahmad, Rafiq and Tehsin, Zara and Tanveer Malik, et al., "Phytoremediation Potential of Hemp (Cannabis sativa L.): Identification and Characterization of Heavy Metals Responsive Genes," *CLEAN—Soil Air Water*. Vol. 44 (2), August 2015, https://www.researchgate.net/publication/281651509_Phytoremediation_Potential_of_Hemp_Cannabis_sativa_L_Identification_and_Characterization_of_Heavy_Metals_Responsive_Genes.
86. Ahmad, R., Tehsin, Z., Malik, S. T., et al., "Phytoremediation Potential of Hemp (Cannabis sativa L.): Identification and Characterization of Heavy Metals Responsive Genes," *Clean Soil Air Water,* Vol 44, Nov 13, 2015, pp. 195–201. https://onlinelibrary.wiley.com/doi/abs/10.1002/clen.201500117.

9

THE EPA IS NOT YOUR FRIEND

So if hemp is essentially nature's vacuum cleaner for toxic substances, why hasn't the EPA gotten on board? After all, the Environmental Protection Agency is tasked with protecting the public from toxic chemical exposure, and the agency estimates that one in four Americans lives within *three* miles of a hazardous waste site.[1] While hemp might not be able to absorb all the toxins at those sites overnight, it can at the very least reduce our exposure. Plus, while it's cleaning up our mess, hemp consumes four times the carbon dioxide than trees do, so it's giving us *more* oxygen as it grows.

Although the EPA was founded to *protect* the environment (and its inhabitants) from the harmful effects of industrial chemicals, it has been failing in its core mission for generations. Granted, I was not a fan when Trump vowed to dismantle the EPA, and I'm not a fan of Andrew Wheeler, whose only interest seems to be in rolling back environmental regulations (which is to be expected, of course, since he was a lobbyist for coal-mining company Murray Energy Corp. prior to being confirmed as the EPA's deputy administrator). However, the irony here is just optics.

EPA officials have a long history of telling the public they're protecting the environment from major corporations, including Monsanto, DuPont, and Dow, while simultaneously working with chemical companies to allow them to accomplish whatever they want. Today, the EPA is acting as it always has in this regard, but the ruse is gone. Unfortunately, when it mattered most, the EPA let American industry do incredible damage to our land, and the majority of Superfund sites (including the very first site) are a result of that negligence.

THE POISON PAPERS

EPA as a whole has never truly put the well-being of the environment over the demands of corporations. The proof of the collusion between the EPA and major chemical companies is now readily available online for anyone to read at PoisonPapers.org. This is a collection of over 200,000 pages—letters, emails, and all sorts of written correspondence between the EPA and chemical corporations. The documents were collected by Carol Van Strum through Freedom of Information Act requests, and they have been instrumental in many lawsuits against chemical corporations, including Dewayne Johnson's.

In the 1970s, Carol Van Strum, who lived in the Oregon Siuslaw National Forest, discovered that the Forest Service was spraying 2,4,5-T and 2,4-D—the chemical herbicides that also happen to be the active ingredients in Agent Orange. These chemicals are known as dioxins; they are man-made and do not degrade easily over time. We now know dioxins cause cancer, birth defects, and serious harms to the environment, but from 1972 to 1977, the US Forest Service was using them as herbicides to kill weeds. On one occasion, the chemicals were actually sprayed from a Forest Service aircraft directly onto Carol Van Strum's house and onto her four children, who were fishing in a nearby river. Immediately following a flyby spraying, her children "developed nosebleeds, bloody diarrhea, and headaches, and many of their neighbors fell sick, too."[2] Van Strum lived near the town of Alsea, where residents were noticing:

- Women had miscarriages shortly after the Forest Service sprayed.
- Local animals were dying with deformities: ducks had backward-facing feet, elk were being born blind, birds had misshapen beaks.
- Cats and dogs were bleeding from their eyes and ears.

Much to the townspeople's surprise, the US Forest Service refused to stop the spraying after learning of these side effects. The chemical 2,4-D (2,4,-Dichlorophenoxyacetic acid) that was causing all of this is still used as a weed killer and it continues to be sold to the public at Walmart, Amazon, Tractor Supply, and just about anywhere herbicides are sold. It was invented during World War II and is one of the oldest pesticides that's still on the market today. Despite the fact that it was one of the two active ingredients in Agent Orange *and* despite the fact that scientific studies link it to horrible health concerns including non-Hodgkin's lymphoma in humans and canine malignant lymphoma in household dogs,[3] it's still used to kill weeds in public spaces such as athletic fields and golf courses.

The only reason 2,4,5-T has been phased out is because of Van Strum; she filed a lawsuit that led to a temporary ban in 1977, and in 1983, the EPA banned its use as a herbicide in the United States. However, the agency continued to allow

2,4-D, despite the fact that the dioxin is just as dangerous to human health and to the environment.

Carol Van Strum has continued to be involved in lawsuits against chemical industries for decades. She became the go-to source for lawsuits against pesticide companies due to the amount of firsthand material she collected.

"People would call up and say, 'Do you have such and such?' And I'd go clawing through my boxes," explained Van Strum, who'd often acquire even more documents through these requests, and she held onto those as well.[4]

The accumulation of these documents was stored (and deteriorating) in a barn on Van Strum's property until a team of volunteers and activists scanned and organized them in a free online database dubbed the Poison Papers. What is most shocking about the documents is what the EPA knew and when the EPA knew it and what little it did to change the environmental impact of toxins.

Dr. Jonathan Latham, the executive director of the Bioscience Resource Project, the organization that orchestrated the compilation of the Poison Papers, explained:[5]

> *These documents represent a tremendous trove of previously hidden or lost evidence on chemical regulatory activity and chemical safety. What is most striking about these documents is their heavy focus on the activities of [EPA] regulators. Time and time again regulators went to the extreme lengths of setting up secret committees, deceiving the media and the public, and covering up evidence of human exposure and human harm. These secret activities extended and increased human exposure to chemicals they knew to be toxic.*

For the most part, the Poison Papers prove: The EPA wasn't deceived by corporations. The EPA wasn't bribed by corporations. The EPA was complicit.

For instance, the Clean Air Act (1970) and Clean Water Act (1972) are certainly crowned the major achievements of the agency, and while these acts have greatly helped to limit our exposure to dioxins and other toxic substances, the EPA mainly protected major chemical industries by granting them more time to phase out toxic substances with no consequences for the damage caused.

Lead-based paint and leaded gasoline are prime examples. While the agency vowed to lower hydrocarbons, carbon monoxide, and nitrogen oxides by 90 percent a few years following the passage of the 1970 Clean Air Act, the EPA didn't actually ban leaded gasoline until the early 1990s.

Meanwhile, Richard Nixon signed the Lead-Based Paint Poisoning Prevention Act in 1971 and subsequent legislation in 1976 that effectively banned leaded paint entirely. However, that legalization was created by the Consumer Product Safety Commission—not by the EPA.[6] Even though the government fully acknowledged the irreversable damage of lead back in 1971, the EPA still gave corporations approximately twenty-five years to stop producing leaded gas.

When the Senate wanted to ban the use of lead in gasoline in 1984, the CDC's Center for Environmental Health fully supported it, and the EPA was tasked with regulating the industry to decrease the lead content in gasoline by 1986. *The New York Times* reported that the agency vowed to ban leaded gas completely by 1988.[7] However, industries were given an even longer extension. The nationwide ban didn't occur until January 1, 1996, when amendments to the Clean Air Act took effect to specifically outlaw leaded gas.

There are many reasons why the EPA isn't as effective as it should be. Ever since the agency was founded in December 1970, it has implemented standards and guidelines for chemical products; companies must submit tests to the EPA to prove a product is safe when used as directed, but many of the chemicals that regulators were tasked to investigate were already in existence for decades prior to the EPA's existence, and, like lead, these chemicals were embedded in nearly every industry.

CEOs had been lobbying politicians and donating to political campaigns for decades, so putting restrictions on these companies—such as how many dioxins they release into the air—wasn't entirely an easy task. Neither was ensuring the companies were in compliance. Even today, research from a nonprofit watchdog group or whistleblower is typically the catalyst for an EPA investigation.

WHAT DOES THE EPA DO?

As a regulatory agency, the EPA enforces the environmental laws that Congress passes by putting together federal guidelines. So, for example, the Safe Drinking Water Act put the EPA in charge of ensuring safe drinking water for the public. Yet, the agency doesn't necessarily follow through to verify rules are being followed. The drinking-water crises in Flint, Michigan, and Newark, New Jersey, weren't discovered by the EPA; the whistle was blown by independent activist groups and independent studies.

In November 2018, CNN reported that HaloSan, a chemical used to treat pools and spas, was being added to the drinking water of Denmark, South Carolina, for ten years without EPA approval. The state of South Carolina actually approved the use of the chemical, and assured people that the water was safe, even though officials had no basis of knowledge to make those statements, since the chemical wasn't evaluated by the EPA as a disinfectant for drinking water.[8] If the EPA was serious about enforcing the Safe Drinking Water Act, this chemical would've been discovered ten years ago.

It's tough to say what the benefit of the EPA is if one of its primary functions—to ensure tap water is indeed safe to drink—isn't a priority.

As it pertains to lead, in December 2018, the EPA put out a memo titled "The Federal Action Plan to Reduce Lead Exposures and Associated Health Impacts,"[9] which was not well-received by independent researchers or activists.

"Here at EPA, we are combating lead exposure on all fronts: in homes, schools, consumer products and drinking water," Wheeler said of his new action plan.[10] "We are updating the Lead and Copper Rule for the first time in over two decades, we are strengthening the dust-lead hazard standards and we are using our grants and financing problems to help communities test for lead, replace lead pipes and upgrade water infrastructure."

While that may be the overall goal, the plan doesn't actually introduce new regulations to physically remove lead in water or soil. It also doesn't outline what the EPA is changing in the Lead and Copper Rule (which hasn't been thoroughly updated since it was first written in 1991). Wheeler stated the new regulations will be announced in spring 2019, but activists are skeptical, as the Federal Action Plan to Reduce Lead Exposures and Associated Health Impacts didn't reveal anything new overall.

Erik Olson, senior director of Health and Food for the Natural Resources Defense Council, wasn't impressed:[11] "This plan does not actually promise to take specific regulatory or enforcement action within any specific time. Feel-good promises to 'consider' and 'evaluate' actions without time frames or commitments . . . won't protect children."

So again, while it appears that the EPA is on our side, the role they actually play in protecting the American people isn't always clear, and this isn't a problem that began recently.

EPA DECEPTION AND COLLUSION

The Poison Papers show that the companies that produce hazardous chemicals knew of the harm, but continued to make them nonetheless for a number of reasons:

1. This is what they made, sold, and how they amassed profits (so why change the formula?).
2. There wasn't a foreseeable alternative that was safer and also cheaper to make. (They weren't interested in investing money to develop alternatives, either.)
3. If there was a safer alternative (and even if it was cheaper), they'd then have to restructure their entire operations (expensive undertaking) and that was out of the question.

4. These chemicals were the industry standard, so if they were to be phased out, then companies would have to disclose the hazards, thus opening them up to lawsuits.
5. As long as no one knew the truth about the harm, then no one was there to stop it.

When someone sues Monsanto or another chemical corporation, the EPA isn't necessarily helpful, since FOIA documents reveal how regulators work with chemical companies instead of standing up to them to demanding change or slap them with fines. Here are some specific examples:

Chemical companies knew dioxins in Agent Orange were harmful for decades

Before getting into what the EPA knew about dioxins and when, the Poison Papers establish what the chemical companies knew and how early they knew it. In January 1979, Vietnam veterans took Dow Chemical Company to trial in a class-action lawsuit due to the effects of Agent Orange exposure. In Dow's annual Securities and Exchange Commission report, the company chose to lie rather than come clean about the toxic effects:[12]

> *Dow believes it has not been scientifically demonstrated that the injuries claimed by the plaintiffs were caused or could have been caused by exposure to Agent Orange.*

However, there is a chain of documents dating back to the early 1950s that speak to Dow's knowledge and the company's role in covering up that knowledge. Prior to using this contaminant in the Vietnam War as part of the "Agent Orange" recipe, Monsanto physicians corresponded with Dow and German chemical manufacturer Boehringer Ingelheim. One particular letter stamped "confidential" from 1956 showed that all three companies knew "the extraordinary danger of the tetrachlorobenzodioxin" to humans, but they agreed to keep the information private.[13] The German manufacturer of 2,3,7,8-TCDD (a chemical used in Agent Orange) had found that in the 1950s, workers developed chloracne welts (blister-like infections) on their faces, neck, arms, and upper half of their bodies as a result of exposure to the dioxin. The company placed caged rabbits into the manufacturing facility for periods of twenty-four to forty-eight hours; within two weeks, the animals were dead, and autopsies revealed they died from acute liver failure.

In 1983, the *New York Times* reported that interoffice memos from 1965 showed Dow Chemical Company met with their competitors to talk about "toxicological problems" in certain samples of dioxin, including 2,4,5-T (also used in Agent Orange). After a meeting at Dow laboratory, chemist C. L. Dunn (the regulatory affairs manager for Hercules, a Dow competitor), summarized in writing that the test results revealed "surprisingly high" amounts of toxic impurities in 2,4,5-T:[14]

> *In addition to the skin effect. . . liver damage is severe . . . even vigorous washing of the skin 15 minutes after application will not prevent damage and may possibly enhance the absorption of the material. There is some evidence it is systemic.*

These memos (which are also included in the Poison Papers database) show that while the chemical companies were conducting these tests, they chose to keep the knowledge of harm confidential. This choice continues to this day, which is why the jury in Dewayne Johnson's case found Monsanto acted with malice and a disregard for human life.

The EPA was aware of dioxin harm and covered it up

From 1981 to 1983, Anne McGill Burford was the head of the EPA and made repeated trips to Dow. While she was only the chief of the EPA for twenty-two months, she still found time to collude with the chemical manufacturer. In 1981, she gave Dow executives a copy of a draft of a 1981 EPA report that placed blame on the company for contaminating Michigan rivers with dioxins, and after doing so, she edited the report's findings to state that the company *wasn't* to blame. While Burford was under investigation for this, she had two paper shredders installed in the EPA's hazardous-waste division (right around the time when the EPA was subpoenaed for its Superfund records).[15]

Internal EPA documents also show that officials knew as early as 1980 that 60 percent of deer and elk in western Oregon had abnormal increases in birth defects and reproductive problems due to the Forest Service spraying 2,4,5-T. The agency also knew the chemical was in the water supply of Lane County (and the town of Alsea, where Carol Van Strum lived). The EPA had completed testing to prove the chemical had entered the breast milk of nursing mothers in Alsea, and that the chemical caused birth defects. However, when the results of those tests were requested by Van Strum's legal team, the agency went so far as to try to discredit the results, stating the samples "may have been contaminated" with dioxin when they were being tested in the lab (a claim a lab technician whistleblower confirmed was false).[16]

The EPA denies its knowledge of dioxin harm

In the early 1980s, the CDC was investigating severe health concerns in Times Beach, Missouri. The researchers determined that the town was essentially being poisoned by high exposures to dioxins (specifically 2,4-D and TCDD). When a flood forced Times Beach residents to evacuate, the CDC recommended the town not be reinhabited for fear that subsequent flooding would spread the contamination farther.[17] By 1985, the entire town was relocated (at a total cost of approximately $36.7 million).[18]

Today, Times Beach, which is a mere thirty miles southwest of St. Louis, is the site of an incinerator, one the federal government paid over $200 million to build, which is used to burn tons of toxin-contaminated materials.[19]

EPA briefing notes from meetings regarding Times Beach dated from 1980 and 1982 show that senior EPA officials met with the "Dioxin Working Group" (leaders and lobbyists in the chemical industry) to discuss how to assess Times Beach. The meeting notes revealed the real problem the EPA faced was twofold:[20] (1) Removing the dioxin was determined to be too expensive an undertaking (because the chemical company certainly wasn't going to do it), and (2) Having the EPA admit the town was contaminated with something health-hazardous would directly conflict with the "previous guidance" the public had received from the agency when it stated TCDD and other dioxins aren't carcinogenic.[21]

Meanwhile, similar dioxin concerns had already occurred in Love Canal, New York. The Hooker Chemical Company buried more than eighty industrial chemicals underneath the town from 1947 to 1952, and chemicals starting oozing up through the ground in the late 1970s. The cleanup effort took twenty-one years and cost close to $400 million.[22] The towns of Love Canal, Times Beach, and Jacksonville—the Arkansas home of Vertac Chemical Corporation, producer of herbicides, including the compounds in Agent Orange—were declared the first three Superfund sites.

Congress realized the EPA hadn't done enough to regulate the amount of dioxins produced by chemical companies or inform the public about health concerns, so the agency was mandated to conduct a National Dioxin Study. While this report was certainly necessary, EPA regulators took the opportunity to avoid announcing what they knew about the harm of dioxins. The Poison Papers documents disclose how the EPA derailed the release of its own report. When Congress initially instructed the EPA to carry out the study, agency regulators told consumer paper product companies to "suppress, modify or delay" their studies and findings.[23] When the National Dioxin Study was finally released in 1987, it showed there were high levels of dioxins in everyday products from baby diapers to paper products, including coffee filters. In the 1980s, Americans were suffering the effects of long long-term exposure to dioxins; yet here we were, bringing dioxins into our homes and exposing newborns to them every day with diapers, and the EPA was clearly aware. Why else would regulators tell companies to suppress, modify, or delay scientific evidence?

In a November 1987 Greenpeace press release titled "EPA & Paper Industry May Have Suppressed Dioxin Findings, Judge Rules" (where Carol Van Strum is listed as the media contact), Federal Judge Owen M. Panner was quoted as stating there was a "valid concern regarding the credibility" of EPA officials, especially since the EPA documents were leaked by an anonymous informer inside the paper industry. Carol Van Strum also noted that the EPA's collusion with these companies led her to question "the integrity of the National Dioxin Study itself" and the EPA as a whole. The leaked documents "show an explicit agreement between EPA and industry officials to forestall regulation of dioxin pollution and to forestall the adverse impact of publicity on the industry by withholding and downplaying [what was already known by the agency]," Van Strum explained.[24]

Further Poison Paper documents reveal:

- In 1990, the EPA allowed hazardous waste to be "recycled" into the ingredients for pesticides and other chemical products.[25]

- Though there hasn't been a particular document detailing what was "recycled" into chemical products, there are hints to suggest that the EPA allowed this because regulators might've known it was already happening. A document from 1985 shows Monsanto sold a chemical to the company that produces Lysol to use in a disinfectant spray that contained TCDD. This disinfectant was sold mainly to hospitals, and was used for twenty-three years. When Monsanto sold the chemical, the company was aware that TCDD increased the risk of cancer in humans and animals,[26] but no one divulged this information to the Lysol manufacturer.

INDUSTRIAL BIO-TEST LABS (IBT) SCANDAL

One of the most widespread American industrial product safety testing scandals happened in 1983 at Industrial Bio-Test Labs (IBT). In an undercover investigation, the FDA discovered that the "independent" lab—which performed more than one-third of all toxicology testing in the United States for pharmaceutical companies and chemical manufacturers[27]—was engaging in scientific fraud.

FDA demanded an investigation into what scientific studies were compromised, and since those studies were what chemical companies gave to the EPA to prove the products were safe for use, the EPA was charged with determining what studies were fraudulent.

One of the many problems with IBT was that it allowed Monsanto to conduct its own tests on its products, including glyphosate.[28] This allowed the chemical giant to provide misleading (or entirely false) data to the EPA, and this fraudulent scientific literature was used to prove that the products could pass regulatory measures. These publications were then used to defend Monsanto in lawsuits because they were passed off as data from an independent lab.

For instance, Aroclor and other PCBs were once used as coolants and lubricants in electrical equipment because the chemicals didn't burn easily and were good insulators. Manufacturers stopped making them in 1977 due to their effects on the environment. According to the EPA, because PCBs never fully degrade, particles can still be found in milk, eggs, dairy products, fish, poultry, and red meat.[29] While manufacturers appeared to do the right thing when they voluntarily decided to stop making them, documents show that Monsanto continued to produce and sell PCBs for *eight* years after their own IBT research had determined the chemicals were hazardous to human health.

The Poison Papers reveal that Monsanto was fully aware of the environmental impact from PCBs because the company had put together a "PCB Environmental Pollution Abatement Plan" in November 1969. The report referenced studies from 1966 that found PCB contamination in fish, birds, and eggs. Monsanto also admitted in the report that the residues of PCB compounds were of global concern because "evidence of contamination" had been found "in some of the very remote parts of the world."[30] Even so, the courses of action were to do nothing (don't tell the public of the harm), phase out PCBs, then "respond responsibly" once *new* scientific research came out to further prove what Monsanto already knew to be true.

In the early 1970s, Dr. Paul Wright (a Monsanto employee who was briefly employed at IBT to conduct Aroclor studies and then rehired by Monsanto as a manager of toxicology once the research was finished) falsified Aroclor studies at IBT after test results showed rodents developed tumors from exposure.[31] The test results were then given to the EPA as a response to the Clean Air Act. The agency was in the process of determining what quantity of PCBs could be discharged into the air, and they were relying on tests from IBT to determine the overall toxic effects of the chemical. According to court documents from a September 1987 lawsuit against Monsanto, the prosecutor asked Dr. Wright if he contacted EPA officials to get them to stall on regulations that limited the discharge of PCBs into the environment. Dr. Wright chose to invoke the Fifth and Fourteenth Amendments rather than answer this question.[32]

By 1972, Monsanto had voluntarily stopped selling PCBs for all uses except electrical applications, but even when executives had the chance to quietly address the health issue with those clients, they didn't come clean. In 1975, a Monsanto manager wrote to a staff supervisor at Westinghouse Electric Corporation: "no human harm has resulted" from over forty years of use of PCBs, but then added, *only* in repeated and prolonged exposure to concentrated levels of PCBs that are above and beyond the accepted "Threshold Limit Levels" can PCBs "have permanent effects on the human body."[33]

When the EPA determined only 16 percent of the tests conducted by IBT were actually valid, the agency announced to the public that the makers of pesticides had ninety days to "offer new test data or make a commitment to do further work to obtain data."[34] (Yes, studies concerning glyphosate's safety were included in this bundle of fraudulent tests; IBT test data was relied upon by the EPA when it approved Roundup for public use in 1974.)[35]

Even though so few IBT tests were accurate, the EPA never publicly disclosed *which specific* products were required to be retested, or which products were potentially affected by IBT's unsound tests, so products remained on the shelves. Poison Papers documents also show that the EPA encouraged the chemical industry and IBT *not to disclose* which consumer products might not be safe. There weren't any recall notices because the EPA never had any intention of recalling anything.

Behind closed doors, the Special Pesticide Reviews Division of the EPA worked with IBT to salvage as many of the original tests as possible. In an audit meeting hosted by the EPA on October 3, 1978, Fred Arnold, the acting branch chief for the Regulatory Analysis and Lab Audits in the EPA's Office of Pesticide Programs, discussed his plan in detail with EPA attorneys and the members of the agency's pesticide and toxicology divisions. David Clegg, the head of Pesticide Section for Canada's Health and Welfare, was also present, as Canada relied on the same IBT studies to determine if a chemical was safe.

At the meeting, Fred Arnold explained that the EPA *didn't* want to purge all files containing IBT data. He also didn't want to impose "data requirements" on the chemical companies by requesting that they resubmit studies, because he found this to be an "extreme action."[36] This directly contradicted the statements the EPA gave to the media regarding the situation.

"We determined that this was neither in the EPA's interest or the public interest or the registrants' [chemical companies] interest because a large number of studies, which were performed at IBT, were performed satisfactorily," Arnold explained.

He went on to say:[37]

We haven't found a single study, to date, where the microfiche is fully consistent with the valuation report. . . . The kinds of problems that we are seeing are, I think, random error in reviewing records, transcribing tables, and re-performing certain kinds of numerical, statistical analyses.

So essentially, the EPA acknowledged *all of the IBT* studies weren't consistent (therefore, they're inaccurate at best). So instead of challenging the results, the agency gave everyone involved the benefit of the doubt that the outcome of the trials (which concluded the chemicals were safe) are all correct.

From here, Arnold stated that the EPA will work with IBT to review the material and do whatever it takes to salvage the studies. If the data from a study conducted at IBT is not entirely complete, then the chemical companies can locate another study that produced the same results found in the IBT study. However, if the EPA and IBT can come up with any logical reasons as to why data is incomplete (such as "random errors" in the reporting process), then the EPA will keep relying on the study.

"If we're satisfied on the safety of the chemical, we'll proceed . . . but we have not established the policy to [reject a study] merely because it's IBT," Arnold further clarified. "In this re-registration process, we are going to come up with a lot of reasons for data gaps and uncertainties . . . and proceed in an orderly fashion and fill data gaps and not interfere with the ability to control pests and market pesticides."

So while the EPA is essentially agreeing to correct and finish IBT's math homework, and whenever IBT made a mistake or didn't complete a proof, the EPA will come up with an excuse—any excuse—to avoid creating even more work for itself and the pesticide industry, it is simultaneously avoiding any interference or potential conflict with the pesticide industry—the very industry the agency was founded to regulate.

While the public was under the impression that EPA forced the chemical industries to redo all pesticide tests and start fresh, the actual evidence suggests that the vast majority of IBT studies were never tossed. Of course, since the EPA willingly *kept* those studies as proof of a pesticide's safety, then how are we to really know if any chemicals that pass the EPA's regulatory process are actually safe?

THE GOOD LABORATORY PRACTICE

After the IBT scandal, the EPA created a system called "Good Laboratory Practice" to put specific guidelines in place to avoid future fraudulent studies.[38] For the EPA to accept a study for review, it must be done in a lab that has the good laboratory practice system in place.

What's interesting about this arrangement is that chemical companies are typically the bread-and-butter customers for labs that adopt the EPA's good laboratory practice, so it is possible that the issues at IBT are just a stone's throw away from reoccurring, especially since the standards and compliance monitoring program is set up as follows:

1. A chemical company pays an independent testing lab to conduct experiments on a chemical (typically by exposing a rat to it in various ways to test for toxicity).
2. The results are then given back to the chemical manufacturer to review.
3. The company presents the findings to the EPA, but the company is able to massage the results by writing justifications for anything alarming.

"What we find is that the tests are faulty. And [the EPA is] allowing the companies to miss experiments, or just not submit them, and the chemical still passes," states Poison Papers director Dr. Jonathan Latham during an NPR interview. "The EPA shouldn't be accepting the data and they are. The Poison Papers are showing this."[39] Latham also explains that even if there is a level of toxicity in the reports, the EPA can decide that the evidence is abnormal, or an aberration, and treat the findings as a reporting error so that the chemical can pass the inspection process and enter the marketplace.

"Their presumption is all chemicals are safe until they're seen as unsafe," says Latham. "If there's cancer in a male rat, but not female, there must have been some kind of error, so therefore the evidence of harm is gone. The final report says there's no evidence—and that's where we get this problem. This is very compartmentalized and inconsistent with how science works. Any argument will do at the end of the day."

THE ACADEMIC COMMUNITY'S RECOMMENDATIONS FOR THE EPA

Prior to the formation of the Poison Papers database in 2017, there were several academic publications detailing the lack of scientific process as it pertains to the EPA's decision making. "Strengthening Science at the U.S. Environmental Protection Agency: Research-Management and Peer-Review Practices" was published in 2000 by several academic committees. For brevity's sake, this was a meeting of the minds, but the many authors included members of the National Academy Press, the National Academy of Sciences, the National Academy of Engineering, the Institute of Medicine, the National Research Council, the Committee on Research and Peer Review in EPA, the Board of Environmental Studies and Toxicology, the Commission on Life Sciences, and the Commission on Geosciences, Environment, and Resources—among many other organizations and academic institutions.

This publication was presented to Congress and focused on the EPA's scientific methods from 1990 to 2000. The materials are of course outdated in regards to today's EPA, but since the report covers nearly the entire presidency of Bill Clinton (with environmentalist Al Gore by his side as vice president), the fact that the EPA struggled to live up to expectations from so many experts is worth exploring.

Clinton enacted ambitious environmental accomplishments during his two terms in office, including:[40]

- Cleaning up six hundred Superfund sites
- Protecting more than 4 million acres of public land through national parks, monuments, and wilderness, including the "roadside rule," which allowed the US Forest Service to put roughly 60 million acres (a third of national forests) off-limits to development
- Creating the Office of Children's Health with the EPA to focus specifically on environmental concerns for children

However, the academic report focused on how the EPA came to its conclusions as to whether a chemical is safe and at what quantity, among other concerns pertaining to the agency's regulatory process. The panel that put together these findings had also written several other reports during the Clinton administration to direct the EPA on a course of effective, scientific decision making, but found that regardless of their recommendations, the EPA did little to actually adopt the policies. Yes, management updated their handbooks, but what the employees actually *did* when evaluating the safety of chemicals didn't change all that much.

The panel concluded that part of this problem is due to the culture at the EPA, since its focus isn't particularly scientific. The EPA was formed "to protect human health and to safeguard the natural environment—air, water, and land—upon which life depends."[41] There aren't any scientific mission statements attached to the agency, like there are at the National Institutes of Health or the National Science Foundation, so although we're under the impression that the agency is filled with scientists, the majority of decision makers have notoriously been lawyers and those who have experience in the industries the EPA is supposed to be keeping tabs on.

In their 1992 "Safeguarding the Future" report, the authors stated that although there were scientists employed at all levels throughout the EPA, the scientific method employed by the EPA was "of uneven quality" and the agency didn't use their "scientific knowledge and resources effectively."[42] The report went on to say that there was "a widely held perception" among people both outside and inside the agency that EPA science was "adjusted" by EPA scientists or decision makers, "consciously or unconsciously, to fit policy" so that products could pass the regulatory system.[43] After this report was published, the EPA promised to achieve several scientific accomplishments over the next decade. In 1997, the agency vowed that its research program:[44]

> *Will measurably increase our understanding of environmental processes and our capability to respond to and solve environmental problems. During the past decade, significant concerns have been expressed about the adequacy of the Agency's ability to assess risks—not only to human health, but also to ecosystems. . . . Our aim is to reduce major areas of uncertainty in our analyses of risk and to minimize reliance on default assumptions.*

However, the academic report published in 2000 found the agency had yet to make good on these promises. The academic panel stated the EPA hadn't evolved into a true watchdog agency for a number of reasons:

- **Refusal to Adapt Peer-Review Practices**

In February 1998, a new peer-review handbook was given to EPA employees to outline what research can be peer-reviewed and when outside experts or independent reviewers can be called upon to improve the quality of scientific data. However, when the academic community reviewed the handbook in 1999, there was concern for conflicts of interest because the EPA allowed "a decision-maker on a particular work product to be the peer-review leader."[45] *How can a project manager be objective enough to also be the peer-reviewer for his or her own findings?*

To increase transparency and accuracy, the academic panel suggested EPA research "should be published in peer-reviewed journals that are open to scientific and public scrutiny."[46] This would also help ensure that the EPA's science is of the same quality and standard of other researchers.

- **Refusal to Hire Scientists in Leadership Positions**

After the 1992 "Safeguarding the Future" report, the EPA hired additional scientists, including a science adviser to the administrator of the EPA. The adviser was to be "a key player when the EPA makes a policy decision," as well as the point person to ensure activities conducted by EPA scientists were professional, and also reach out to the scientific community at large when appropriate for additional counsel.[47] (Essentially, the science adviser was responsible for conducting all of the things the EPA should be doing.)

While the adviser who was hired was extremely qualified, he was never "given the authority that would be required" for his position, and once there was a change in administration in 1993, he was not only let go, but the position was never filled again, leaving the top echelon of the decision-making body without objective scientific guidance.

Since EPA administrators "have typically been trained in law, not science," and the "EPA's regulatory offices are not required to follow scientific advice" from the Office of Research and Development, this lack of input from an actual scientist deeply undermined scientific credibility at the EPA.[48]

- **Refusal to Engage in Graduate Fellowship and Postdoctoral Programs**

In 2000, the workforce at the EPA was aging; more than 47 percent of employees in the Office of Research and Development were fifty years old or older, and more than 550 employees were eligible for retirement within the next five years. Since the EPA has had periodic hiring freezes (and budget cuts), this makes it difficult for the agency to compete for new talent, especially when private-sector and academic institutions provided a more secure work environment.

The academic panel recommended the EPA start a fellowship program to open research analysis to four to six "senior research scientists and engineers with world-class reputations in areas vital to the EPA's long-term strategy and direction."[49]

In 2000, the Office of Research and Development had eight fellows, but there weren't any present in upper management or administration (where EPA regulatory decisions were actually made), so it was suggested that the EPA expand the program to include endowed academic research chairs, mimicking programs offered by the National Institutes of Health.

- **Vague Research Accountability**

Since the EPA chief is appointed by the president, the agency's mission, goals, priorities, and practices often change every four to eight years. Theoretically, the scientific evidence behind environmental policies should remain constant, regardless who is president, but when the EPA changes a regulation or updates a policy, the academic community questions if science is taken into account because the research that led the EPA to make a decision isn't accessible. At the time, the EPA didn't provide much documentation or transparency, which can obviously be concerning if the outside academic community's thought process doesn't mirror what the EPA is suggesting.

The academic panel recommended that the EPA provide documentation—particularly regarding what research studies were used to analyze product safety and why that research in particular was relied upon when compared to other studies.

While there were many course corrections at the EPA under President Clinton, including lowering the quantity of naturally occurring arsenic in drinking water, his main objective was to prove that EPA regulators didn't have to make an either/or decision as it pertained to either siding with American industry or saving the environment (thus harming industry profits). On January 11, 2000, President Clinton stated:[50]

> *From our inner cities to our pristine wild lands, we have worked hard to ensure that every American has a clean and healthy environment. We've rid hundreds of neighborhoods of toxic waste dumps, and taken the most dramatic steps in a generation to clean the air we breathe. We have made record investments in science and technology to protect future generations from the threat of global warming. We've worked to protect and restore our most glorious natural resources, from the Florida Everglades to California's redwoods to Yellowstone. And we have, I hope, finally put to rest the false choice between the economy and the environment, for we have the strongest economy perhaps in our history, with a cleaner environment.*

Yet, the concerns mentioned in the academic report continued under George W. Bush, thus confirming how quickly the EPA is subject to the whims of whatever political party wins the presidency:

- In October 2004, NASA scientist James Hansen claimed that data collected by NASA showing the acceleration in global warming was being withheld from the public by the Bush administration.

- In July 2008, Jason Burnett, a former official at the EPA, wrote a letter to the Senate stating Vice President Dick Cheney and the White House Council on Environmental Quality were censoring discussions on climate change.

- Key sections of the Clean Water and Clean Air Acts were gutted, the Endangered Species Act was dismantled, millions of acres of

> wilderness were open to mining, drilling, and logging, and the EPA was told to reduce its enforcement of regulations.[51]

According to the *Encyclopaedia Britannica*:[52]

> *The Bush administration was frequently accused of politically motivated interference in government scientific research. Critics charged that political appointees at various agencies, many of whom had little or no relevant expertise, altered or suppressed scientific reports that did not promote administration policies, restricted the ability of government experts to speak publicly on certain scientific issues, and limited access to scientific information by policy makers and the public.*

While President Obama's two terms seemed to recapture the idea that American industry can get along with the environment and still be profitable, his first choice for EPA administrator, Lisa Perez Jackson, authorized and defended BP's choice to use Corexit to clean up the *Deepwater Horizon* drilling rig explosion in 2010.[53] Corexit is a dispersant, a toxic chemical marketed to break oil down into smaller particles and remove it from the environment. The chemical, which was never used in a large-scale oceanic oil spill previously, was later found to be extremely harmful to human health, and to the health of wildlife and the environment as a whole, because it breaks oil down into tiny particles that are then absorbed by cell walls at a faster rate.[54]

At a Senate hearing on June 15, 2010, Jackson stated, "In the use of dispersants we are faced with environmental tradeoffs."[55] Since the EPA has never stated the wide-ranging, irreversible damage caused by Corexit is enough of a hazard to remove it from the list of substances used to clean up oil spills, the chemical continues to be used in ocean spills around the world.

Lisa Jackson spent twenty years as an environmental officer at the state and national levels prior to heading up the EPA, and was widely accepted by scientists and environmental activists[56] when she was appointed by President Obama. In December 2012, she announced that she'd leave the EPA as a form of protest, claiming the Obama administration would move to support the Keystone pipeline and she didn't want to be a part of it. However, her critics stated the timing was a little too coincidental, given the damning reports that had surfaced on Corexit by independent studies. She also had a secondary email account at the EPA, not under her real name, but under the alias "Richard Windsor." After digging further into her dual email accounts, there was a correspondence with Alison Taylor, a vice president at Siemens (and lobbyist). Jackson told Taylor to contact her on her personal email "when you need to contact me directly."[57] Although Jackson claimed her second email was innocuous, it left many unanswered questions.

"I don't know any other agency that does this," stated Anne Weismann, the chief counsel of the watchdog group Citizens for Responsibility and Ethics in Washington. "Why would you pick a fictitious name of someone of different gender? To me it smacks of . . . trying to hide."[58]

From 2009 to 2016, the EPA published approximately 3,900 rules in the Federal Register, many of them initiated by environmental health and air-quality expert Gina McCarthy, whom Obama chose to replace Lisa Jackson as the head of the EPA. McCarthy structured additions to the Clean Water Act and Clean Air Act, including the Clean Power Plan, which decreased coal plant emissions. Yes, overall, the Obama administration changed the course of the EPA by following what the "Strengthening Science at the U.S. Environmental Protection Agency" report recommended: if scientists and environmentalists are in charge of the EPA, then the EPA as a whole will employ those standards. However, if you recall from chapter 5, there was still collusion between Monsanto and the EPA. Monsanto was effective in manipulating the agency into delaying their review of glyphosate, a review that ultimately concluded the chemical isn't carcinogenic, which is at odds with the World Health Organization and California's Prop 65 ruling. The EPA was also criticized for not acting sooner in the Flint water crisis.

In 2015, the Natural Resources Defense Council wrote a letter to Gina McCarthy informing her of the intention to sue the EPA under the Clean Water Act for failing to issue regulations "to prevent and contain the discharge of hazardous substances" from above-ground storage tanks.[59] Then in February 2016, the council did in fact sue under the Clean Water Act to "force it to set limits on perchlorate," a toxic chemical used as an oxidizer in rocket propellants, fireworks, and signal flares that is also widely used by the military and defense industry.[60] According to the Natural Resource Defense Council, "even at low levels, it can present health risks to children and pregnant women," including brain damage in babies. The council detected perchlorate in the drinking-water systems of 16.6 million Americans, including the residents of Flint.

"The situation in Flint has highlighted the importance of acting swiftly to protect kids from toxic chemicals in our drinking water," stated Erik Olsen, Director of the Health and Environment Program at the Natural Resources Defense Council. "Five years after EPA promised to protect millions of Americans—including children and pregnant women—who drink perchlorate-contaminated water, the agency still hasn't even proposed a standard. We have an obligation to ensure that every American has access to safe drinking water. It is high time for EPA to get toxic perchlorate out of our kitchen taps."[61]

While the EPA has admitted since 2011 that it needed to place limits on the chemical, as of an August 30, 2018, court filing, it claimed it needed another six months to review research to determine what those limits should be.[62] In this case, the EPA under the Trump administration and the Obama administration are both to blame for stalling on regulating perchlorate. The EPA should've already addressed the issue because it pertains to the Clean Water Act, but Congress is just as much to blame. In March 2017, Senator Kirsten Gillibrand introduced a bill to force the EPA's hand by requiring the agency to set drinking water standards for certain contaminants, including 1,4-dioxane (a stabilizer used to manufacture and process paper, cotton, textiles, cosmetics, automotive coolant, and other products[63] that affects the immune system) and perchlorate—the bill has yet to pass.[64]

Meanwhile, the FDA has been conducting testing to discover that "virtually all types of food" have "measurable levels" of perchlorate.[65] An FDA study published in 2016 in the *Journal of Exposure Science & Environmental Epidemiology* noted that the highest concentrations were found in bologna, salami, and rice cereal for babies.[66] As a note for those who don't know, rice cereal is recommended as the first solid food to feed to infants (at around four to five months of age).

For some reason, food manufacturers are also able to add perchlorate as an anti-static agent to plastic packaging for dry food. The toxin is either added to the packaging of the final products or to the packages surrounding the raw materials prior to processing. Either way, contamination via food could be entirely avoidable if the FDA removed the approval for its use, as it did in October 2018 for seven artificial flavors that cause cancer in animals.[67] The Natural Resource Defense Council is currently suing the FDA as well, hoping the lawsuit will push the agency to ban the toxin.

Perhaps the FDA should stop worrying about how to regulate CBD and start regulating this egregiously harmful contaminant in baby food.

THE EPA'S FUTURE ROLE IN REGULATING TOXINS

In April 2018, several university academics published a paper titled "The Environmental Protection Agency in the Early Trump Administration: Prelude to Regulatory Capture" in the *American Journal of Public Health*. The paper details how new EPA leadership under the Trump administration has "explicitly sought to reorient the EPA toward industrial and industry-friendly interests, often with little or no acknowledgment of the agency's health and environmental missions."[68]

According to the findings, the administration is aimed at "deconstructing, rather than reconstructing" the EPA "by comprehensively undermining many of the agency's rules, programs, and policies while also severely undercutting its budget, work capacity, internal operations, and morale." Interestingly, the authors note in the acknowledgments that the research was supported by the National Institute of Environmental Health Sciences of the National Institutes of Health (under Award Number T32ES023769), and the study was "conceptualized, designed, analyzed, written, and supported by the Environmental Data and Governance Initiative."[69] While the authors admit that their content does not represent the views of the NIH, they relied on interviews from members of the EPA and the Occupational Safety and Health Administration to compile the information.

Now that Andrew Wheeler is the current EPA chief (a former coal production lobbyist who is neither an environmentalist nor a scientist), there are several reports that speak to the agency's plans to roll back regulations imposed under the Obama administration, since most Republicans viewed them as government overreach. As President Bill Clinton referenced near the end of his two terms in office, there's a false theory that EPA regulations either help industry (which harms the environment) or harm industry (by bettering the environment). Historically speaking, the EPA seems to choose to help industry more often than not. Until this *industry versus environment* perception changes, we can't count on the EPA to be effective.

NOTES

1. United States Government Accountability Office, "Superfund: Funding and Reported Costs of Enforcement and Administration Activities," July 18, 2008, https://books.google.com/books?id=1tOqgipAX7oC.
2. Sharon Lerner, "100,000 Pages of Chemical Industry Secrets Gathered Dust in an Oregon Barn for Decades," *The Intercept,* July 26, 2017, https://theintercept.com/2017/07/26/chemical-industry-herbicide-poison-papers/.
3. Ibid.
4. Ibid.
5. Center for Media and Democracy, "The Poison Papers: Secret Concerns of Industry and Regulators on the Hazards of Pesticides and Other Chemicals," *EcoWatch,* July 27, 2017, https://www.ecowatch.com/the-poison-papers-2465841261.html.
6. Tristan Fowler, "A Brief History of Lead Regulation," *Science Progress,* Oct 21, 2008,https://scienceprogress.org/2008/10/a-brief-history-of-lead-regulation/.
7. Philip Shabecoff, "EPA Orders 90 Percent Cut in Lead Content of Gasoline by 1986," *The New York Times,* March 5, 1985, https://www.nytimes.com/1985/03/05/us/epa-orders-90-percent-cut-in-lead-content-of-gasoline-by-1986.html.
8. Sara Ganim, "For 10 years, a chemical not EPA approved was in their drinking water," Health, *CNN,* Nov 28, 2018, https://www.cnn.com/2018/11/11/health/denmark-sc-water-chemical-not-epa-approved/index.html.

9. EPA, "Federal Action Plan to Reduce Childhood Lead Exposures and Associated Health Impacts," *President's Task Force on Environmental Health Risks and Safety Risks to Children,* Dec 2018, https://www.epa.gov/sites/production/files/2018-12/documents/fedactionplan_lead_final.pdf.
10. Timothy Cama, "Trump admin lays out plan to confront lead poisoning 'head-on,'" *The Hill,* Dec 19, 2018, https://thehill.com/policy/energy-environment/422096-trump-admin-lays-out-plan-to-confront-lead-poisoning-head-on.
11. Brakkton Booker, "Trump Administration Reveals Plan To Limit Lead Exposure, Critics Say It's Not Enough," National, *NPR,* Dec 19, 2018, https://www.npr.org/2018/12/19/678270138/critics-call-trump-administration-plan-to-reduce-lead-exposure-toothless.
12. David Burnham, "1965 Memos Show Dow's Anxiety on Dioxin," *The New York Times,* April 19, 1983, https://www.nytimes.com/1983/04/19/us/1965-memos-show-dow-s-anxiety-on-dioxin.html.
13. Letter from Monsanto Chemical Company to Dr. R. Emmet Kelly, "Chloracne cases at Badischen Anilin due to trichlorophenol," June 12, 1956, pp 509–156, https://www.documentcloud.org/documents/3418422-1A-1-563.html.
14. David Burnham, "1965 Memos Show Dow's Anxiety on Dioxin," *The New York Times,* April 19, 1983, https://www.nytimes.com/1983/04/19/us/1965-memos-show-dow-s-anxiety-on-dioxin.html.
15. UPI, "Acting EPA Chief let Dow edit dioxin study," March 1983, p 76, https://assets.documentcloud.org/documents/3418996/33B-19636-20098part2.pdf.
16. Letter from Peter DeFazio to House of Representatives Speaker Thomas O'Neill regarding EPA suppressing/delaying test results, Feb 20, 1985, https://www.documentcloud.org/documents/3720462-PP-B0450.html.
17. EPA press release, "Join Federal/State Action Taken to Relocate Times Beach Residents," Feb 22, 1983, last modified Sept 22, 2016, https://archive.epa.gov/epa/aboutepa/1983-press-release-joint-federalstate-action-taken-relocate-times-beach-residents.html.
18. William Powell, "Remember Times Beach: The Dioxin Disaster, 30 Years Later," *St Louis Magazine,* Dec 3, 2012,https://www.stlmag.com/Remember-Times-Beach-The-Dioxin-Disaster-30-Years-Later./
19. Patricia M. Szymczak, "Incinerator Burns Up Times Beach," *Chicago Tribune,* Feb 18, 1987, https://www.chicagotribune.com/news/ct-xpm-1987-02-18-8701130202-story.html.
20. EPA Briefing Document from Sept 27, 1982 Dioxin Working Group Meeting Summary, Dioxin Issueshttps://www.documentcloud.org/documents/3892977-PP-D0504b-Dioxin-Working-Group-Meeting-Summary.html. .
21. Ibid.

22. Anthony DePalma, "Love Canal Declared Clean, Ending Toxic Horror," *The New York Times,* March 18, 2004, https://www.nytimes.com/2004/03/18/nyregion/love-canal-declared-clean-ending-toxic-horror.html.
23. "The Papers," The Poison Papers, accessed Jan 25, 2019, https://www.poisonpapers.org/the-poison-papers/.
24. Greenpeace Press Release, "EPA & Paper Industry May Have Suppressed Dioxin Findings, Judge Rules," Nov 11, 1987, https://www.documentcloud.org/documents/3705611-PP-D0173.html.
25. EPA Letter to Rep. James McClure Clarke regarding FOIA request for pesticide ingredients, Dec 11, 1990, https://www.documentcloud.org/documents/3223271-Scan-20111114-171409.html.
26. Elizabeth Fay transcript regarding TCDD in Santophen, Aug 14, 1985, https://www.documentcloud.org/documents/3253654-Poison-Papers-B-1435.html.
27. Carson, Phillip A.; Dent, N. (2007-06-01). Good Clinical, Laboratory and Manufacturing Practices. Royal Society of Chemistry. pp. 172–174.
28. Phillip S. Smith testimony related to IBT lab fraud, Sept 3, 1987, Cecil Scott, et. al. vs. Monsanto Company, C. A. No. 84-1103-CA, p. 2, https://assets.documentcloud.org/documents/3253579/Poison-Papers-B-1360.pdf.
29. "Toxic Substances Portal–Polychlorinated Biphenyls (PCBs)," Agency for Toxic Substances & Disease Registry, *CDC*, July 2014, last modified Aug 27, 2014, https://www.atsdr.cdc.gov/toxfaqs/tf.asp?id=140&tid=26.
30. Monsanto, "PCB Environmental Pollution Abatement Plan," Nov 10, 1969, p. 15-17, https://www.documentcloud.org/documents/3032105-Monsanto-PCB-Pollution-Abatement-Plan.html#document/p1.
31. Phillip S. Smith testimony related to IBT lab fraud, Cecil Scott, et al. vs Monsanto Company, C. A. No. 84-1103-CA, Sept 3, 1987, https://www.documentcloud.org/documents/3253579-Poison-Papers-B-1360.html#search/p2/IBT.
32. Dr. Paul Wright of Monsanto/IBT testimony, Cecil Scott, et al. vs Monsanto Company, C. A. No. 84-1103-CA, Sept 3, 1987, https://www.documentcloud.org/documents/3116395-Poison-Papers-0188.html#search/p8/IBT.
33. Letter from Monsanto to Dan A. Albert Westinghouse Electric Corporation regarding health impacts of PCBs, March 18, 1975, https://www.documentcloud.org/documents/3131805-Poison-Papers-0424.html.
34. Philip Shabecoff, "EPA Threatens to Suspend Approval of Pesticides Over Test Flaws," *The New York Times,* July 12, 1983, https://www.nytimes.com/1983/07/12/us/epa-threatens-to-suspend-approval-of-pesticides-over-test-flaws.html.

35. Helen Christophi, "Monsanto Accused of Fraudulent Data in Roundup Cancer Trial," *Courthouse News Service,* July 27, 2018, https://www.courthousenews.com/monsanto-accused-of-fraudulent-data-in-roundup-cancer-trial/.
36. EPA IBT Audit Meeting transcript, Oct 3, 1978, https://assets.documentcloud.org/documents/3417915/IBT-Howard-Johnson-Transcript.pdf.
37. Ibid.
38. EPA, "Good Laboratory Practices Standards Compliance Monitoring Program," Compliance, US Environmental Protection Agency, accessed Jan 25, 2019, https://www.epa.gov/compliance/good-laboratory-practices-standards-compliance-monitoring-program.
39. Against the Grain, "The EPA and the Chemical Industry: A Cosy Alliance," *KPFA*, Oct 10, 2018, https://kpfa.org/episode/against-the-grain-october-10-2018/.
40. Juliet Eilperlin, "Does Bill Clinton Deserve to have EPA Named In His Honor? Actually, Yes He Does," *The Washington Post,* July 17 2013, https://www.washingtonpost.com/news/the-fix/wp/2013/07/17/does-bill-clinton-deserve-to-have-epa-named-in-his-honor/.
41. Committee on Research and Peer Review in EPA, et al, "Strengthening Science at the US Environmental Protection Agency," The National Academies of Sciences Engineering Medicine, *The National Academies Press,* 2000, p 25, https://www.nap.edu/read/9882/chapter/3#25.
42. Committee on Research and Peer Review in EPA, et al, "Strengthening Science at the US Environmental Protection Agency," The National Academies of Sciences Engineering Medicine, *The National Academies Press,* 2000, p 29, https://www.nap.edu/read/9882/chapter/3#29.
43. Ibid.
44. Committee on Research and Peer Review in EPA, et al, "Strengthening Science at the US Environmental Protection Agency," The National Academies of Sciences Engineering Medicine, *The National Academies Press,* 2000, p 26, https://www.nap.edu/read/9882/chapter/3#26.
45. Committee on Research and Peer Review in EPA, et al, "Strengthening Science at the US Environmental Protection Agency," The National Academies of Sciences Engineering Medicine, *The National Academies Press,* 2000, p 124, https://www.nap.edu/read/9882/chapter/5#124.
46. Ibid.
47. Committee on Research and Peer Review in EPA, et al, "Strengthening Science at the US Environmental Protection Agency," The National Academies of Sciences Engineering Medicine, *The National Academies Press,* 2000, p 127, https://www.nap.edu/read/9882/chapter/6#127.
48. Committee on Research and Peer Review in EPA, et al, "Strengthening Science at the US Environmental Protection Agency," The National Academies of

Sciences Engineering Medicine, *The National Academies Press,* 2000, p 129, https://www.nap.edu/read/9882/chapter/6#129.

49. "Bill Clinton on Environment," *On the Issues*, last modified Sept 11, 2018, http://www.ontheissues.org/Celeb/Bill_Clinton_Environment.htm.
50. Suzanne Goldenberg, "The worst of times: Bush's environmental legacy examined," Environment, *The Guardian,* Jan 16, 2009, https://www.theguardian.com/politics/2009/jan/16/greenpolitics-georgebush.
51. Brian Duignan, "Geroge W. Bush," *Encyclopaedia Britannica*, last modified Jan 2, 2019, https://www.britannica.com/biography/George-W-Bush.
52. Renee Schoof and Anita Lee, "Researchers worry about oil dispersants' impact, too," National, *McClatchy Newspapers,* Sept 18, 2013, https://www.mcclatchydc.com/news/nation-world/national/article24582025.html.
53. "What are Dispersants and How Do They Work?" *Beachapedia,* Aug 23, 2015, http://www.beachapedia.org/Dispersants.
54. Senate Hearing before the Committee on Small Business and Entrepreneurship, 111 Congress, Second Session, June 17, 2010, https://www.govinfo.gov/content/pkg/CHRG-111shrg73969/html/CHRG-111shrg73969.htm.
55. Lila Guterman, "Obama's choice to direct EPA is Applauded," News of the Week, *Science Magazine,* Vol. 322, Issue 5909, Dec 19, 2008, pp. 1775, http://science.sciencemag.org/content/322/5909/1775.
56. Erica Martinson, "Jackson denies secrecy at hearing," *Politico*, Sept 10, 2013, https://www.politico.com/story/2013/09/lisa-jackson-email-secrecy-096550.
57. Erica Martinson, "Lisa Jackson's 'Windsor' knot," *Politico,* Nov 20, 2012,https://www.politico.com/story/2012/11/lisa-jacksons-windsor-knot-084112.
58. Letter from the Natural Resources Defense Council to Gina McCarthy EPA regarding intent to sue, May 12, 2015, https://assets.documentcloud.org/documents/2170094/1-1.pdf.
59. "NRDC Sues EPA to Force it to Limit Toxic Chemical in Drinking Water," *National Resources Defense Council,* Feb 18, 2016, https://www.nrdc.org/media/2016/160218.
60. Ibid.
61. David Schultz, "It's Rocket Science: EPA Needs Six More Months to Study Perchlorate," News, Bloomberg BNA, Aug 31, 2018,https://www.bna.com/rocket-science-epa-n73014482439/.
62. The American Water Works Association, "What is dioxane?," What's in my Water?, *Drink Tap,* accessed Jan 25, 2019,https://drinktap.org/Water-Info/Whats-in-My-Water/Dioxane.
63. Emily Dooley, "Gillibrand bill would require EPA to set drinking water standards," Long Island, *Newsday,* March 3, 2017, https://www.newsday.com/long-island

/gillibrand-bill-would-require-epa-to-set-drinking-water-standards-1.13207460.
64. Tom Neltner, "FDA finds more perchlorate in more food, especially bologna, salami and rice cereal," Environmental Defense Fund, Jan 9, 2017, http://blogs.edf.org/health/2017/01/09/fda-finds-more-perchlorate-in-more-food/.
65. Abet E., Spungen J., Pouillot R., et al., "Update on dietary intake of perchlorate and iodine from U.S. food and drug administration's total diet study: 2008–2012," *Journal of Exposure Science and Environmental Epidemiology,* Vol. 28, Dec 14, 2016, pp. 21–30, https://www.nature.com/articles/jes201678.
66. FDA, "Food Additive Regulations; Synthetic Flavoring Agents and Adjuvants," *Federal Register,* Oct 9, 2018, https://www.federalregister.gov/documents/2018/10/09/2018-21807/food-additive-regulations-synthetic-flavoring-agents-and-adjuvants.
67. Dillon L, Sellers C, Underhill V, et al. "The Environmental Protection Agency in the Early Trump Administration: Prelude to Regulatory Capture," *Am J Public Health,*Vol.108(S2):S89-S94, April 2018, https://www.ncbi.nlm.nih.gov/pmc/articles/PMC5922212/.
68. Ibid.

10

CLEAN UP WITH HEMP

Okay, so we're on our own (for the most part). The EPA won't advocate for carbon-neutral means of cleaning up the environment. Chemical companies responsible for the pollution have a curious way of making bank, putting out a propaganda campaign stating their products are harmless, and then, when it's discovered that the chemicals truly are damaging, they're already out of business, or filing for bankruptcy, or finding some other way for the EPA and FEMA and the CDC to foot the bill and fund cleanup efforts with taxpayer money.

In general, the EPA isn't going to start cracking down on industrial pollution because the federal government itself (or more specifically, the military industrial complex) is one of the worst, yet often unacknowledged, causes of climate change—here are just a few examples:

- According to *Newsweek*, the DOD has a carbon footprint that dwarfs any private corporation. About 10 percent of all of America's Superfund sites are Pentagon installations (which equates to nearly nine hundred domestic military facilities). In 2014, Maureen Sullivan, the head of the Pentagon's environmental program, stated it would cost taxpayers $27 billion to clean up all of the Pentagon's pollution—which at the time included 39,000 contaminated sites on American soil.[1]

- As of 2018, the Defense Environmental Restoration Program (DENIX), which is tasked with cleaning up these 39,000 sites, has a budget of $70 billion and spends over one billion dollars per year cleaning up military sites that contain toxic waste and explosives. In May 2018, *ProPublica* gathered this data and mapped it out to discover that contaminated DOD sites are in nearly every American's backyard; they're next to schools, rivers, lakes, and residential neighborhoods.[2]

- In May 2017, the *Washington Post* reported that the DOD is the largest consumer of fuel in the world. Each year, the Pentagon purchases about 100 million barrels (or 4.2 billion gallons) of refined petroleum for machines of war.[3]

- A 2003 article in *The Guardian* outlined the environmental damage of the first Gulf War. Oil refineries were targeted, which led to massive chemical pollution. Iraqi forces in Kuwait destroyed more than seven hundred oil wells, spilling 60 million barrels of oil. Wells burned for nine months, which resulted in a reduced amount of sunlight; the air temperature decreased by fifty degrees Fahrenheit and sea temperatures also fell. Sewage treatment plants were also destroyed to contaminate the water supply. The groundwater aquifer that contains Kuwait's entire freshwater reserve is still contaminated.[4]

- The Defense Advanced Research Projects Agency (known as DARPA) developed napalm and the formula for Agent Orange. Under the US War Powers Act, the government contracted nine manufacturers to produce Agent Orange and other herbicides with specific manufacturing specifications (Dow Chemical and Monsanto were the two largest producers). Through a program called Operation Ranch Hand, more than 20 million gallons of herbicide were sprayed over Vietnam, Cambodia, and Laos from 1961 to 1971.[5] In Vietnam, more than 4.5 million acres of forest cover were destroyed; crops died and water sources were heavily contaminated. Today, over half a million Vietnamese children have been born with serious birth defects, and many have developed cancer from exposure to the dioxins. The erosion and deforestation that occurred as a result of Operation Ranch Hand also negatively impacted environment; animals that inhabited the forests and jungles were threatened with extinction, and crop fields were displaced.

- In the Iraq and Afghanistan wars, the US government had more than 250 open-air burn pits at US bases in both countries to burn garbage and massive amounts of waste. Hundreds of tons of waste were burned per day. There weren't any standards in place for what could or couldn't be burned, and toxins from petroleum, oil, rubber, plastic, Styrofoam, batteries, explosives, medical waste, animal and human carcasses, pesticides, to aerosol cans entered the atmosphere and were inhaled by the US troops deployed on the bases—those service members are now suffering from cancers and brain tumors.[6]

- A 2018 DOD records release indicated that there are more than fifty active open-air burn pit sites in the United States. Regardless of how close they are to schools, homes, and water supplies, these sites burn raw explosives. According to *ProPublica*, these actions "have led to 54 separate federal Superfund declarations and have exposed the people who live near them to dangers that will persist for generations."[7]

While phytoremediation can help clean up military-grade toxins including TNT, there are limitations. Research is still ongoing, but existing studies examining various plants and their ability to remediate dioxins (such as PCBs) have shown that the process of degrading these toxins is "extremely slow" in the plants and microbial communities as a whole.[8] However, hemp can still be a valuable part of the cleanup team, just as it is currently battling lead contaminations in Italy. Even though the process is effective, naysayers are quick to point out three phytoremediation challenges:

1. The cleanup effort can only go as deep as the plant's root system.

While the topsoil may in fact become decontaminated within one or two growing cycles, the toxins will continue to leach up through the ground, and can eventually recontaminate the soil. This is technically true of any cleanup system currently used. Superfund sites don't become pristine after a dirt removal effort. They're just less contaminated. Phytoremediation allows us to continually clean the soil, continually remove the toxins. Yes, it's a process, and a long one at that, but remember it took twenty-one years and $400 million dollars to clean up Love Canal, New York.[9] And there are problems with contamination sites years later (sometimes sooner), which is why most sites are repurposed into something other than housing communities (but that doesn't mean they aren't also affecting nearby cities). And why can't nearby communities use phytoremediation in their backyards as a way to minimize their risk of exposure?

As of 2018, there are still 22,000 tons of chemicals buried underneath Love Canal, which the EPA claims to be monitoring for leaching through several sophisticated means. While the EPA says the land is perfectly safe for humans, nonprofit and environmental groups have tested the soil to find that dioxins and PCBs *are still there;* locals also claim this is the cause of unusual birth defects and miscarriages. In August 2018, PBS *NewsHour* spoke to one woman who had a total of *eleven* miscarriages after moving to Love Canal.[10] Since the EPA won't admit the land actually *isn't* safe, what can the locals do? Move? Sure. With what money? Who's going to buy these people's homes after learning they've all been built over the first toxic waste dump that started the Superfund program?

2. Wildlife will just eat the contaminated plants and spread toxins around.
Luckily, most animals don't eat hemp, but birds do like to snatch hempseed from the plant as it grows. It's of course conceivable for wild animals to eat contaminated plants that are being used for phytoremediation. Of course it's even more likely that they're already eating or drinking from their contaminated environment anyway, so where's the solution for that problem? We already know from the Poison Papers that the EPA is well aware of how toxins are already affecting wildlife by causing blindness and severe deformities.

3. Once the plants have absorbed the toxins, what do we then do with the toxic plants?
Excellent question. How do we dispose of toxic plants? Incineration? Won't that just bring the toxins back into the air? Luckily, science has come up with some solutions known as post-harvest management techniques.

POST-HARVEST MANAGEMENT TECHNIQUES

Dry-ashing, wet oxidation, and microwave digestion are methods that researchers use to extract heavy metals from plants. In a 2003 study published in *Microchemical Journal,* researchers tested all three methods of extraction. They took mushrooms that had absorbed heavy metals (cadmium, zinc, and copper) and were able to recover 95 to 103 percent of all elements the plant samples accumulated.[11] Essentially, these processes use high heat and chemical reactions to extract the trace elements in plant matter. (I was able to find academic papers dating back to 1957 that use dry-ashing methods to extract lead and copper from plants.[12]) Researchers suggest that the heavy metals and toxins that accumulate in plants from phytoremediation can then be collected, stored, or even recycled.[13]

Incineration is the easiest method of discarding hemp after the phytoremediation process because we already have incinerators for dioxins, but there are other uses for hemp's biomass.

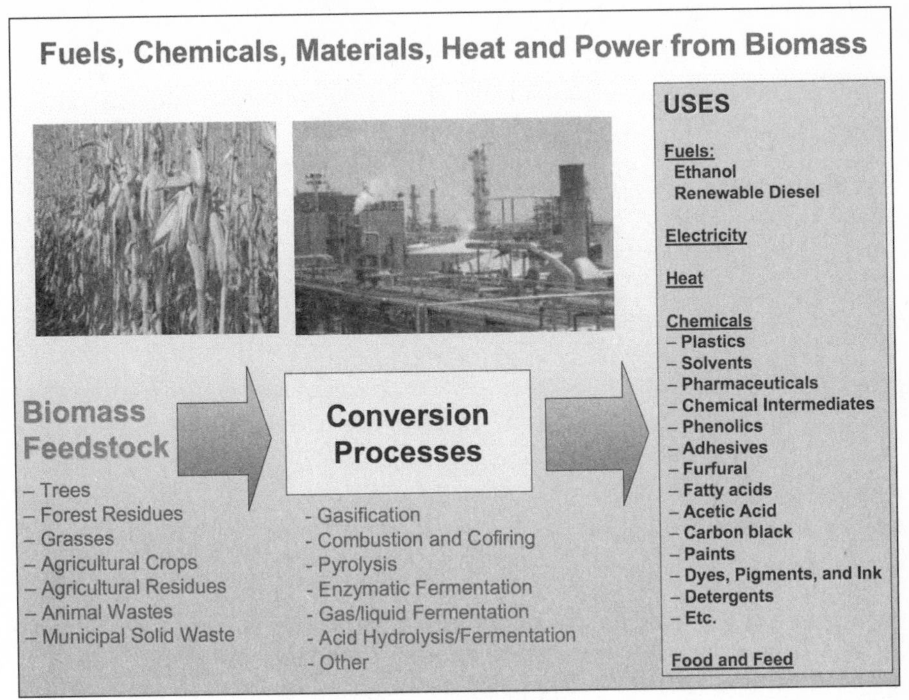

Current technology makes it possible to convert plants into many different uses, and the conversion process discards the toxins. *Source: The National Renewable Energy Laboratory; a national laboratory of the US Department of Energy, Office of Energy Efficiency & Renewable Energy, 2004 (https://www.nrel.gov/docs/gen/fy04/36831e.pdf)*

- **Biomass Thermochemical Conversion Process**

Biomass is plant matter that is converted into a solid fuel, or converted into a liquid or gaseous form to produce electric power, heat, or fuels.

Once the contaminated hemp is plucked from the ground, there are several ways to use thermochemical conversion to turn it into energy. Essentially this method can be considered a type of incinerator that still allows us to use hemp's biomass for other purposes.

The process used to generate electricity is known as gasification, a fermentation method where methane is siphoned off and burned. This creates heat that can then be used to produce electricity.

In 2004, the National Renewable Energy Laboratory factored in the environmental impact of a biomass power plant operation creating electricity and determined it would actually result in *negative* greenhouse gas emissions. A hemp biomass power plant would not just be carbon neutral, but theoretically, the entire process (growing the plants, transporting them, building and firing up the power plant, creating the electricity) would actually decreases greenhouse gas because the plants themselves are decreasing carbon emissions *and producing oxygen*, which

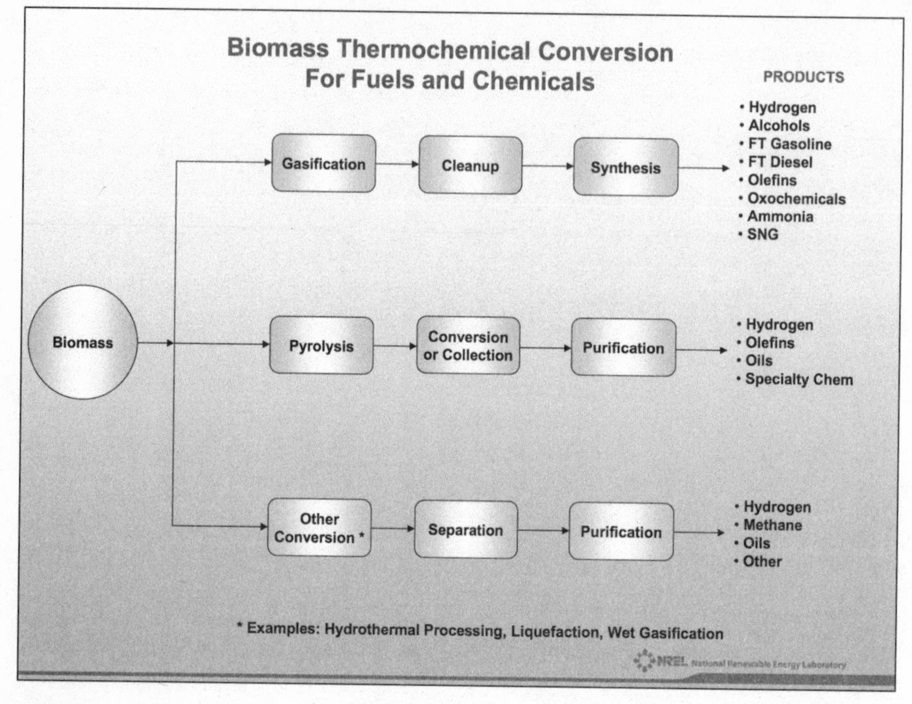

A simple diagram to illustrate how biomass thermochemical conversion is used to create fuels or chemicals that can be found in everyday products. *Source: The National Renewable Energy Laboratory; a national laboratory of the US Department of Energy, Office of Energy Efficiency & Renewable Energy, 2004 (https://www.nrel.gov/docs/gen/fy04/36831e.pdf)*

cancels out the carbon emissions that are created when their biomass is converted into electricity.

The US Energy Information Administration states that as of 2017, the US electric power sector is responsible for 34 percent of the total carbon dioxide emissions in the United States.[14] In 2004, the National Renewable Energy Laboratory compiled data to show the future, widespread potential of using a biomass power system to produce electricity. They estimated that the entire operation could produce only 49 g of carbon emissions per kilowatt hour (hWH)[15]—49 grams is equal to 0.108027 pounds.

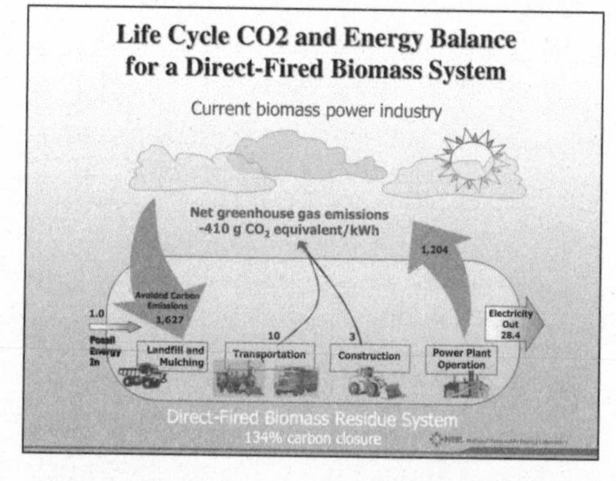

A simplified model of a current biomass power plant produces -410 g of carbon emissions per kWh (kilowatt hour). *Source: The National Renewable Energy Laboratory; a national laboratory of the US Department of Energy, Office of Energy Efficiency & Renewable Energy, 2004 (https://www.nrel.gov/docs/gen/fy04/36831e.pdf)*

Currently most biomass power plants utilize wood waste solids that are by-products of making paper-related products—such as black liquor, a by-product of the kraft pulping process—and this accounted for 27 percent of all biomass and waste-generated electricity. According to the US Energy Information Administration, as of 2016, this small yet significant biomass industry generated 21,813,231 megawatt hours (or 21,813,231,000 kWh) of electricity and created *zero CO_2 emissions.*[16] When biomass *includes* waste fuels, such as landfill gas and biogenic municipal solid waste produced from wastewater treatment plants, that brings the total to 71.4 billion kWh (or 2 percent of the total generation).[17]

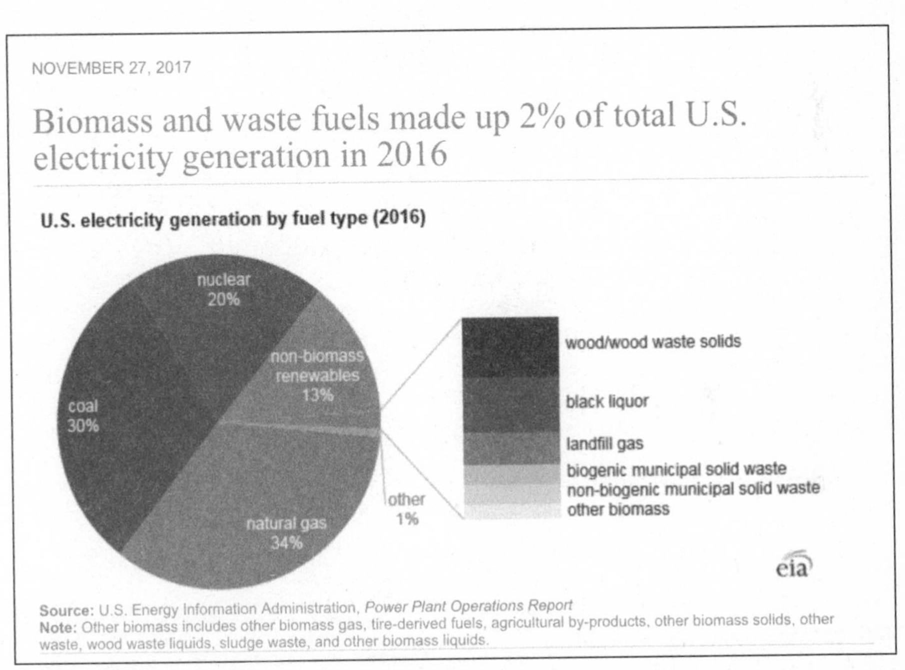

Only 2 percent of electricity generation in the United States came from biomass and waste fuels. *Source: US Energy Information Administration, 2017 (https://www.eia.gov/todayinenergy/detail.php?id=33872)*

According to the US Energy Information Administration, the total CO_2 emissions for 2016 from the United States' electric power industry as a whole was 1,928,400,912 metric tons (multiply by 1,000,000 to get that number in grams: 1,928,400,912,000,000—this equates to roughly 4,251,396,274,589 pounds). Those carbon emissions generated a net amount of 4,076,674,984 megawatt hours (or 4,076,674,984,000 kWh) of electricity.

Year			Energy Source	CO2 Emissions (Metric Tons)
2016	US-TOTAL	Total Electric Power Industry	All Sources	1,928,400,912
2016	US-TOTAL	Total Electric Power Industry	Coal	1,265,500,155
2016	US-TOTAL	Total Electric Power Industry	Geothermal	416,326
2016	US-TOTAL	Total Electric Power Industry	Natural Gas	621,148,811
2016	US-TOTAL	Total Electric Power Industry	Other Biomass	0
2016	US-TOTAL	Total Electric Power Industry	Other Gases	0
2016	US-TOTAL	Total Electric Power Industry	Other	14,590,727
2016	US-TOTAL	Total Electric Power Industry	Petroleum	26,744,893
2016	US-TOTAL	Total Electric Power Industry	Wood and Wood Derived Fuels	0
				Net Generation (Megawatt hours)
2016	US-Total	Total Electric Power Industry	Total	4,076,674,984
2016	US-Total	Total Electric Power Industry	Coal	1,239,148,654
2016	US-Total	Total Electric Power Industry	Geothermal	15,825,807
2016	US-Total	Total Electric Power Industry	Pumped Storage	-6,686,127
2016	US-Total	Total Electric Power Industry	Hydroelectric Conventional	267,812,153
2016	US-Total	Total Electric Power Industry	Natural Gas	1,378,306,934
2016	US-Total	Total Electric Power Industry	Nuclear	805,693,948
2016	US-Total	Total Electric Power Industry	Other Gases	12,807,432
2016	US-Total	Total Electric Power Industry	Other	13,754,235
2016	US-Total	Total Electric Power Industry	Petroleum	24,204,806
2016	US-Total	Total Electric Power Industry	Solar Thermal and Photovoltaic	36,054,121
2016	US-Total	Total Electric Power Industry	Other Biomass	21,813,231
2016	US-Total	Total Electric Power Industry	Wind	226,992,562
2016	US-Total	Total Electric Power Industry	Wood and Wood Derived Fuels	40,947,227

Total carbon emissions and electricity generated in 2016 by the power industry. *Source: US Energy Information Administration (https://www.eia.gov/tools/faqs/faq.php?id=74&t=11)*

Regardless, this data is stating that the vast majority of our current methods of creating electricity (coal, petroleum, natural gas, etc.) give out *way more emissions* than biomass power.

Here's the kicker though: Contrary to what these government agencies are claiming, using wood waste products for biomass *isn't* actually carbon neutral. First and foremost, it takes decades for one tree to grow back. There's nothing carbon neutral about that. Secondly, when compared to coal, wood is a low density fuel, which means to generate the same amount of power as coal, we need a lot more of it. So to keep up with biomass demands, that means we have to keep pulp and paper mills pumping—and these facilities also produce dioxins as a by-product, so that's again not carbon neutral when we're *poisoning* the environment to make wood waste for electricity.

To make matters worse, the Partnership for Policy Integrity came out with a report in 2014 titled *Trees, Trash, and Toxins: How Biomass Energy Has Become the New Coal* which details all the ways that biomass companies are exploiting loopholes in renewable energy policy.[18] One problem is that biomass facilities are producing just as much pollution as coal facilities to manufacture wood pellets. So while wood pellets themselves "burn clean" and are considered carbon neutral, the process to make them is as dirty as coal energy. In fact, the *Trees, Trash, and Toxins* report claimed biomass plants can emit "almost 50 percent more CO_2 than coal per unit of energy produced."[19] These companies often go unmonitored by any outside, third-party inspection or environmental agency, so there isn't much of a consequence if the facilities produce more pollution than what is allowed by law.

Currently, biomass companies receive subsidies based on the assumption that the fuel created is renewable and clean, and southern states like Virginia, Florida, and Georgia are leading the way in generating it.[20] The subsidies are creating a steady expansion in the biomass industry; however, it seems the companies that are getting into the field are already cutting corners and/or running into wood waste supply problems. Southern forests are becoming the source of material for wood pellets, and in Georgia, the world's largest biomass pellet facility exports 750,000 tons of processed wood pellets a year to Europe, so nations can produce "clean" electricity in accordance with the European Union's clean energy laws.[21]

The forest and climate activist group Dogwood Alliance has documented how Enviva Partners, a corporation that has biomass plants in North Carolina, Florida, and Virginia, has been taking mature trees in the wetlands of North Carolina as fuel for their power plant facilities.[22] In 2015, the Partnership for Policy Integrity petitioned the EPA about the Piedmont Green Power biomass plant in Barnesville, Georgia. The plant received a $49 million taxpayer-funded stimulus; this grant was intended to promote clean energy.[23] However, the company had such broadly written air permits that the permits didn't actually require the company to comply with any state air guidelines, and the state itself didn't enforce safe emissions limits, either. Mary Booth, the founder of the Partnership for Policy Integrity, explained that the EPA recognized the problem and demanded a new permit be issued from the state to the company, but no changes were actually made on the new permit.

"So the EPA looked at the permit and they agreed with us, but the new permit Georgia issued was like déjà vu all over again, it still didn't ensure this was a safe plant," Booth explained in 2017 to the Energy News Network. "We had to submit a whole new set of comments."[24]

But instead of solely relying on the EPA, Booth wrote reports to the Federal Trade Commission (FTC) and the Securities and Exchange Commission (SEC) documenting how biomass power producers are misleading customers and shareholders about the process being green and carbon neutral. She also requested that the SEC "give concrete guidance to companies" on how to account for all their carbon emissions.

"Biomass itself is really an offsetting scheme," states Booth. "You say you're burning these trees and emitting this carbon but new tree growth will offset it . . . But who is actually checking whether the foresters are planting new trees or just depending on natural regeneration?"[25]

Even worse is the bottom line to residents: people who live in a state that is giving a subsidy to a biomass company are directly paying that subsidy (known as renewable energy credits) every time they pay their electric bill. For example, Gainesville, Florida's 2009 contract with the Gainesville Renewable Energy Center biomass plant required the city to pay $70 million annually (or $2.1 billion over thirty years) to the biomass plant company; power generated from the plant provided electricity for 70,000 homes.[26] Anyone with an electric bill in Gainesville directly contributed to this cost every month. However, this was just the cost to have the privilege of a biomass plant in town; this fee does not include the cost of purchasing electricity from the plant, and the residents had to pay this cost even if the city decided at some point not to purchase power from the plant. While the plant employed about three hundred locals,[27] the city ultimately *didn't* purchase power from it because they were able to buy cheaper power from other utility facilities. So residents of Gainesville were stuck paying *more* for their electricity even though they weren't using the biomass facility.

In September 2017, Gainesville city commissioners negotiated to purchase the facility for $754 million, which was estimated to save electric customers 8 to 10 percent on monthly electric bills, even though the buyout will cost the city $1.2 billion over 30 years due to interest.[28]

In October 2018, the *Gainesville Sun* reported that Laura Haight, the Partnership for Policy Integrity's group policy director, called the Gainesville plant "the poster child of everything that could go wrong with biomass subsidies,"[29] and one of the worst biomass deals in the country. Regardless, the Gainesville City Commission set a goal to run the city on 100 percent renewable energy by 2045, with the Gainesville Renewable Energy Center—now known as the Gainesville Deerhaven Renewable Generating Station—reportedly producing about 25 percent of the city's energy as of late 2018.

As of 2018, biomass plants have received $10 million or more in grants for construction from the U.S. Department of the Treasury, and according to Mary Booth, the plants "tend to over promise and under deliver." Booth examined twenty-five biomass plants that have received federal funding; the Gainesville plant received a $116.8 million grant from the federal government, marking the largest grant funding ever received—it is also the largest biomass facility *and* was the costliest to electric customers.[30]

However, if we can convert all these wood biomass plants into hemp biomass plants, and then we'd actually see a true reduction in carbon emissions. Not only does hemp grow *much* faster than trees, it converts CO_2 into oxygen at a faster rate than trees.

According to the College of Agriculture & Life Sciences at North Carolina State University's Department of Horticultural Science, the average tree "can absorb as much as 48 pounds of carbon dioxide per year and can sequester 1 ton of carbon dioxide by the time it reaches 40 years old."[31] A tree can take fifteen to thirty years to reach its full size (and maximize its carbon dioxide intake), depending on the climate and type of tree.

As a comparison, the National Hemp Association states one ton of hemp removes 1.63 tons of CO_2—or 3,260 pounds.[32] Since hemp can be harvested every four to six months, and it's possible to grow five tons of hemp per acre each growing cycle, then conservatively, if five tons of hemp are grown twice a year on a one-acre plot of land, then that one acre can remove 40.75 tons of CO_2 (or 81,500 pounds) per year.

As James Vosper, the CEO of Good Earth Resources, explained to Australia's House of Representatives Standing Committee on Climate Change, Water, Environment and the Arts in 2011:[33]

> *One hectare [2.47 acres] of industrial hemp can absorb 22 tonnes of CO_2. . . . It is possible to grow to 2 crops per year so absorption is doubled. Hemp's rapid growth (grows to 4 metres in 100 days) makes it one of the fastest CO_2-to-biomass conversion tools available, more efficient than agro-forestry. . . . For a crop, hemp is very environmentally friendly, as it is naturally insect resistant, and uses no herbicides. Hemp grows rapidly in Australia and matures in 90 days compared to traditional forestry taking 20 years. It therefore starts absorbing CO_2 from almost from the day it is planted.*

Hemp is classified as a C4 plant, while trees are a C3 plant. Hemp and other C4 plants are able to capture more CO2 inside their cells and produce higher levels of oxygen than C3 plants. Although C4 plants represent about 5 percent of the Earth's biomass and 3 percent of its known plant species,[34] they account for about 23 percent of terrestrial carbon fixation (the process where plants convert carbon dioxide into organic carbon compounds).[35] If the proportion of C4 plants is increased, then that means *more* carbon dioxide can be captured, and researchers have noted this would be a simple, cheap, and effective way of combating climate change.

So to recap, we started growing hemp to clean the soil via phytoremediation, then took the biomass from the contaminated hemp and turned it into fuel for electricity—and stopped using wood biomass—we'd make a significant dent in decreasing greenhouse gases.

Currently, Sweden has a small commercial hemp production for hemp briquettes,[36] which are essentially a competitor of the wood pellets or compressed coal that are used to start a fire. While hemp briquettes burn faster and at a higher heat than wood pellets, hemp briquettes are made from hemp hurds and *no chemical additives* are used to produce them. In a 2008 article, Poland's Institute of Natural Fibres and Medicinal Plants stated:[37]

> *[Hemp] Briquettes and pellets make a fuel obtained by compressing dry waste materials. No chemical additives are used in the production process. Shives get stuck together as a result of the action of steam and high pressure. The material utilized during shaping of briquettes and pellets is lignin.*

Researchers noted that one hectare (which is roughly 2.47 acres) of hemp produces ten to fifteen tons of biomass that can be used for manufacturing hemp briquettes. As three acres of land produces up to fifteen tons of hemp briquettes, the plant is simultaneously absorbing 2.5 tons of CO_2 as it grows.

We could absolutely use hemp to power our homes, clean our soil, *and make a significant impact on climate change* all at the same time. But wait! There's more.

- **Hemp as Biofuel**

Transesterification is how biomass such as vegetable oils, animal fats, and waste cooking oils can be converted into biofuel. The process exchanges the organic group R of an ester (carboxylic acid ester) with the organic group R of an alcohol (a different carboxylic acid ester)[38]—this is the same process for producing biofuel from hemp.

Through transesterification, oil from hempseed can be turned into biofuel or it can be blended with petroleum diesel to create biodiesel fuel. According to a 2015 Swedish research paper from the University of Gävle titled "Advantages and Challenges of Hemp Biodiesel Production: A comparison of Hemp vs. Other Crops Commonly used for biodiesel production," unlike other forms of biodiesel fuel, hemp biodiesel fuel doesn't require any modifications to diesel engines.[39] The paper also notes that the most common hiccup with using biomass for fuel is the problem of food versus fuel. Current biodiesel plants include food crops such as soybeans, olives, peanuts, and rapeseed.

"For sustainable fuels, often it comes down to a question of food versus fuel," echoed Richard Parnas, a professor of chemical, materials, and biomolecular engineering who led a hemp biofuel study at the University of Connecticut. "It's equally important to make fuel from plants that are not food, but also won't need the high-quality land."[40]

Crops that can be used for food and fuel are typically only grown on agricultural land, therefore farmers must decide if they'll use their lands to produce food or fuel, and this can affect the price of food and the quantity of certain foods. However, since hemp can be grown on infertile soil (and contaminated soil), this means hemp used for biofuel won't encroach on fertile land already reserved for food crops. Hemp also doesn't require as much water or fertilizers or pesticides as other crops, and the average harvest yield can be calculated and predicted. Plus, hemp can be grown in a variety of climates, all over the world.

The 2015 University of Gävle paper also notes that hemp is a carbon-neutral replacement for diesel fuel:

> *During the three month life cycle of the plant, the cannabis ingests carbon dioxide at a rapid rate much greater compared to that of trees, which makes hemp a very effective scrubber of carbon dioxide. Effectively, hemp could provide the means to by which we are not introducing additional carbon into the environment. Therefore, offering another alternative fuel source to offset our reliance on fossil fuels.*

In Southeast Asia, the most commonly produced biodiesel fuel is made from palm oil. Unfortunately, palm oil has a huge environmental impact because the land that is being cleared to farm palm trees is primarily tropical forests. Indonesia has the third largest tropical forest in the world, but it is also one of the largest greenhouse gas emitters because of this deforestation. From 1990 to 2012, 60 percent of the tropical forest of Kalimantan (the Indonesian portion of the island of Borneo) was cleared for palm tree cultivation for palm oil. While this greatly displaced animals and wildlife in Kalimantan, the land-clearing process in 2010 alone "emitted more than 140 million metric tons of carbon dioxide—an amount equivalent to annual emissions from about 28 million vehicles."[41]

Although palm oil produces more biofuel than hemp oil, hemp does have a high efficiency of conversion. A 2010 study at the University of Connecticut found that "97 percent of the hemp oil was converted to biodiesel—and it passed all the laboratory's tests, even showing properties that suggest it could be used at lower temperatures than any biodiesel currently on the market."[42] Plus, "as a rule of thumb, the amount of emission reduction corresponds roughly to the biodiesel blend rating of the fuel."[43] Meaning a hemp biodiesel blend would be given a higher rating than a palm oil blend because it requires fewer carbon emissions to produce.

A 2010 study published in *Bioresource Technology* also noted that because hemp has high yields of both oil and biomass, it can be used to produce both biodiesel and bioethanol:[44]

> *Hemp seeds have high oil content, ranging from 26% to 38%, making the plant an ideal candidate for producing biodiesel fuel. . . . Furthermore, hemp biodiesel meets the standards for biodiesel fuel set by ASTM 6751–09 [the American Society for Testing and Materials standards for biodiesel fuel blend stock B100]. The distinct properties of hemp biodiesel are its low cloud point and low kinematic viscosity. These promising cold flow properties make hemp biodiesel attractive and competitive. Its biomass content can be fermented to create low carbon fuels, such as bioethanol or biobutanol.*

Remember when Lyster Dewey wrote that hemp hurds—the waste product of industrial hemp—were highly absorbent and could be used for animal bedding? Hemp hurds, also known as shives or hemp core fiber, are the woody inner part of the hemp stalk. The bark (or bast) of the hemp is separated from the rest of the plant and the fibers from the bark are what is used to make clothing, plastics, and basically all commercial products. The inner part of the hemp plant (hemp hurds), and all the plant's stems and leaves, are the remaining biomass that can be used to make low-carbon fuels including bioethanol and biobutanol (butanol produces less carbon dioxide than gasoline for the same amount of energy). Currently, feedstocks such as corn grain biomass produce bioethanol and biobutanol, which are then blended with gasoline to decrease overall gas emissions.[45]

Rudolf Diesel, inventor of the diesel engine, warned that since fossil fuels aren't a renewable resource, it was not wise to rely on them to power machines. Here's what he had to say about renewable resources in 1912:[46]

> *The fact that fat oils from vegetable sources can be used may seem insignificant today, but such oils may perhaps become in course of time of the same importance as some natural mineral oils and the tar products now. In any case, they make it certain that motor power can still be produced from the heat of the sun, which is always available for agricultural purposes, even when all our natural stores of solid and liquid fuels are exhausted.*

The concept and practice of diverging from fossil fuels was also alive and well in Europe in the late 1800s. Prior to World War I, the Germans had an extensive biofuels program. From 1890 to 1916, they researched the possibility of fueling trucks, cars, and trains with ethanol. The government placed tariffs on imported oil, which prompted the emergence of other fuel sources, including using potato alcohol. Ethanol-fueled household appliances such as lamps, water heaters, laundry irons, hair curlers, coffee roasters, and cooking stoves were soon invented; there are estimates of approximately 95,000 alcohol-fueled stoves and 37,000 lamps that were produced in Germany by 1902.[47]

According to a January 1945 article published in *Industrial & Engineering Chemistry* by the federal government's War Production Board's Office of Production Research and Development, in the early 1930s, the Germans developed and utilized the Scholler Process, a means to turn wood waste into ethyl alcohol, which was being used as an additive for gasoline, a fuel source, and an alternative to petroleum.[48] By 1941, the Nazis had built an estimated twenty plants that turned sawdust and wood chips into this industrial alcohol.[49] The US government briefly tested the Scholler Process in the 1940s to consider ethanol as a possible "wartime necessity" in the event that our oil supply chain collapsed.

The point to this history lesson is that creating biofuel isn't a new or revolutionary concept. Again, if we're already using hemp to clean the soil, or if we're using the bast fiber for all of its wonderful commercial potential, then we still have plenty of hemp hurds and hemp biomass left for other uses. We can easily create our own locally sourced form of biofuel—the technology is already there, so why not expand it?

In 2007, the United States implemented the Energy Independence and Security Act, which forced the transportation industry to start integrating biofuels at the gas pump. This is why you see signs at gas stations such as "fuel may contain up to 10 percent ethanol." Today's ethanol is sourced from corn, soybeans, or milo.[50] However, the majority of American ethanol is produced from corn, and the demand for it has farmers caught in the middle of the food-versus-fuel debate.

According to *Scientific American,* "only a tiny fraction of the national corn crop is directly used for food for Americans, much of that for high-fructose corn syrup."[51] While this demand for corn-based ethanol has also decreased the amount of wheat, oats, sorghum, barley, alfalfa, sunflower, and other crops grown in the United States, it has also infringed on waterways, wetlands, and wildlife. The *Proceedings of the National Academy of Sciences* states that as of 2013, approximately 1.3 million acres of grassland and prairie were converted to either grow or process corn.[52]

According to a 2017 article in *Modern Farmer,* between 2006 and 2011, the amount of cropland devoted to corn increased by more than 13 million acres, with roughly 40 percent of America's total harvest used for ethanol production and 36 percent used to feed livestock.[52] In addition to this, corn that isn't used for food purposes still requires the use of fertilizers, pesticides, and an abundance of water. It also depletes the soil of nutrients, and many farmers aren't able to rotate crops on their land to keep up with the demand.[53] *Scientific American* states the total landmass of all domestic corn production is 97 million acres—which is roughly the size of California[54]—and even with all that corn production, we're *still* importing corn for food.

Instead of depleting natural resources to create biofuel from corn, why aren't we planting hemp? Yes, the USDA claims ethanol biofuel decreases carbon emissions by about 43 percent,[55] but compromising the environment to decrease emissions doesn't do us any good. The ends don't justify the means.

In 2017, *Smithsonian* magazine noted the top five biofuels that can replace corn. Hemp was the first one, followed by switchgrass, carrizo cane, jatropha, and algae.[56] Why was hemp favored? Because it doesn't have to be grown on agricultural land. Hemp requires little to no irrigation, fertilizer, or pesticides, and it absorbs more CO_2 than corn, *and* it cleans the soil.

- **Supercapacitor Hemp Bast Fiber**

The energy storage industry has found a revolutionary use for hemp bast fiber. Researchers discovered that bast (the outer bark of the hemp stalk) can be turned into a nanocomposite for supercapacitors to conduct electrodes.

According to a 2013 Canadian research paper published in *ACS Nano,* bast fiber has "good electrical conductivity"[57] and can store energy the same way graphene-based, ultrafast supercapacitors do.

Dr. David Miltin, the chemical and materials engineering professor who led the study, told the American Society of Mechanical Engineers that "we were delighted at how well this material performed as super capacitor electrodes."[58]

To uncomplicate that as much as possible: Graphene is a single sheet of carbon atoms that is incredibly thin, strong but bendable, and able to conduct electricity better than copper. It is the most expensive nanomaterial used in supercapacitors and batteries. The material can cost as much as $2,000 per gram to manufacture.[59] However, devices made with graphene are able to store more energy and discharge the energy faster.

The supercapacitor is the *thing* that stores and releases the energy that powers fast-charging flashlights, solar cells, medical devices, hybrid electric vehicles, and high-tech portable electronics that require an uninterrupted power source. (It's essentially a high-tech battery.)

The regenerative braking system on electric vehicles is one application of how quickly a supercapacitor can convert energy. When a car goes downhill and brakes are applied, the energy used isn't lost but instead stored and converted into more energy to power the batteries that operate the car.

Mitlin and his research team discovered that by separating hemp's bast fiber into nanosheets, the fiber can be used the same way graphene is used in a supercapacitor, and the fiber is actually able to outperform the graphene. The best part is that the bast nanosheet costs less than $500 per ton to manufacture.

The hemp-based device has an electrochemical performance that is "on par with or better than graphene-based devices," Mitlin explained. "The key advantage is that our electrodes are made from biowaste using a simple process, and therefore, are much cheaper than graphene."

Scientists at the University of Manchester in England started developing graphene in 2004, and won a Nobel Prize for the discovery in 2010. To date, graphene is also the strongest mineral ever discovered, with forty times the strength of diamond. Graham Stuart, the UK Minister for Investment at the Department for International Trade, has stated graphene has the potential "to make Internet connections faster, filter salt water, and make phone screens unbreakable."[60] In 2013, researchers discovered that graphene oxide even has the ability to "quickly remove radioactive material from contaminated water" by soaking it up like a sponge; if the graphene oxide is then burned, it'll leave behind a "cake of radioactive material you can then reuse."[61]

Graphene sounds a bit like snake oil, doesn't it? Almost like a man-made version of hemp.

According to MiningTechnology.com, "graphene is capable of transferring electricity 140 times faster than lithium, while being 100 times lighter than aluminum. This means it could increase the power density of a standard Li-ion [lithium-ion] battery by 45%."[62] However, all of that is fairly irrelevant, now that Canadian scientists have concluded that hemp can be fashioned to do the same thing at a cheaper cost.

Since its discovery in 2010, graphene researchers have been advancing the technology, and they're aiming to accomplish some pretty amazing advances, such as making it possible for electric vehicles to charge fully depleted car batteries within ten minutes.[63] However, it's still an extremely expensive process to make graphene, and *Popular Mechanics* notes that the process can be "absurdly complex."[64] Articles also suggest that since graphene is made from carbon, it can be recycled, and basic products such as used cooking oil can be converted into graphene.[65] In May 2018, researchers were able to manually remove graphite waste from batteries and turn them into graphene products,[66] but overall most articles that state graphene is recyclable don't give sources to actual studies or detail the process (so it's difficult to say how carbon neutral the process actually is).

Graphene and graphite aren't just spelled similarly; on an atomic level, there's one major difference. Graphene is *one layer of carbon atoms.* When multiple layers are stacked on top of one another, the product becomes graphite, which is also the same carbon-based substance found in number 2 pencils.

In September 2018, a paper was published in the journal *Advanced Materials* that analyzed the purity of graphene produced from sixty different manufacturers to find that *not one product* contained more than 50 percent graphene, even though every product was being marketed as a pure source.[67] The authors write:[68]

> *It is worrisome that producers are labeling black powders as graphene and selling for top dollar, while in reality they contain mostly cheap graphite. This kind of activity gives a bad reputation to the whole industry and has a negative impact on serious developers of graphene applications. Only through standardization and following protocols for characterization as proposed here, the graphene industry can evolve reliably.*

The National University of Singapore researchers also noted in their paper that manufacturers might not be completely aware of their "graphene" shortcomings. Whether or not these findings lead to any industry standards for graphene, it is clear that a graphene product might very well be plain old graphite. If the recent—shall we say pitfalls—of the American auto industry has taught us anything, Ford Motor Co. had better figure out how to get their graphene formula right. The company announced that it is partnering with Eagle Industries Inc and XG Sciences Inc to add graphene to engine components for the F-150 pickup and Mustang as early as 2019.[69]

Whatever lies ahead for graphene, Dr. David Milton's research on hemp bast fiber is truly revolutionary. Tech industry giants are pumped about graphene's potential, and here we have *a plant* that is *carbon-neutral* and able to accomplish *the same thing* that the man-made product can. Companies from Samsung to IBM have graphene patents,[70] and it makes no sense to me that all these corporations can be so enamored with graphene yet so dismissive of hemp.

Graphene aside, we have some serious environmental issues overall with today's high-tech electronics and electric vehicles because they are powered by lithium-ion batteries. Yes, it's great that these batteries are rechargeable, and in many cases, they are moving us away from fossil fuels. However, mining the lithium—and cobalt and nickel, the other two key metals used to manufacture batteries—requires a lot of water, and the areas of the world (like Bolivia) where there are rich deposits of these metals, water is scarce. Not only does this lead to conflict between miners and locals, but the deposits left from the mining process are toxic to humans, livestock, and crops.[71] Many of the mines, like those in the Democratic Republic of Congo, also use child labor to find new sources of cobalt.[72]

What makes matters even worse is that when we're ready to upgrade our laptop, cell phone, or whatever, the lithium-ion batteries typically end up in landfills, where the chemicals leach into the soil and create even more problems. Many environmentalists concede that our newfound dependence on lithium, cobalt, and nickel is spiraling out of control. Since the power and reliability of lithium batteries also decreases over time with each use, we'll continue in this cyclical loop (unless another more renewable power source for batteries is implimented).

According to the *Los Angeles Times*, less than 3 percent of all lithium-ion batteries worldwide are recycled—which is a huge problem considering there's a limited amount of lithium, cobalt, and nickel (just as there's a limited amount of fossil fuels), and mining them isn't exactly carbon-neutral.[73]

While British and French governments have already committed to outlawing the sale of gas-powered cars by 2040, and Volvo has pledged to produce only electric or hybrid vehicles starting in 2019,[74] *and* there are processes in the works to start recycling lithium batteries economically on a worldwide scale,[75] we're still driving cars that aren't "clean" energy because the process to create them is impacting the environment at a rate that offsets their value from an emissions standpoint. Yet, science has already shown us we can create biofuel from hemp's biomass and electrical components from its fiber to solve the vast majority of these problems.

In 1941, *Popular Mechanics* reported that after twelve years of research, the Ford Motor Company was successful in creating a car with a plastic body—made entirely from plant-based substances, including hemp:[76]

> *When Henry Ford recently unveiled his plastic car, result of 12 years of research, he gave the world a glimpse of the automobile of tomorrow, its tough panels molded under hydraulic pressure of 1,500 pounds per square inch (psi) from a recipe that calls for 70 percent of cellulose fibers from wheat straw, hemp and sisal plus 30 percent resin binder. The only steel in the car is its tubular welded frame. The plastic car weighs a ton, 1,000 pounds lighter than a comparable steel car.*

In 2016, Bruce Michael Dietzen created Renew Sports Cars—a car company that took the ideas of Henry Ford's plant-based car and modernized them. This car is truly the closest we've come to a carbon-neutral automobile since Henry Ford's prototype. The Renew car is made primarily from hemp fibers and plant biomass:[77]

- The body is made of three layers of woven cannabis hemp from China, making it lighter than a car made from fiberglass.
- The hemp that is woven into fabric for the interior is ten times more dent resistant than steel.
- The car itself runs on recycled agricultural waste.

On his blog, Bruce Dietzen stated his reasoning to start the Renew car company:[78]

> *Electric Vehicles aren't expected to make all that much of an impact on climate change between 2020 and 2050. . . . Don't get me wrong. I love electric vehicles. But making a mere 1% impact on climate change after altering 16% of a significant market like cars demonstrates a very scary fact.* **We are not thinking big enough.** *Minuscule 1% changes won't cut it. We need 10% changes and we need them fast. And I have one to propose:* **Make everything we can from plants.**

And what better way to use hemp for phytoremediation and then use hemp's biomass to create nearly every aspect of our technology—everything from car parts to supercapacitors?

- **Phytomining Resources from Hemp**

As far as the environmental impact of lithium mining is concerned, hemp and other hyper-accumulator plants absorb cobalt, copper, nickel, and other metals that are used to create, say, lithium-ion batteries. We can use hemp to absorb these metals from the soil, then extract the metals from the plants post-harvest. This process is called *phytomining.*

Theoretically, if we can use hemp to not only clean the soil but also *mine* the soil for its resources,[79] then we can further decrease our carbon footprint. A 1998 research study published in *Trends in Plant Science* states that the process of phytomining metals *after* phytoremediation is certainly possible,[80] and a 2013 *Live Science* article states that fast-growing plants such as mustard, sunflowers, and tobacco can be used to mine gold to offset the current environmental impact.[81]

Chris Anderson, an environmental geochemist and gold phytomining expert at Massey University in New Zealand who was quoted in the *Live Science* article, is currently working with researchers in Indonesia to develop a more environmentally sustainable system for small-scale gold miners. The goal is to decrease the pollutants used in traditional mining operations—including mercury, arsenic, and copper.

The geochemist explains that he doesn't think retrieving gold through phytomining will ever take the place of current gold-mining techniques, but "the value of it is in the remediation of polluted mine sites."[82]

Anderson believes "if we can generate revenue by cropping gold while remediating the soil, then that is a good outcome."

Yes, unfortunately, mining the Earth for resources is here to stay, but as the expert stated, one way we can *limit the impact* on the environment is through bioharvesting (or phytomining) the metals. Since scientists are already able to take these metals out of plants post-harvest,[83] why don't we offset the environmental impact of mining for lithium, copper, cobalt, and nickel by growing hemp at mining sites?

- **Oil Spill Cleanup Methods**

As reported by *USA Today,* the US Department of Transportation lists the largest and most damaging oil spills from 2010 to 2017, and includes an estimate for each spill. The total costs for the property and environmental damages are anywhere from $345,554 to $927,270,213 per spill depending on the location, type of oil, and quantity.[84]

The 2010 BP *Deepwater Horizon* spill, also known as the Gulf Oil Spill, is recognized as the worst oil spill in US history, and as of January 2018, the company claimed its total costs were approximately $65 billion—which accounted for total cleanup costs, legal fees, lawsuits, and settlement fees.[85] Over 1,000 miles of shoreline on the Gulf of Mexico—from Texas to Florida—was impacted when over 130 million gallons of oil leaked from the tanker into the ocean.[86] To get the oil out of the ocean, cleanup crews used three techniques:

- Physical barriers known as **containment booms** were used to control the spread of the oil by containing and diverting it along a desired path.[87]

- The oil was then removed by skimmers or **sorbents** that are used to pick up the oil and remove it from the water. Sorbents can be natural/organic materials such as peat moss, straw, hay, or sawdust. Clay, volcanic ash, sand, vermiculite (natural inorganic materials), and synthetic sorbents such as polyurethane, polypropylene, and polyethylene were also used. According to the EPA, synthetic sorbents can absorb up to seventy times their own weight in oil, while most organic sorbents can absorb only three to fifteen times their weight in oil.[88] However, as long as natural sorbents aren't pretreated, they are biodegradable, while synthetic sorbents are not.

- Chemicals called **dispersants** were also used to break up the oil.

In theory, if oil is broken down by dispersants, the smaller particles mix with water more easily so that evaporation and bacteria can degrade oil more quickly. However, the combination of oil and Corexit—the dispersant that was approved by the EPA and used in the 2010 BP cleanup—was found to be toxic to sea life, and even caused the bacteria that consumes oil molecules to die. Natural sorbents contain these microbes, which are used as a natural form of bioremediation in oil spills, and although there was a tenfold increase in those microbes soon after the spill occurred, researchers at the University of Tennessee–Knoxville determined that they were killed off once Corexit was added to the water.[89]

According to the *Smithsonian* and the *National Museum of Natural History*, roughly 58 percent of the total 1.84 million gallons of Corexit was sprayed from the air into the Gulf,[90] and about 771,000 gallons was injected into the leaking oil column to dilute the oil at the source of the rupture.[91] In 2014, researchers from the University of Texas Marine Science Institute discovered that 50 percent of mesozooplankton communities died after 48 hours of exposure to Corexit alone.[92] In 2015, a study by the University of Alabama at Birmingham showed that Corexit can damage human lung tissue *and* cause respiratory injuries among wildlife, including the cells in fish gills, sea turtles, and crabs.[93] While oil is extremely harmful to coral because it reduces its ability to reproduce and grow, exposure to Corexit is even worse as the dispersant causes coral health to decline at a much faster rate. The University of California's School of Natural Sciences in Merced published research in December 2017 that included photos of coral *literally* disintegrating after being exposed to Corexit 9500 and Corexit 9527.[94]

In the 2010 BP oil spill, Corexit was sprayed for ninety days. C-130 airplanes sprayed it onto the ocean, but it frequently hit cleanup crews on the ground—many of whom were not given protective respirators. Wilma Subra, the scientific adviser for the Louisiana Environmental Action Network, told state and federal authorities that workers exposed to crude oil and Corexit have serious short-term health concerns including "acute respiratory problems, skin rashes, cardiovascular impacts, gastrointestinal impacts, and short-term loss of memory" while long-term impacts included "cancer, decreased lung function, liver damage, and kidney damage."[95] A scientific study published in the peer-reviewed *Environmental Pollution* journal a mere nineteen months after the BP explosion confirmed Subra's findings by noting that when crude oil is mixed with Corexit, toxicity increases up to 52 percent.[96] The study determined that when the chemical is combined with crude oil, it becomes fifty times more toxic to small marine life because the oil is now broken down into more easily consumable parts. Since the chemical also killed off all the oil-eating bacteria, there's no way for wildlife to escape the full effects of oil *and Corexit* contamination.

In 2011, the National Institutes of Health announced it would study the 33,000 cleanup workers' long-term health issues,[97] and by September 2017, researchers concluded that workers exposed to dispersants still experienced common symptoms, such as "cough, wheeze, tightness in the chest, and burning in the eyes, nose, throat, or lungs."[98] A 2018 study on the Coast Guard personnel found those who were exposed to oil dispersants during their deployment to the *Deepwater Horizon* crisis were also experiencing the same symptoms.

Over 8,500 Coast Guard responders were deployed in 2010, but only 4,855 completed the respiratory survey conducted by the Uniformed Services University. Out of this pool, more than half (54.6 percent) were exposed to crude oil, and nearly one-fourth (22 percent) were exposed to oil dispersants. Out of the total who participated, nearly two thousand Coast Guard members experienced side effects: 19.4 percent experienced coughing, 5.5 percent experienced shortness of breath, and 3.6 percent experienced wheezing. According to the report:

> *Increasing frequency of inhalation of oil was associated with increased likelihood of all three respiratory symptoms. A similar pattern was observed for contact with oil dispersants for coughing and shortness of breath. The combination of both oil and oil dispersants presented associations that were much greater in magnitude than oil alone for coughing.*

Keep in mind, these results are being compiled *eight years* after the spill occurred, and if they're *still* experiencing these side effects, I'm going to take a logical leap here and conclude that these "symptoms" are actually signs of permanent damage.

These long-term *symptoms* also correlate to a John Hopkins study, published in the March 2019 issue of *Science of the Total Environment.* Researchers took a closer look at the human health risk for workers who inhaled dispersants at the 2010 BP oil spill. The study concluded the toxic chemicals in Corexit can cause serious pulmonary diseases, including lung cancer. After dispersants are added to a spill, they can separate the oil particles until they're so fine that the oil can be carried by ocean waves and breeze. Reporters covering the study referred to the phenomenon as a "toxic mist" because it can travel up to fifty miles, and if inhaled, the toxic mist is "drawn to the innermost parts of the lungs, where they are quickly absorbed into the body."[99]

According to the British newspaper *The Telegraph,* Corexit was banned in the UK in 1998,[100] but after reviewing what is currently approved by the government's Marine Management Organisation to treat oil spills, this is no longer true. As of November 2018, Corexit EC9500B is in fact approved for sea-spills in the UK through July 2020.

As for the EPA? In 2011, the agency recommended that the Office of Solid Waste and Emergency Response revise the National Contingency Plan to "clarify roles and responsibilities for Spills of National Significance" and "capture" more testing on dispersants; the EPA also recommended that the Office of Research and Development study the long-term health and environmental effects of dispersants.[102] Yet, even after the many independent studies that have been conducted over the past eight years, the EPA's website lists Corexit as a product that can be used for oil-spill cleanup.[103]

NALCO, the manufacturer of Corexit, states that the dispersant "is a highly concentrated, bio-degradable formulation."[104] I've literally not seen one shred of independent scientific documentation that proves the product is biodegradable. I *have seen* plenty of verifiable research that shows Corexit is an environmental hazard and a toxin that literally kills life. So I'm perplexed. Mainly by how the satanic darlings who create and sell Corexit can sleep at night.

According to Center for Responsive Politics and OpenSecrets.org, Nalco Holding Company spent $440,000[105] in 2010 to lobby Congress, the EPA, and several other government agencies on bills relating to the BP oil spill, including H.R. 5626, the Blowout Prevention Act of 2010; S. 3516, the Outer Continental Shelf Reform Act of 2010; H.R. 5629, the Oil Spill Accountability and Environmental Protection Act of 2010; H.R. 3534, the Consolidated Land, Energy, and Aquatic Resources Act of 2009; and S. 3661/H.R. 6119, the Safe Dispersants Act. Which of course leads me to logically conclude that they have something to hide . . . such as perhaps the product *isn't* biodegradable.

In 2011, Ecolab Inc. (a chemical industry parent company for cleaning products) bought Nalco. I suppose those two go together better than peanut butter and jelly because since 2000, Ecolab has paid out $118,345,231 in violations, including workplace safety or health violations, fraud, and environmental violations.[106] And, hey big spender, their most extravagant violation expenses are doled out on Nalco.

Nalco was founded in 1928. The company's name has changed slightly over the years—Nalco Environmental Solutions, LLC is the current subsidiary that sells Corexit. (It was founded in 2010, ain't that a BP coincidence?)[107] A simple search in the Poison Papers for "Nalco" brings up some truly alarming information. I'm not going to go into every detail, but once I read certain documents, I felt compelled to share what hasn't been reported in relation to Corexit and the company that makes it, so here we go . . .

Known as Nalco Chemical Company at the time, Nalco acquired Industrial Bio-Test Laboratories (IBT) in 1966. While I won't go so far as to say Nalco created IBT's culture of fraudulent laboratory testing, the federal investigation into IBT occurred under Nalco's watch—*and at the time,* the company was producing chemicals that required testing by a lab before they could be approved by the EPA. According to the Nalco's 1980 annual report, the chemical company was named in four lawsuits by former IBT clients for "misrepresenting" test data,[108] and a 1983 *Business Week* article states that Nalco lost $20 million when IBT was finally shut down.[109]

But IBT was just one of many sketchy ventures for Nalco—ventures that seemed to define the company's core wheelhouse and values (or lack thereof). Even after IBT fraud was exposed, a clear pattern of *we don't give a shit about the environment* mentality continued to flourish at this phenomenal corporation.

In 1985, Nalco Chemical Company wrote a letter to the EPA to request more time to comply with new federal dioxin regulations, as they were using the dioxin TCDD (2,4,5-trichlorophenate) at two wastewater treatment facilities (facilities that were worth over $90 million, according to Nalco).[110] While the letter, written by the company's Environmental Control Manager, stated the "wastewater treatment sludges" were being safely accumulated in tanks until the toxins could be incinerated—and the Nalco treatment plants had a "removal efficiency of 99.9999%"—documents compiled by the EPA for Congress in July 1986[111] stated that the Bedford Park, Illinois location sent the sludge to be "disposed" of in off-site landfills,[112] which is clearly not the same as an incinerator. In the 1980s, EPA referred the Bedford Park site to their Superfund department after ten soil samples were found to contain the TCDD dioxin.[113] Nalco's site was also described as "extensively paved." Paving over soil to eliminate the immediate possibility of it being sampled was of course the universal indication that on-site dumping of dioxins was in fact occurring. And in fact they *definitely* did, because the Bedford Park location was eventually declared a Superfund site by the EPA.[114] The declaration was lifted in the 1990s once Nalco agreed to abide by guidelines—including keeping accurate records regarding the amount of chemicals on site and their storage.[115]

In June 2000, Nalco paid a settlement of $300,000 for the costs associated with the Byron Salvage Superfund site in Ogle County, Illinois.[116] The site is also on the National Priorities List because it is "one of the worst hazardous waste sites identified by the EPA."[117] "Industrial wastes were reportedly dumped directly on the ground" *and* when there was heavy rain in Ogle County, those wastes were carried off-site into nearby waterways.[118]

According to HomeFacts.com, Nalco's Brookhaven, Mississippi facility is another (archived) Nalco Superfund site,[119] as is their Odessa, Texas location,[120] among several other facilities in America. So while "Environmental Solutions" may in fact be in the title of the company that manufactures Corexit, the company's history is one that presents nothing but environmental *problems*. I'd simply adore an actual environmental solution from them, considering the immense damage they've caused.

Prior to the Ecolab acquisition, Nalco formed a "joint venture" with Exxon in 1994 to develop products to combat oil spills. If that isn't the definition of double dipping, well, Exxon, *yes it certainly is, so we'll move on.*

While Corexit might seem the likely outcome of such a venture with an oil company, the formula existed generations prior *and* was actually used in lab research during the IBT days. A report compiled in 1976 titled "Some Biological Effects of PCBs"[121] examined the effects of PCBs and Aroclor 1254 on just about every aspect of life. As far as reproduction is concerned, Aroclor was shown to decrease the amount of spermatic fluid in brook trout; it also decreased the hatchability of eggs. The analysis of the reproductive section of the report concluded: "the significance of this study is difficult to evaluate because the solubilizing agent used, Corexit, seemed to have some effect on testicular pathology and on vitro steroid metabolism."[122] After the 2010 BP oil spill, studies examining reproductive harm showed that when capelin—a smelt fish (aka dinner for whales, seals, Atlantic cod/mackerel, squid, and birds)—are exposed to Corexit that has been added to crude oil, the combination affects the sperm's ability to fertilize eggs.[123] With fewer capelin in the sea, there's also less food for all other marine life.

But wait a minute, what concentration of Corexit are we talking about here?

Well, there's a 1977 study was published in *Marine Pollution Bulletin* that showed concentrations down to .0003 ppm (parts per million) will affect the sea urchin's sperm and decrease its ability to fertilize eggs.[124] That's *less than* the amount of glyphosate allowed in food by the EPA.

While some people might not care if sea urchin or capelin can reproduce, *something is affecting* sperm count and motility (sperm's ability to move) in humans. According to the American Society for Reproductive Medicine, as of October 2018, two new studies show that "sperm quality has declined noticeably this century in infertility patients and sperm donors alike."[125] While researchers don't want to place total blame on environmental or lifestyle-related issues, they found the low sperm numbers unprecedented and concerning.

From 2007 to 2017, a total of 2,586 sperm donor specimens in the United States were analyzed, with donors ranging from nineteen to thirty-eight years of age. Researchers looked at the total motile sperm count (the total amount of sperm capable of moving/swimming), the total amount of sperm, and the average concentration of sperm in a particular specimen for the past ten years in six cities: Los Angeles, Palo Alto, Houston, Boston, Indianapolis, and New York City. All aspects of sperm production and mobility *declined as a whole* in every region *except* New York City. While researchers weren't sure why the results were different in Manhattan, I'm going to go out on a very sturdy limb here and suggest that if chemicals like Corexit affect wildlife's ability to breathe and reproduce, those chemicals will eventually affect mankind's ability to do the same. Since Corexit isn't going away any time soon, as a species, if we can't breathe or reproduce, then the very chemicals we've manufactured will play a hand in our own extinction.

According to scienceline.org, the National Oceanic and Atmospheric Administration responds to about 100 oil spills each year.[126] While oil spills do seem to happen on a fairly regular basis, we're probably going to see even more of them if the US Interior Department opens "nearly the entire U.S. Outer Continental Shelf for potential oil and gas lease sales."[127] Currently, 94 percent of the Outer Continental Shelf (OSC) is off-limits to offshore drilling, but according to a press release from the former US secretary of the interior, Ryan Zinke, the agency intends to update the Natural Outer Continental Shelf Oil and Gas Leasing Program. Once approved, the changes will take effect from 2019 to 2024 to make "over 90 percent of the total OCS acreage and more than 98 percent of undiscovered, technically recoverable oil and gas resources in federal offshore areas available," marking the largest number of available lease sales in US history.[128]

While safety and environmental regulations increased after the *Deepwater Horizon* oil spill, the effort to dial them back again occurred soon after President Trump was elected. The Department of the Interior's Bureau of Safety and Environmental Enforcement (BSEE) developed a proposal to essentially *reverse* the measures that were put in place. According to *The Wall Street Journal,* these rollbacks will "reduce the role of government in offshore oil production and return more responsibility to private companies," which is estimated to save the oil and gas industry more than $900 million over the next ten years.[129] Of course, if they have that kind of money to spend, they'll certainly be inclined to "drill baby drill" on 90 percent of the OCS, so I suppose now is the time to freeze your partner's sperm, while he's still somewhat fertile?

Although the Department of the Interior's plans haven't come to fruition yet, environmentalists have already stated we need better and more sustainable, environmentally safe methods of cleaning up oil spills. Or at the very least, we need to start implementing more of the natural/organic methods that already exist and are proven to be effective.

In 1999, the US Navy compiled a research paper on kenaf (rhymes with giraffe) that highlighted the plant's ability to clean up oil spills. Kenaf (known as *Hibiscus cannabinus* L) is related to cotton, hemp, and okra (the USDA refers to kenaf as a cousin to these crops).[130] Since kenaf doesn't contain any THC, it is completely legal to grow, and at the time of the Navy's study, the largest US manufacturer of kenaf was KenGro Corporation in Charleston, Mississippi. The company grew 2,000–2,500 tons of kenaf on approximately five hundred acres (so four or five tons per acre), and reported using very little herbicide and pesticide on the crop.[131]

According to the 1999 report compiled by the Naval Facilities Engineering Service Center in Port Hueneme, California:[132]

- Kenaf grows quickly—twelve to fourteen feet within five or six months.

- The outer stalk is called bast, and it is often used for manufacturing fiber-based products in Africa, including rope, writing paper, cigarette paper, and even a fiberglass-like product (with the addition of polypropylene).

- The rest of the stalk is the white inner core (woody, like hemp hurds) that can be used to make absorbent products, including composite panels, animal bedding, oil-absorbent materials, and building materials such as insulation.

- The stalks can be burned for fuel, and the leaves can be consumed as a vegetable.

- Kenaf absorbent made for oil-spill cleanups is a sustainable, renewable resource that is all-natural, bio-based, and biodegradable in approximately twenty-four months.

- Kenaf absorbent is nontoxic to animals and has no adverse effects on plants or animals.

The similarities between hemp and kenaf may seem uncanny at first, but they have different physical appearances and characteristics:

- White, yellow, or purple hibiscus flowers bloom on the kenaf plant.

- While hemp seeds are round, kenaf seeds are pointy and look more like a shark's tooth.

- Kenaf also has thorny stalks whereas hemp does not.
- Although kenaf is considered a vegetable, it doesn't contain the nutritional value of hemp as a superfood, and it also doesn't have medicinal properties, since it doesn't contain CBD, THC or any other cannabinoids.

The Navy's 1999 research paper included third-party testing (conducted by Mississippi State University and Millsaps College in Mississippi) which put kenaf absorbent through three phases to verify that KenGro Corporation's claims were indeed accurate: the kenaf product absorbed up to *six* times its own weight and it could be used to clean up oil spills on soil, water, and hard surfaces (such as factory floors or highway pavement).[133]

Kenaf also outperformed the other methods of absorbing oil—both organic and synthetic sorbents—including peat moss, kitty litter, and polypropylene. The Navy recommended kenaf be used for Navy/DOD operations that involve petroleum, oil, and hydrocarbon products, and concluded it is "cost effective" when compared to similar products already used: "procuring agencies servicing the Federal government should implement purchase preference for kenaf absorbent where feasible."[134] Not only did the Navy determine that kenaf is an excellent sorbent of oil, but they also noted that it carries microorganisms (bacteria) which can be used for bioremediation of petroleum wastes (the bacteria eat the oil). So relying on dispersants to break down the oil in a large spill could potentially be avoided as well if kenaf is used.

In 2004, a study published in *Marine Pollution Bulletin* examined a new oil adsorption method called adsorption filtration. The technology cleans oil from water by running it through "a simple filter made from freeze treated, dried, milled and then fragmented plant material."[135] The adsorption filtration method works best to gather spilled oil in shallow coastal waters *before* the oil reaches the shore (which is what dispersants like Corexit are used for).

The study took fiber from reed canary grass, flax, and hemp to determine what is the best size for the plant material fragments, and found that "fine spring harvested hemp fibre (diameter less than 1 mm) and reed canary grass fragments adsorb 2–4 g of oil per gram of adsorption material compared to 1–3 g of water."

The best part about this research is that it shows how the biomass from hemp *and* other *legal* plants can absorb oil at a reliable rate, so if we start developing this technology now, hemp hurds can be used as a sorbent to clean up oil spills or any other chemical or toxic liquid spill.

Hemp, Inc., an American-based company focused on revamping the industrial hemp industry, has already been doing the legwork on this technology. According to Reuters, the company's subsidiary, Industrial Hemp Manufacturing LLC, produces DrillWall (a nontoxic, biodegradable drilling fluid additive used in oilfield drilling applications), SpillSuck, and Spill-Be-Gone—which are made from the inner core particles (hurd) of kenaf and hemp plants:[136]

> *Spill-Be-Gone [and SpillSuck] can be used for emergency oil spill cleanup. It can also be used in automotive repair shops to keep garage floors free of oil,*

or manufacturing plants to keep oil spills from machinery posing a danger to workers on the job.

The only difference between SpillSuck and Spill-Be-Gone is the size of the packages the product is sold in, and overall, all three biodegradable products (DrillWall, SpillSuck, and Spill-Be-Gone) are less expensive and much better for the environment than synthetic competitors.

According to David Schmitt, COO of Industrial Hemp Manufacturing LLC, SpillSuck was experimentally used in 2010 to combat the BP *Deepwater Horizon* oil spill in the Gulf of Mexico.

"It was totally amazing to watch all the birds and other small animals that were completely covered in oil walk in and through the pile of kenaf. And how easily it removed the oil from them," Schmitt recalled. "It was like they instinctively knew that this mound of kenaf would help save their lives."[137]

As of January 2018, Hemp, Inc. claimed it has the largest hemp mill in the western hemisphere;[138] most likely because in 2008, it purchased the only large-scale hemp decorticator in the United States.[139] (The decorticator is a machine used to separate fiber from the plant's stalks.)

Since the decorticator works the same on either hemp or kenaf, the company is proof positive of how easily a farmer can transition from growing kenaf to hemp and integrate both of them into the biomass production industry. The oil cleanup industry itself is unfortunately going to continue to grow and expand. Why else would Ecolab purchase Nalco for $5.9 billion in 2011, after the initial findings on the harm of Corexit were made public?[140] At the time, the *New York Times* reported that Nalco had operations in 150 countries, so unless of course we start using more biofuel and less fossil fuel, expect to see more Corexit being dumped into the ocean to "mitigate" oil spills.

While we might not be able to stop drilling operations from expanding, companies can take it upon themselves to make more environmentally sound choices so that drilling and mining make less of an impact. In August 2018, the *Colorado Springs Independent* reported that Industrial Hemp Manufacturing's nontoxic, biodegradable drilling fluid (DrillWall) is being sold exclusively to Quadco, LLC—a corporation providing products, equipment, and services for the Alaskan oil and gas industry.[141] Certainly a step in the right direction. And for those oil companies who don't care about stepping just one little toe onto the right path, they might want to consider how Corexit can impact their bottom line once the inevitable oil spill occurs and lawsuits ensue (because their shareholders quite frankly *will* give a damn).

According to bloomberglaw.com, as of November 2018, there is a "surge of back-end litigation" from 2010 BP cleanup workers who were exposed to Corexit.[142] Again, this is *eight years later* and the *Deepwater Horizon* Medical Settlement Claims Administrator states that BP has received 6,389 notices to sue, and more than 62 percent of those claims have been approved.[143] In 2016 alone, BP paid $5.5 billion in civil suits related to the oil spill. The *Maritime Executive* reported that BP paid:[144]

- A Clean Water Act penalty of $5.5 billion (plus interest)

- $8.1 billion in natural resource damages, up to an additional $700 million to address injuries to natural resources that are presently unknown, and $600 million for other claims, including claims under the False Claims Act

- Royalties and reimbursement of natural resource damage assessment costs and other expenses due to this incident

- A total of over $20 billion, the largest settlement with a single entity in the history of federal law enforcement

So unless oil companies prefer to continue to pay, and pay, and pay, and pay, and pay for exposing cleanup crews to oil and Corexit, they might want to consider some more natural alternatives—like hemp and kenaf absorbents.

- **Hemp for Paper**

This chapter has presented some novel uses for hemp's biomass to combat climate change and environmental pollution, so I'm going to end on a classic: taking hemp's biomass and turning it into paper.

Producing paper from hemp dates back more than two thousand years. According to a 2016 conference paper from the 1st International Mediterranean Science and Engineering Congress in Turkey, the oldest surviving piece of hemp paper dates back to between 140 and 87 BC; it was discovered in 1957 in a tomb near Sian in Shensi province, China.[145]

As previously stated in chapter 1, in 1916, Lyster H. Dewey and Jason L. Merrill published "Hemp Hurds as Paper-Making Material" in the USDA's Bulletin No. 404, which documents a variety of methods used turn hemp hurds into a No. 1 machine-finish printing paper.[146] Since that time, there have been research articles that use traditional wood-pulp papermaking methods, swapping wood pulp for hemp hurds, with similar success. Traditionally, the process involved in taking wood and turning it into pulp for paper is a chemical-intensive process, yet manufacturers are able to use less chemicals with hemp fiber.[147]

According to a 2012 article titled "Environmental Impact of Pulp and Paper Mills," published in the *Environmental Engineering and Management Journal*, the three main pollutants in today's paper industry are:[148]

1. The chemicals used in the wood-pulping and pulp-bleaching process, which includes sulfur compounds and nitrogen oxides that are emitted into the air. Nitrogen dioxide and sulfur dioxide contribute to acid rain while chlorinated and organic compounds (used to bleach the paper fibers) contain dioxins and are discharged as wastewater that then has to be properly disposed. Chlorine has been largely replaced by chlorine dioxide, which in turn greatly reduces the amount of dioxins.
2. The amount of water needed to manufacture paper is more than any other industry. Water is turned into wastewater (contaminated with chlorine/dioxins or chlorine dioxide) at an average rate of 20–100 cubic meters per ton of paper product.
3. The remaining wood wastes (such as black liquor) and sludge are usually incinerated or disposed of in landfills; however, a bark boiler can turn them into energy and fuel for the mill to operate.

There are also Elemental Chlorine Free (ECF) and Totally Chlorine Free (TCF) bleaching processes,[149] but they are not universally used worldwide. Even if the entire pulp-paper industry were able to produce paper *without* dioxins, we're still cutting down trees, so we aren't doing much to lessen deforestation and our overall carbon footprint.

And as much as paper mills push their PR campaign that the manufacturing process isn't as environmentally harmful as it once was, in December 2018, a Georgia-Pacific plant in Crossett, Arkansas, was charged $600,000 in fines and ordered to spend $1.8 million on environmental projects and facility upgrades. An independent investigation by the *Arkansas Democrat-Gazette* found the plant emitted more hydrogen sulfide than its permit allowed, and in 2015, an EPA inspection found that leaks in the company's air-management systems caused hazardous chemicals, such as formaldehyde, into the air, which was in violation of the Clean Air Act and the Resource Conservation and Recovery Act. Aside from the fines, Georgia-Pacific must also:[150]

- Commit to roughly $4.7 million in environmental projects to reduce hydrogen sulfide emissions and monitor the local concentration of the toxic gas in the air.

- Install a $2.9 million filter to prevent "dregs," or compounds that don't dissolve, from entering the plant's wastewater treatment system.

- Spend $1.8 million on three "supplemental environmental projects," including a $1 million pulp mill collection tank to make the plant's wastewater system more stable.

- Reduce the hydrogen sulfide emissions (and the smell of the emissions) by injecting more oxygen into the wastewater treatment system, which costs $500,000.

A 2015 paper written by the Institute of Papermaking and Printing at the Technical University of Lodz, Poland, looked at hemp as a comparison and solution to wood-pulp paper to determine:[151]

1. Hemp paper *doesn't* require the use of toxic bleaching substances because it can be whitened with hydrogen peroxide. This means the hemp-hurd-pulp mills won't be at risk for poisoning waterways with chlorine or dioxins.
2. Hemp paper can be recycled up to seven to eight times, compared with only three times for wood-pulp paper.

Much of the hemp paper market is considered a specialty fiber, or high-priced niche market, including rolling papers and cigarette papers, banknotes made from hemp fibers, and fine art paper. Currently, the Colorado Hemp Company is producing paper through their company "Tree Free Hemp." The company states it manufactures "100% sustainable, hemp-blended, post-consumer recycled paper products."[152] However, eco-friendly paper products can also be made from hemp's cousin kenaf.

In 1981, a headline for a *New York Times* article on kenaf stated "You Don't Need Trees to Make Paper," and quoted a 1979 Bulletin from the ANPA (American Newspaper Publishers Association) that stated "an average acre of kenaf will likely yield over nine times the pulp per acre of forest land"—or seven to fourteen tons an acre each year, compared to one-and-a-half to three tons for wood.[153]

The article quoted Dr. Marvin Bagby of the U.S. Department of Agriculture as stating: "Kenaf can be a supplement, extender, or alternative to wood in making newsprint." He admitted that "every major pulp and paper company has shown interest from time to time in Kenaf,"' but paper is an "inert industry geared to working with wood," meaning the paper industry itself isn't interested in changing anything. While some newspapers reportedly experimented with kenaf paper or adding kenaf fibers to the wood-based pulp, the fad didn't catch on to a substantial degree.

In 1982, the USDA published a paper titled "Kenaf—Promising New Fiber Crop," which explained that the methods for producing paper from kenaf were more sustainable, less toxic, and "appeared promising on all counts."[154] Since kenaf fibers are naturally whiter than tree pulp, less bleaching is required, and it is possible to use hydrogen peroxide in place of chlorine, which is of course means kenaf paper can be devoid of dioxins, just like hemp paper.[155]

According to research from Vision Paper, the first company to produce paper sourced entirely from kenaf and process it without the use of chlorine bleaching chemicals, kenaf contains 25 to 50 percent less lignin than wood pulp, making it easier to pulp. They estimate that the energy requirements for producing pulp from kenaf are about 20 percent less than from wood, *and* paper mills can easily convert their wood-pulping equipment to pulp kenaf.[156]

Unfortunately, there aren't as many studies on hemp's papermaking potential as there are on kenaf's due to the fact that kenaf has always been legal and hemp hasn't, but the kenaf studies are certainly relevant due to the similarities between the two plants. Both have two regions of fibers in their stalks—an outer bast core and inner woody core.

Whether hemp hurds are added to current means of papermaking or we make a switch to phase out wood-pulp products, one aspect of everyday life that isn't going away is our dependence on paper. According to the EPA, the United States uses about 69 million tons of paper and paperboard on a yearly basis, and each year, more than 2 billion books, 350 million magazines, and 24 billion newspapers are published.[157] As far as recycling goes, in 2013 half of all recyclables collected in the United States by weight were paper. The EPA averaged that out to 275 pounds for each person living in the United States. At the time, we were recycling a total of 43 million tons of paper and paperboard, which is a recycling rate of 63 percent. However, the majority of our paper products are sent to China for recycling, and as of January 2018, China stopped taking the majority of our "foreign garbage"—including paper products.[158]

Since August 2018, China has enacted a 25 percent import tariff on the remaining recyclable materials that they accept from the United States including OCC (Old Corrugated Cardboard) and scrap plastic.[159] Prior to the tariff, in the first half of 2018 alone, China imported 2.73 million short tons of OCC, 1.4 million short tons of all other US recovered fiber (paper products), and 30 million pounds of scrap plastic.[160]

Between China's tariffs and new regulations, the majority of these items are now winding up in American landfills. While paper products such as newspapers and cardboard can decompose faster than most other waste products, we still aren't saving any landfill space by throwing them away. Since hemp is also used for composting, recycling hemp paper products into biodegradable landscape paper or mulch could be another solution worth exploring.

NOTES

1. Alexander Nazaryan, "The US Department of Defense Is One of the World's Biggest Polluters," *Newsweek* Magazine, July 17, 2014, https://www.newsweek.com/2014/07/25/us-department-defence-one-worlds-biggest-polluters-259456.html.
2. Defense Environmental Restoration Program, "Defense Environmental Restoration Program Updates Munitions Response Dialogue (MRD)," *DENIX*, accessed Jan 27, 2019, https://www.denix.osd.mil/derp/home/.
3. Abrahm Lustgarten, "Get an Inside Look at the Department of Defense's Struggle to Fix Pollution at More Than 39,000 Sites," Bombs in our Backyard, *ProPublica,* May 7, 2018, https://www.propublica.org/nerds/data-get-an-inside-look-at-the-department-of-defense-struggle-to-fix-pollution.
4. Craig Witlock and Bob Woodward, "At the Pentagon, overpriced fuel sparks allegations—and denials—of a slush fund," Investigations, *The Washington Post,* May 20, 2017, https://www.washingtonpost.com/investigations/at-the-pentagon-overpriced-fuel-sparks-allegations—and-denials—of-a-slush-fund/2017/05/20/c5ff4bf4-31b2-11e7-9dec-764dc781686f_story.html.
5. Duncan McLauren and Ian Willmore, "The environmental damage of war in Iraq," Iraq, *The Guardian*, Jan 18, 2003, https://www.theguardian.com/world/2003/jan/19/iraq5.
6. "Agent Orange," *History.com,* last modified Aug 21, 2018, https://www.history.com/topics/vietnam-war/agent-orange-1.
7. Abrahm Lustgarten, "Open Burns, Ill Winds," Bombs in our Backyard, *ProPublica,* July 20, 2017, https://www.propublica.org/article/military-pollution-open-burns-radford-virginia.
8. Marjorie Cohn and Jeanne Mirer, "The Toxic Effects of Agent Orange Persist 51 Years After the Vietnam War," *Truthout.org,* https://truthout.org/articles/the-toxic-effects-of-agent-orange-persist-51-years-after-the-vietnam-war/.

9. Lauren Walker, "US military burn pits built on chemical weapons facilities tied to soldiers' illness," US military, *The Guardian,* Feb 16, 2016, https://www.theguardian.com/us-news/2016/feb/16/us-military-burn-pits-chemical-weapons-cancer-illness-iraq-afghanistan-veterans.
10. F Campanella, Bruno & Bock, Claudia & Schröder, Peter, "Phytoremediation to increase the degradation of PCBs and PCDD/Fs," *Environmental science and pollution research international*, Vol 9, Feb 2002, pp 73-85, https://www.researchgate.net/publication/11478552_Phytoremediation_to_increase_the_degradation_of_PCBs_and_PCDDFs.
11. Anthony DePalma, "Love Canal Declared Clean, Ending Toxic Horror," *The New York Times,* March 18, 2004, https://www.nytimes.com/2004/03/18/nyregion/love-canal-declared-clean-ending-toxic-horror.html.
12. Thompson M., Rothman M., Regan M, "Residents say Love Canal Chemicals Continue to Make Them Sick," *PBS News Hour Weekend,* Aug 5, 2018, https://www.pbs.org/newshour/show/residents-say-love-canal-chemicals-continue-to-make-them-sick.
13. Mustafa Tuzen, "Determination of heavy metals in soil, mushroom and plant samples by atomic absorption spectrometry," *Microchemical Journal,* Vol 74, Issue 3, June 2003, pp 289-297, https://www.sciencedirect.com/science/article/pii/S0026265X03000353.
14. Abson D., Lipscomb A.G., "The determination of lead and copper in organic materials (foodstuffs) by a dry-ashing procedure," *Analyst,* Vol 82, Issue 972, 1957, pp 152–160, https://pubs.rsc.org/en/content/articlelanding/1957/an/an9578200152/unauth#!divAbstract.
15. P. B. A. Nanda. Kumar, Viatcheslav. Dushenkov, Harry. Motto, and Ilya. Raskin, "Phytoextraction: The Use of Plants To Remove Heavy Metals from Soils," *Environmental Science & Technology,* Vol 29(5), May 1995, pp 1232-1238, https://pubs.acs.org/doi/abs/10.1021/es00005a014.
16. "Frequently asked questions," Independent Statistics & Analysis, *US Energy Information Administration,* last modified June 8, 2018, https://www.eia.gov/tools/faqs/faq.php?id=77&t=11.
17. National Renewable Energy Laboratory powerpoint, "Innovation for Our Energy Future," Midwest Research Institute, *Office of Energy Efficiency & Renewable Energy*, August 2004, https://www.nrel.gov/docs/gen/fy04/36831e.pdf.
18. "Frequently asked questions," Independent Statistics & Analysis, *US Energy Information Administration,* last modified April 5, 2018, https://www.eia.gov/tools/faqs/faq.php?id=74&t=11.

19. "Biomass and waste fuels made up 2% of total U.S. electricity generation in 2016," Today In Energy, Independent Statistics & Analysis, *US Energy Information Administration,* last modified Nov 27, 2017, https://www.eia.gov/todayinenergy/detail.php?id=33872.
20. Mary Booth and Edward Miller, "Trees, Trash, and Toxics: How Biomass Energy Has Become the New Coal," *Partnership for Policy Integrity,* April 2, 2014, http://www.pfpi.net/trees-trash-and-toxics-how-biomass-energy-has-become-the-new-coal.
21. Ibid.
22. "Southern states lead growth in biomass electricity generation," Today in Energy, Independent Statistics & Analysis, *US Energy Information Administration,* last modified May 25, 2016, https://www.eia.gov/todayinenergy/detail.php?id=26392.
23. Gillian Neimark, "Q&A: Advocate says biomass energy is the new coal," *The Energy News Network,* Aug 7, 2017, https://energynews.us/2017/08/07/southeast/qa-advocate-says-biomass-energy-is-the-new-coal/.
24. Ibid.
25. "EPA Holds Piedmont Green Power Accountable for Air Pollution," Partnership for policy integrity, Dec 16, 2016, http://www.pfpi.net/epa-holds-piedmont-green-power-accountable-for-air-pollution.
26. Gillian Neimark, "Q&A: Advocate says biomass energy is the new coal," *The Energy News Network,* Aug 7, 2017, http://biomassmagazine.com/articles/5634/gainesville-biomass-plant-completes-construction-financing.
27. Ed Bielarski, "Ed Bielarski: Biomass plant purchase produced long-term savings," Opinion, *Gainesville Sun,* Oct 19, 2018, https://www.gainesville.com/opinion/20181019/ed-bielarski-biomass-plant-purchase-produced-long-term-saving.s
28. Andrew Caplan, "Report: GREC contract among country's worst," *Gainesville Sun,*Oct 24, 2018, https://www.gainesville.com/news/20181024/report-grec-contract-among-countrys-worst.
29. Ibid.
30. Ibid.
31. Erv Evans, "Tree Facts," Trees of Strength, *NC State University*, accessed Jan 25, 3019, https://projects.ncsu.edu/project/treesofstrength/treefact.htm.
32. "How Industrial Hemp Can Reduce Our Carbon Footprint," Hemp news, *National Hemp Association,* July 13, 2016, http://nationalhempassociation.org/cannabis-and-climate-change-how-industrial-hemp-can-reduce-our-carbon-footprint/.

33. James Vosper, "The Role of Industrial Hemp in Carbon Farming," March 24, 2011, accessed Jan 25, 2019 http://www.aph.gov.au/DocumentStore.ashx?id=ae6e9b56-1d34-4ed3-9851-2b3bf0b6eb4f.
34. Sage, Rowan; Monson, Russell (1999). "7". *C4Plant Biology*. pp. 228–229.
35. Osborne, C. P.; Beerling, D. J. (2006). "Nature's green revolution: the remarkable evolutionary rise of C4 plants". *Philosophical Transactions of the Royal Society B: Biological Sciences*. 361 (1465): 173–194.
36. Thomas Prade, "Is industrial hemp the ultimate energy crop?," Environment +Energy, *The Conversation,* Jan 2, 2014, https://theconversation.com/is-industrial-hemp-the-ultimate-energy-crop-20707.
37. Kolodzieg J., Wladyka-Przybylak M., Mankowski J., Grabowska L., "Heat of Combustion of Hemp and Briquettes Made of Hemp Shives," Institute of Natural Fibres & Medicinal Plants, *Renewable Energy and Energy Efficiency, 2012, Conditioning of the energy crop biomass compositions,* pp 163–166, http://llufb.llu.lv/conference/Renewable_energy_energy_efficiency/Latvia_Univ_Agriculture_REEE_conference_2012-163-166.pdf.
38. Caoline Burges Clifford, "9.2 The Reaction of Biodiesel: Transesterification," Energy Institute, *PennState College of Earth and Mineral Sciences,* accessed Jan 25, 2019, https://www.e-education.psu.edu/egee439/node/684.
39. Ahmad Alcheikh, "Advantages and Challenges of Hemp Biodiesel Production: A comparison of Hemp vs. Other Crops Commonly used for biodiesel production," *University of Gavle,* June 2015, http://hig.diva-portal.org/smash/get/diva2:842842/ATTACHMENT01.pdf.
40. Christine Buckley, "Hemp produces viable biodiesel UConn study finds," Environment, Science, *UConn Today,* Oct 6, 2010, https://today.uconn.edu/2010/10/hemp-produces-viable-biodiesel-uconn-study-finds/#.
41. Harrison, Christine, "Stanford researchers show oil palm plantations are clearing carbon-rich tropical forests in Borneo," Stanford University, October 7, 2012.
42. Christine Buckley, "Hemp produces viable biodiesel UConn study finds," Environment, Science, *UConn Today,* Oct 6, 2010, https://today.uconn.edu/2010/10/hemp-produces-viable-biodiesel-uconn-study-finds/#.
43. Ahmad Alcheikh, "Advantages and Challenges of Hemp Biodiesel Production: A comparison of Hemp vs. Other Crops Commonly used for biodiesel production," *University of Gavle,* June 2015, http://hig.diva-portal.org/smash/get/diva2:842842/ATTACHMENT01.pdf.
44. Li, si-yu, James D. Stuart, and Richard S. Parnas, "The feasibility of converting Cannabis sativa L. oil into biodiesel." *Bioresource Technology* Vol. 101, no. 21 (2010), pp. 8457-8460.

45. "Biobutanol," Alternative Fuels Data Center, Energy Efficiency & Renewable Energy, *US Department of Energy,* accessed Jan 25, 2019, https://afdc.energy.gov/fuels/emerging_biobutanol.html.
46. Joseph Toomey, *An Unworthy Future: The Grim Reality of Obama's Green Energy Delusions,* Archway Publishing 2014, p 312.
47. Bill Kovarik, "History of Biofuels," *Ethanol History,* 2013, http://www.ethyl.environmentalhistory.org/?page_id=58.
48. W.L. Faith, "Development of the Scholler Process in the United States, *Industrial & Engineering Chemistry,* Vol 37 (1), January 1945, pp 9–11, http://pubs.acs.org/doi/abs/10.1021/ie50421a004.
49. Ibid.
50. Brian Barth, "The next generation of biofuels could come from these five crops," Future of Energy, *Smithsonian.com,* Oct 3, 2017, https://www.smithsonianmag.com/innovation/next-generation-biofuels-could-come-from-these-five-crops-180965099/
51. Christopher K. Wright and Michael C. Wimberly, "Recent land use change in the Western Corn Belt threatens grasslands and wetlands," *PNAS* March 5, 2013, Vol 110 (10), pp 4134-4139, http://www.pnas.org/content/110/10/4134.
52. Brian Barth, "A Kernel of Truth: Why Biofuels May Not Be Better," Lifestyle, *Modern Farmer,* Sept 19, 2017, https://modernfarmer.com/2017/09/a-kernel-of-truth-biofuels-ethanol-and-corn/?xid=PS_smithsonian.
53. Ibid.
54. Jonathan Foley, "It's time to rethink America's corn system," Sustainability, *Scientific American,* March 5, 2013, https://www.scientificamerican.com/article/time-to-rethink-corn/.
55. USDA Report, "Corn ethanol significantly reduces greenhouse gases," *Growth Energy,* Jan 12, 2017, https://growthenergy.org/2017/01/12/usda-report-corn-ethanol-significantly-reduces-greenhouse-gas-emissions/
56. Brian Barth, "The next generation of biofuels could come from these five crops," Future of Energy, *Smithsonian.com,* Oct 3, 2017, https://www.smithsonianmag.com/innovation/next-generation-biofuels-could-come-from-these-five-crops-180965099/
57. Wang H., Xu Z., Kohandehghan A., Li Z., "Interconnected Carbon Nanosheets Derived from Hemp for Ultrafast Supercapacitors with High Energy," *ACS Nano*, Vol. 7 (6), May 7, 2013, pp. 5131-5141, https://pubs.acs.org/doi/abs/10.1021/nn400731g.
58. Mark Crawford, "Hemp Carbon Makes Supercapacitors Superfast," *ASME,* July 2013, https://www.asme.org/engineering-topics/articles/energy/hemp-carbon-makes-supercapacitors-superfast.
59. Ibid.

60. Molly Lempriere, "Could graphene batteries change the face of graphite mining?" Analysis, *Mining Technology,* Sept 19, 2018, https://www.mining-technology.com/features/graphene-batteries-change-face-graphite-mining/.
61. Mike Williams, "Another tiny miracle: Graphene oxide soaks up radioactive waste," Current News, *Rice University News & Media,* Jan 8, 2013, http://news.rice.edu/2013/01/08/another-tiny-miracle-graphene-oxide-soaks-up-radioactive-waste-2/.
62. Molly Lempriere, "Could graphene batteries change the face of graphite mining?" Analysis, *Mining Technology,* Sept 19, 2018, https://www.mining-technology.com/features/graphene-batteries-change-face-graphite-mining/.
63. Tim Loh, "Once-hot material graphene could be next battery breakthrough," Climate Changed, *Bloomberg,* April 10, 2018, https://www.bloomberg.com/news/articles/2018-04-10/once-hot-material-graphene-could-be-next-battery-breakthrough.
64. Eric Limer, "The process of making graphene is absurdly complex," *Popular Mechanics,* June 21, 2016, https://www.popularmechanics.com/science/a21436/how-graphene-is-made/.
65. Harry Pettit, "'Miracle material' graphene can now be made with cooking oil (and it's 200 times stronger than steel)," Science, *Daily Mail,* https://www.dailymail.co.uk/sciencetech/article-4179670/British-material-graphene-COOKING-OIL.html.
66. Ahmed, T.M., Asmatulu, E., Rahman, M.M., "Recycling of Graphite Waste into High Quality Graphene Products," *TechConnect Briefs*, Vol 2, May 13, 2018, pp 207-210, https://briefs.techconnect.org/papers/recycling-of-graphite-waste-into-high-quality-graphene-products/
67. Kauling A., Seefeldt A., Pisoni D., et al., "The worldwide graphene flake production," *Advanced Materials,* Vol 30, Issue 44, Nov 2, 2018, https://onlinelibrary.wiley.com/doi/abs/10.1002/adma.201803784.
68. Ryan F. Mandelbaum, "Most commercial graphene is just expensive pencil lead, new study finds," *Gizmodo,* Oct 9, 2018, https://www.extremetech.com/computing/278818-most-commercially-available-graphene-is-actually-pencil-lead.
69. "Ford develops car parts made out of graphene," *Assembly Magazine,* Oct 29, 2018, https://www.assemblymag.com/articles/94540-ford-develops-car-parts-made-out-of-graphene.
70. Rich Smith, "Should You Invest In Graphene Stocks?" *The Motley Fool,* last modified April 6, 2018, https://www.fool.com/investing/2016/12/08/should-you-invest-in-graphene-stocks.aspx.

71. Amit Katwala, "The spiralling environmental cost of our lithium battery addiction," Energy, *Wired,* Aug 5, 2018, https://www.wired.co.uk/article/lithium-batteries-environment-impact.
72. Anna Hirtenstein, "How to Mine Cobalt Without Going to Congo," Business, *Bloomberg,* Nov 30, 2017, https://www.bloomberg.com/news/articles/2017-12-01/the-cobalt-crunch-for-electric-cars-could-be-solved-in-suburbia.
73. Rob Nikolewski, "New way to recycle lithium-ion batteries could be a lifeline for electric cars and the environment," Technology Business, *LA Times,* March 16, 2018, https://www.latimes.com/business/technology/la-fi-lithium-ion-battery-recycling-20180316-story.html.
74. Joey Gardiner, "The rise of electric cars could leave us with a big battery waste problem," Energy, *The Guardian,* Aug 10, 2017, https://www.theguardian.com/sustainable-business/2017/aug/10/electric-cars-big-battery-waste-problem-lithium-recycling.
75. Rob Nikolewski, "New way to recycle lithium-ion batteries could be a lifeline for electric cars and the environment," Technology Business, *LA Times,* March 16, 2018, https://www.latimes.com/business/technology/la-fi-lithium-ion-battery-recycling-20180316-story.html.
76. "Auto body made of plastics resists denting under hard blows," *Popular Mechanics Magazine,* Vol. 76, No. 6, December 1941, https://www.ukcia.org/research/PopularMechanics/index.php.
77. Darren Boyle, "That's a smoking hot car, Florida businessman builds environmentally friendly sports car made from cannabis hemp," May 6, 2016, https://www.dailymail.co.uk/news/article-3576981/That-s-SMOKING-hot-car-Florida-businessman-builds-environmentally-friendly-sports-car-cannabis-hemp.html.
78. Bruce Michael Dietzen, "How Plants Like Hemp Will Save The Planet," *Bruce Dietzen Blog Post,* June 13, 2018, https://brucedietzen.wordpress.com/2018/06/13/how-plants-like-hemp-will-save-the-planet/.
79. Monalisa Mohanty, "Post-harvest management of phytoremediation technology," *J Environ Anal Toxicol,* Vol 6(5), p 398, Aug 31, 2016, https://www.omicsonline.org/open-access/postharvest-management-of-phytoremediation-technology-2161-0525-1000398.php?aid=80025.
80. Brooks R., Chambers M., Nicks L., et al., "Phytomining," *Trends in Plant Science,* Vol 3 (9), Sept 1, 1998, pp 359–362, https://www.sciencedirect.com/science/article/pii/S1360138598012837.
81. Lindsey Konkel, "There's Gold in Them Thar Plants," Strange News, *Live Science,* April 12, 2013, https://www.livescience.com/28676-plants-grow-gold.html.
82. Ibid.

83. Nicks L, Chambers MF (1994) Nickel farm. Discover, p: 19.
84. Associated Press, "Top 20 onshore US oil and gas spills since 2010," *USA Today,* Nov 17, 2017, https://www.usatoday.com/story/news/nation/2017/11/17/top-20-onshore-oil-and-gas-spills/876390001/.
85. Ron Bousso, "BP Deepwater Horizon costs balloon to $65 billion," Environment, *Reuters,* Jan 16, 2018, https://www.reuters.com/article/us-bp-deepwaterhorizon/bp-deepwater-horizon-costs-balloon-to-65-billion-idUSKBN1F50NL.
86. The Ocean Portal Team, "Introduction," Gulf Oil Spill, *Smithsonian: Ocean Find your Blue*, accessed Jan 25, 2019, https://ocean.si.edu/conservation/pollution/gulf-oil-spill.
87. "Booms," Emergency Response, *EPA*, accessed Jan 25, 2019, https://www.epa.gov/emergency-response/booms.
88. "Sorbents," Emergency Management, *EPA,* last modified Feb 20, 2016, https://archive.epa.gov/emergencies/content/learning/web/html/sorbents.html.
89. Mark Schleifstein, "An explosion in oil-munching bacteria made fast work of BP oil spill, scientists says," Louisiana Environment and Flood Control, *Nola.com,* April 8, 2013, https://www.nola.com/environment/index.ssf/2013/04/an_explosion_in_oil-munching_b.html.
90. The Ocean Portal Team, "Introduction," Gulf Oil Spill, *Smithsonian: Ocean Find your Blue*, accessed Jan 25, 2019, https://ocean.si.edu/conservation/pollution/gulf-oil-spill.
91. Mark Schleifstein, "Dispersant used in BP spill might cause damage to human lungs, fish, crab gills, new study says," Louisiana Environment and Flood Control, *Nola.com,* April 3, 2015, https://www.nola.com/environment/index.ssf/2015/04/dispersant_used_in_bp_spill_mi.htm.l
92. "Study: Dispersant, UV Radiation Increase Oil Spill Impacts on Zooplankton But Food Web Interactions may Reduce Them," *Gulf of Mexico Research Initiative,* July 8, 2014, http://gulfresearchinitiative.org/study-dispersant-uv-radiation-increase-oil-spill-impacts-zooplankton-food-web-interactions-may-reduce/.
93. Li FJ., Duggal R., Oliva O., Karki S., Surolia R., et al., "Heme Oxygenase-1 Protects Corexit 9500A-Induced Respiratory Epithelial Injury across Species," *PLOS One,* Vol 10(4), April 2, 2015, https://journals.plos.org/plosone/article?id=10.1371/journal.pone.0122275.
94. Ruiz-Ramos, D. V., Fisher, C. R., & Baums, I. B., "Stress response of the black coral Leiopathes glaberrima when exposed to sub-lethal amounts of crude oil and dispersant," *Elem Sci Anth*, Vol 5, p. 77, Dec 14, 2017, https://www.elementascience.org/articles/10.1525/elementa.261/.

95. Mark Hertsgaard, "The worst part about BP's oil-spill cover-up: it worked," Business&Technology, *Grist,* April 22, 2013, https://grist.org/business-technology/what-bp-doesnt-want-you-to-know-about-the-2010-gulf-of-mexico-spill.
96. Rico-Martinez, R., Snell, T., Shearer, T., et al., "Synergistic toxicity of Macondo crude oil and dispersant Corexit 9500A® to the Brachionus plicatilis species complex (Rotifera)," *Environmental Pollution,* Vol 173, Feb 2013, pp 5–10, https://www.sciencedirect.com/science/article/pii/S0269749112004344.
97. National Institutes of Health, "Gulf Study," *US Department of Health and Human Services,* accessed Jan 25, 2019, https://gulfstudy.nih.gov/en/index.html.
98. NIH, "Gulf spill oil dispersants associated with health symptoms in cleanup workers," News Releases, Sept 19, 2017, https://www.nih.gov/news-events/news-releases/gulf-spill-oil-dispersants-associated-health-symptoms-cleanup-workers.
99. Afshar-Mohajer N., Fox M., Koehler K., "The human health risk estimation of inhaled oil spill emissions with and without adding dispersant," *Science of the total Environment,* Vol 654, March 1, 2019, pp 924-932, https://www.sciencedirect.com/science/article/pii/S0048969718344656.
100. Rowena Mason, "BP sued over dispersant used on spill," Oil and Gas, *The Telegraph,* Aug 7, 2010, https://www.telegraph.co.uk/finance/newsbysector/energy/oilandgas/7931891/BP-oil-spill-law-suit-looms-over-spill-dispersant.html.
101. "Statutory guidance approved oil spill treatment products," *Gov.UK,* last modified Jan 14, 2019, https://www.gov.uk/government/publications/approved-oil-spill-treatment-products/approved-oil-spill-treatment-products.
102. EPA, "Revisions needed to national contingency plan based on Deepwater Horizon oil spill," EPA Office of Inspector General, Aug 25, 2011, https://nepis.epa.gov/Exe/tiff2png.cgi/P100BYF0.PNG?-r+75+-g+7+D%3A%5CZYFILES%5CINDEX%20DATA%5C11THRU15%5CTIFF%5C00000034%5CP100BYF0.TIF.
103. "Alphabetical list of NCP products schedule (products available for use during an oil spill)," Oil Spills Prevention and Preparedness Regulations, EPA, accessed Jan 25, 2019, https://www.epa.gov/emergency-response/alphabetical-list-ncp-product-schedule-products-available-use-during-oil-spill.
104. "Corexit Technology," NALCO Environmental Solutions LLC, accessed Jan 25, 2019, https://www.nalcoenvironmentalsolutionsllc.com/corexit/.
105. "Nalco Holding Co.: Summary, 2010," *Open Secrets,* accessed Jan 25, 2019, https://www.opensecrets.org/lobby/clientsum.php?id=D000027772&year=2010.
106. "Violation Tracker Parent Company Summary: Ecolab," Violation Tracker, *Good Jobs First,* accessed Jan 25, 2019, https://violationtracker.goodjobsfirst.org/parent/ecolab.

107. "Company Overview of Nalco Environmental Solutions, LLC," *Bloomberg,* accessed Jan 28, 2019, https://www.bloomberg.com/research/stocks/private/snapshot.asp?privcapId=261319662.
108. Paul Merrell, "IBT Officials Indicted for Fraud," *NCAP News,* Vol 2, No 4., Spring-Summer 1981, p. 12, https://www.documentcloud.org/documents/3692959-PoisonPapersC0927.html#search/p14/Nalco.
109. *Business Week,* "Toxicology labs face a test of their own," Industrial Edition, Safety health, Oct 17, 1983, p. 54, https://www.documentcloud.org/documents/3416805-Poison-Papers-B-2587.html#search/p269/Nalco.
110. Nalco chemical company letter to EPA Matthew A. Straus, Regarding dioxin-containing wastes, final rule, April 15, 1985, p. 263, https://www.documentcloud.org/documents/3701864-PP-B0135.html#search/p263/Nalco.
111. EPA, "Work/Quality Assurance Project Plan for the Bioaccumulation Study," Prepared by the Monitoring and Data Support Division Office of Water Regulations and Standards, July 1986, https://www.documentcloud.org/documents/3719997-PP-B0330.html.
112. EPA, "Work/Quality Assurance Project Plan for the Bioaccumulation Study," Prepared by the Monitoring and Data Support Division Office of Water Regulations and Standards, July 1986, p.213, https://assets.documentcloud.org/documents/3719997/PP-B0330.pdf.
113. EPA FOIA requests from 1988, "Region Nalco Chemical Company–Bedford Park, IL," p 55, https://www.documentcloud.org/documents/3702042-PP-B0170.html#search/p55/Nalco.
114. Superfund site, "Nalco Chemical Company," *Homefacts,* accessed Jan 28, 2019, https://www.homefacts.com/environmentalhazards/Illinois/Cook-County/Bedford-Park/Superfund-Nalco-Chemical-Company-Ild005092572.html.
115. Federal Register, "Rules and Regulations", EPA, Vol 60, No. 54, March 21, 1995, p 14899, https://www.gpo.gov/fdsys/pkg/FR-1995-03-21/pdf/FR-1995-03-21.pdf.
116. Federal Register, "Notice of Lodging of Ninth Consent Decree in United States v. Nalco Chemical Company, et al., Under the Comprehensive Environment Response, Compensation and Liability Act," DOJ, July 19, 2000, https://www.federalregister.gov/documents/2000/07/19/00-18214/notice-of-lodging-of-ninth-consent-decree-in-united-states-v-nalco-chemical-company-et-al-under-the.
117. Superfund Site, "Byron Salvage Yard," *Home Facts,* accessed Jan 28, 2019, https://www.homefacts.com/environmentalhazards/Illinois/Ogle-County/Byron/Superfund-Byron-Salvage-Yard-Ild010236230.html.
118. EPA, "Superfund Site: Byron Salvage Yard," accessed Jan 28, 2019, https://cumulis.epa.gov/supercpad/cursites/csitinfo.cfm?id=0500261.

119. Superfund Site, "Nalco Chemical Lab," *Home Facts,* accessed Jan 28, 2019, https://www.homefacts.com/environmentalhazards/Mississippi/Lincoln-County/Brookhaven/Superfund-Nalco-Chemical-Lab-Msd985966647.html.
120. Superfund Site, "Nalco Chemical Co," *Home Facts,* accessed Jan 28, 2019, https://www.homefacts.com/environmentalhazards/Texas/Ector-County/Odessa/Superfund-Nalco-Chemical-Co-Txd095217766.html.
121. Patrick R. Durkin, "Some Biological Effects of PCBs," May 20, 1976, p.152, https://assets.documentcloud.org/documents/3701948/PP-D0074.pdf.
122. Patrick R. Durkin, "Some Biological Effects of PCBs," May 20, 1976, p.195, https://assets.documentcloud.org/documents/3701948/PP-D0074.pdf
123. Beirao J., Litt MA., Purchase CF., "Chemically-dispersed crude oil and dispersant affects sperm fertilizing ability, but not sperm swimming behaviour in capelin (Mallotus villosus)" *Environ Pollut.,* June 5, 2018, p 521-528, https://www.ncbi.nlm.nih.gov/pubmed/29883953.
124. Hagstrom B., Lonning S., "The effects of Esso Corexit 9527 on the fertilizing capacity of spermatozoa," *Marine Pollution Bulletin,* Vol 8, Issue 6, June 1977, pp 136-8, https://www.sciencedirect.com/science/article/pii/0025326X77901539.
125. ASRM, "Sperm Count and Motility in Decline in North America and Europe; Phenomenon Affects Infertility Patients and Sperm Donors Alike – Except for New York's Donors, They're All Right," Press Release, Oct 9, 2018, https://www.asrm.org/news-and-publications/news-and-research/press-releases-and-bulletins/sperm-count-and-motility-in-decline-in-north-america-and-europe-phenomenon-affects-infertility-patients-and-sperm-donors-alike—except-for-new-yorks-donors/.
126. Will Sullivan, "Clean-up chemical at the BP oil spill tied to health problems," Environment, *Science Line*, Nov 8, 2017, https://scienceline.org/2017/11/clean-chemical-bp-oil-spill-tied-health-problems/.
127. US Department of the Interior, "Secretary Zinke Announces Plan For Unleashing America's Offshore Oil and Gas Potential," Jan 4, 2018, https://www.doi.gov/pressreleases/secretary-zinke-announces-plan-unleashing-americas-offshore-oil-and-gas-potential.
128. Ibid.
129. Ted Mann, "Regulators Propose Rollbacks to Offshore Drilling Safety Measures," Politics, *The Wall Street Journal,* Dec 25, 2017,https://www.wsj.com/articles/regulators-propose-rollbacks-to-offshore-drilling-safety-measures-1514206800.
130. Thomas P. Abbott, "New Uses for Kenaf," *AgResearch Magazine,* Aug 2000, https://agresearchmag.ars.usda.gov/2000/aug/kenaf/.
131. Naval Facilities Engineering Service Center, "Evaluation of Bio-Based Industrial Products for Navy and DOD Use, Phase I, Kenaf Absorbent,"

March 1999, https://static1.squarespace.com/static/5ae9dcbe50a54fa7a4012f62/t/5aec7ac688251b5c062f175a/1525447366974/kengroresearch_navykenaf.pdf.
132. Ibid.
133. Ibid.
134. Ibid.
135. Antti Pasila, "A biological oil adsorption filter," *Marine Pollution Bulletin,* Vol 49, Issues 11-12, Dec 2004, pp 1006-1012, https://www.sciencedirect.com/science/article/pii/S0025326X04002681.
136. "Hemp Inc company profile," *Reuters,* accessed Jan 28, 2019, https://www.reuters.com/finance/stocks/companyProfile/HEMP.PQ.
137. Hemp, Inc., "7,000 Pound Sale of Kenaf Kicks Hemp, Inc. Product Sales into Action," May 16, 2016, http://www.marketwired.com/press-release/7000-pound-sale-of-kenaf-kicks-hemp-inc-product-sales-into-action-2124992.htm.
138. Hemp, Inc., "Hemp, Inc. Announces Completion of First Purchase Order for Natural Absorbent Product, Spill-Be-Gone," Jan 25, 2018, https://globenewswire.com/news-release/2018/01/25/1304774/0/en/Hemp-Inc-Announces-Completion-of-First-Purchase-Order-for-Natural-Absorbent-Product-Spill-Be-Gone.html.
139. Steve Bloom, "The Return of the Hemp Decorticator," News, *Freedom Leaf Magazine,* June 2015, p. 8, http://online.fliphtml5.com/wixy/dciy/#p=9
140. Michael J. De La Merced, Chris V. Nicholson, "Ecolab agrees to buy Nalco for $5.4 billion," *New York TImes,* July 20, 2011, https://dealbook.nytimes.com/2011/07/20/ecolab-to-buy-nalco-for-5-4-billion/.
141. Jake Altinger, "Hemp Inc.'s new product makes oil drilling a bit more eco-friendly," *Colorado Springs Independent,* Marijuana news, Aug 22, 2018, https://www.csindy.com/coloradosprings/hemp-incs-new-product-makes-oil-drilling-a-bit-more-eco-friendly/Content?oid=14668202.
142. Fatima Hussein, "Long after BP spill, worker lawsuits still coming," News, *Bloomberg Law,* Nov 14, 2018, https://news.bloomberglaw.com/daily-labor-report/long-after-bp-spill-worker-lawsuits-still-coming.
143. *Plaintiffs v. BP Exploration & Production Inc., et al.*, Nov 1, 2018, no. 12-cv-968, http://src.bna.com/Dcv.
144. Wendy Laursen, "Winners and Losers in Deepwater Horizon Payout," *The Maritime Executive,* April 5, 2016, https://www.maritime-executive.com/article/winners-and-losers-in-deepwater-horizon-payout.
145. Tutuş, Ahmet & Karataş, Barış & Çiçekler, Mustafa, "Pulp and Paper Production from Hemp by Modified Kraft Method," *1st International Mediterranean Science and Engineering Congress*, Oct 2016, p 1036-1042, https://www.researchgate.net/publication/310124359_Pulp_and_Paper_Production_from_Hemp_by_Modified_Kraft_Method.

146. Dewey L., Merrill J., "Hemp Hurds as Paper-Making Material," *US Department of Agriculture Bulletin No. 404*, Oct 14, 1916, http://users.isr.ist.utl.pt/~wurmd/Livros/pg26-table2-1916-USDA-Bulletin-404-ayz-1-Acre-of-HEMP-is-EQUAL-to-4-Acres-of-TREES-Over-20-Year-Rotation-Period.pdf.
147. "Markets for Hemp," *Vote Hemp,* accessed Jan 28, 2019, http://www.votehemp.com/markets_stalk.html.
148. Gavrilescu, Dan & Puitel, Adrian & Dutuc, Gheorghe & Craciun, Grigore, "Environmental impact of pulp and paper mills," *Environmental Engineering and Management Journal*, Vol. 11 (1), Jan 2012, pp. 81-86, https://www.researchgate.net/publication/281761323_Environmental_impact_of_pulp_and_paper_mills.
149. "Frequently asked questions on Kraft pulp mills," *CSIRO, The Joint Forces of CSIRO & Forest Research,* March 4, 2005, https://web.archive.org/web/20071202204438/http://www.gunnspulpmill.com.au/factsheets/BleachingByCSIRO.pdf.
150. Emily Walkenhorst, "Paper mill fined over foul air," *Northwest Arkansas Deomcrat Gazette,* Dec 16, 2018, https://www.nwaonline.com/news/2018/dec/16/paper-mill-fined-over-foul-air-20181216/.
151. Annals of Warsaw University of Life Sciences; SGGW Forestry and Wood Technology No 91, 2015: 134-137
152. "Tree Free Hemp," *CoHempCo,* Jan 28, 2019, https://treefreehemp.com/.
153. Steven Price, "You Don't Need Trees to Make Paper," *New York Times,* Sept 13, 1981, https://www.nytimes.com/1981/09/13/business/you-don-t-need-trees-to-make-paper.html.
154. L.H. Princen, "Kenaf-Promising New Fiber Crop," *The Herbarist*, No 48, pp. 79-83, 1982, https://naldc.nal.usda.gov/download/26222/PDF.
155. "Pulp and Paper: Bleaching and the Environment," Industry Commission, Rept No. 1, May 21, 1990, https://www.pc.gov.au/__data/assets/pdf_file/0004/156712/01pulp.pdf.
156. Thomas A. Rymsza, "Vision Paper," *Journal of Industrial Ecology,* Vol 7, Nov 3–4, 2004, pp 215–18, http://www.visionpaper.com/PDF_speeches_papers/jie.pdf.
157. "Frequent Questions," EPA, Feb 21, 2016, https://archive.epa.gov/wastes/conserve/materials/paper/web/html/faqs.html#sources.
158. Livia Albeck-Ripka, "Your recycling gets recycled, right? Maybe, or maybe not," *New York Times,* May 29, 2018, https://www.nytimes.com/2018/05/29/climate/recycling-landfills-plastic-papers.html.
159. Colin Staub, "China to enact tariffs on OCC and other recycled paper," *Resource Recycling,* Aug 8, 2018, https://resource-recycling.com/recycling/2018/08/08/china-to-enact-tariffs-on-occ-and-other-recycled-paper/.
160. Ibid.

11

BUILDING WITH HEMP

In April 2018, my family and I moved from California to Missouri. At the time, my husband and I made the decision so that our daughter could live closer to family and cousins, but we soon found out just how fortunate we were to leave when we did.

We left California a couple of months prior to the most destructive wildfire season in the state's history. Paradise, a nearby town of nearly 27,000 people, was completely devastated. Over 18,000 homes and businesses were destroyed. Had we waited to move until later in the year, I have no idea how my husband would've been able to drive the U-Haul truck to Missouri. Major highways were affected; people burned to death in their cars in bumper-to-bumper traffic because they couldn't escape how fast the fire was spreading. The Carr Fire, Delta Fire, and Hirz Fire were particularly devastating to where we were living.

From December 2016 to the spring of 2018, we lived in Shasta County (or NorCal as it's often called) in the Redding/Lakehead area to grow medical marijuana at Hobbs Greenery on a plot of forty acres. We were surrounded by the national forest—mountains, redwood trees, lakes, bears, deer, rattlesnakes. The nearest schools and hospitals (and essentially civilization) were forty-five minutes away. When the fires reached our neck of the woods, the heat and winds were so intense that footage from local news crews contained *firenadoes*—the fire was able to go up into a cyclone due to how fast and hot it was burning.[1]

If we had stayed in California, there was a strong likelihood that we would've been trapped by the Hirz Fire. The house we were renting was the last house at the end of a dead-end dirt road cut out of a mountainside by the milling and logging industry. To get to our house, we had to drive about three miles down this road, crossing over two wooden bridges. On the driver's side, there was a cliff—no guardrail—and a rapid, white-water creek miles below. On the passenger's side, there was a steep embankment, covered in mossy trees. The road was the width of a one-way street, with various gravel and dirt driveways along the way. All our neighbors had about forty acres of land, with multiple forty-acre lots between them for sale. There was no cell service from our house all the way to our exit ramp on the highway (what was easily a ten- to fifteen-minute drive).

Our house had its own source of gravity-fed spring water (without any chemicals, softeners, or additives) that came from one of the many waterfalls on the property. The house was built in 1990, and came equipped with a wood-burning stove for heat in the winter. (The thermostat used our propane tank, and refilling that cost much more than a hunk of logs.)

In 2018, the laws that legalized recreational marijuana in California changed the costs associated with medical marijuana farms like ours. The expensive application process and permits made it financially infeasible for us to continue to grow marijuana legally, and gave us the final push to move to the suburbs of St. Louis.

While the fire didn't burn down the house we were living in (and all our neighbors were spared), the path of the Hirz Fire closed our exit ramp on the I-5; within days it reached the base of our dead-end dirt road, and our neighbors were told to evacuate.[2] Keep in mind, our propane tank—and the tanks of our neighbors—could've easily fueled the fire. Our gravity-fed drinking water was compromised by all the ash and debris. The air-quality index (AQI) was listed at 275, "very unhealthy," and a little too close to the AQI of 300, which is deemed hazardous. My daughter's school in Redding was closed due to the air-quality concerns, and the evacuated Paradise residents were told not to move back until the toxic ash and hazardous waste that the fire left behind were cleared away.

According to the California Department of Forestry and Fire Protection (CAL FIRE) and the National Interagency Fire Center (NIFC), as of December 2018, there were a total of 7,994 wildfires in the state that affected 1,824,542.6 acres.[3] While many scientists pointed to global warming as the main culprit because increased temperatures equate to increased dryness and increased wildfire risk,[4] the real cause of concern in the case of the two largest fires in California that year was criminally lax fire-prevention maintenance by utility companies.

Poorly maintained electrical equipment either caused or escalated the Woolsey Fire and the Camp Fire. (Coincidentally, both the Woolsey Fire and Camp Fire began on November 8, 2018.) These fires were so close to residential areas that they wiped out entire neighborhoods. The entire town of Paradise was displaced due to the Camp Fire; residents were given mere hours to evacuate due to how quickly the blaze spread. Eighty five people died in the Camp Fire, making it the most deadly wildfire to-date in California's history.[5]

As of November 2018, authorities investigating the Camp Fire determined that the fire was fueled as a result of Pacific Gas & Electric Co.'s failure to maintain electrical lines; the utility company didn't clear away brush from the lines or update faulty equipment. Fire officials traced the Camp Fire back to "a single steel hook that held up a high voltage line on a nearly hundred-year-old PG&E transmission tower."[6] While PG&E hasn't taken any direct responsibility, the company did disclose that "a transmission line located near the origin of the fire experienced a problem a few minutes before the fire began."[7] As a result, PG&E, the largest utility company in California, is now facing several lawsuits from Camp Fire victims.

This isn't the first time PG&E was to blame for the cause of a wildfire. According to CNN, state fire officials have found that PG&E equipment caused seventeen fires in 2017, and, in eleven of those cases, the company violated codes regarding brush clearance.[8]

Jamie Court, president of the advocacy group Consumer Watchdog, told CNN that these fires were entirely avoidable, but because PG&E won't maintain their power lines, the likelihood of another fire caused by "a failure at gardening" is eminent.

"What we have to do is put a fire chief in charge and make PG&E garden its power lines and make PG&E modernize its equipment," he said.

The *San Francisco Chronicle* reported in December 2018 that federal judge William Alsup asked California's attorney general, Xavier Becerra, to clarify the extent to which any "reckless operation or maintenance of PG&E power lines would constitute a crime under California law."[9] If the utility is found to be guilty of criminal negligence in the Camp Fire, then that also means they violated the terms of their probation from the 2017 San Bruno natural gas pipeline explosion.

In the San Bruno incident, PG&E was found guilty on six felony charges for violating safety regulations; it was also found guilty for obstructing the investigation. The explosion killed eight people and destroyed thirty-eight homes.[10] PG&E was fined $1.6 billion by state regulators and had to pay out civil settlements to the tune of hundreds of millions of dollars (which it is still doing).

The *Sacramento Bee* reported that the Camp Fire's financial damages "could exceed $7.5 billion."[11] Unfortunately, California residents might find their electric and gas rates increasing to cover the utility's payouts, and rates will certainly increase regardless to cover the costs of repairing and maintaining lines. While many of the fire investigations and lawsuits against PG&E are ongoing, there is a similarity between PG&E and the Southern California Edison, which is the utility being sued for the Woolsey Fire. Lawsuits claim Edison's poorly maintained electrical equipment, found near the start of the fire, is to blame.

According to the California Department of Forestry and Fire Protection, the Woolsey Fire burned 96,949 acres and destroyed 1,643 structures.[12] Corelogic Inc, a property data and analytics provider, determined that damages are currently estimated at up to $500 million in commercial losses and some $3.5 billion to $5.5 billion in residential losses.[13] The lawsuits being filed on behalf of the victims of the Woolsey wildfire state that "the wildfire could have been avoided if Edison had properly maintained its overhead power lines and electrical equipment and cleared away vegetation from its wires."[14]

While I haven't lived in an area serviced by Edison, as far as my personal experiences with PG&E are concerned, the company neglects to do pretty much everything. PG&E didn't want to service our area, so instead of checking the readings on the meter regularly, they "estimated" that monthly rates from previous years were the same. In 2017, they charged us the same monthly rates (plus increases in taxes, etc.) as 2016—despite the fact that we (1) had opened a new account with them, (2) weren't living there during those months in 2016, and (3) weren't using the same energy consumption as the people who were renting the house in 2016. There were many months where our PG&E bill was $2,000 and even higher than that (the home was less than 1,900 square feet). We came to find out that the previous renters (who wound up moving to another house down the road from us) were using *lots* of indoor grow lights (something we weren't doing). The only way to get PG&E to recalculate the bill was to constantly call until someone came out and checked the meter. I'm pretty sure they still owe me a pretty significant refund.

PG&E prefers to use smart meters (which we didn't have in our California house). Smart meters are attached to the house and then report back to PG&E on exactly what amount of electricity is being used. The only problem there is that smart meters have been known to catch fire, something unheard of for analog meters. In 2014, Oregon's Portland General Electric had to replace seventy thousand residential smart meters after three of them caused small house fires due to electrical component failures.[15] Since 2014, the technology in smart meters has improved, but the fear of even the possibility of them catching fire is certainly warranted in fire-prone areas like Northern California.

In November 2017, a Fresno resident filed a lawsuit against PG&E for a fire in his home. He stated that PG&E employees arrived at his house "almost as quickly" as the firefighters so that they could remove the smart meter "while the firefighters worked" and take it with them.[16] The local Fresno ABC affiliate reported that there were at least five lawsuits in California making similar fire-hazard claims about smart meters.

"When I asked for PG&E's response to the lawsuits, the company's representatives asked me [to email them] a list of questions," stated the Action News reporter covering the story. "They answered none of them, but they did send me a statement saying smart meters have to meet safety requirements and standards spelled out in the national electric safety code."[17]

In December 2018, Robert Kennedy Jr. and his law firm hosted a workshop in Chico for Camp Fire survivors.

"PG&E needs to step up and face responsibility for the choices it made," stated RFK Jr. at the meeting.[18] "[It needs to] begin being invested in this community in making sure it gets rebuilt and rebuilt in a way that this will never happen again."

THE COST OF LIVING (AND BUILDING) IN CALIFORNIA

After the Camp Fire, the unemployment rate in Butte County (the Chico metropolitan area, which includes Paradise) increased from 4.2 percent in October 2018 to 4.3 percent in November 2018.[19] According to California's Employment Development Department, this increase was higher than the overall unemployment rate in the state (3.9 percent) as well as the unemployment rate in the United States as a whole (3.5 percent).

Meanwhile, in Chico, a house of approximately 1,500 square feet typically sells for $330,000 (or $220 per square foot),[20] and the average annual salary is $42,092.[21]

One of the biggest challenges in rebuilding the area is overcoming the reason for residents to continue to stay and rebuild if their source of income was also wiped out by the fire . . . or if the average person (working or not) can afford a mortgage (even if it's discounted or a zero-interest disaster aid loan). Most residents affected by the Camp Fire say they're committed to rebuilding, but the process is going to have to involve a regrowth of local businesses and work opportunities, given the median cost of a home in California is 2.5 times higher than the rest of the country (and the average salary in NorCal isn't).

In 2015, a run-of-the-mill California home (and fixer-upper) cost $437,000; the cost of constructing a new home in California has been increasing and the state has a lack of public dollars for subsidized housing.[22] While FEMA stepped in to provide trailers to house Camp Fire survivors in November 2018, the executive director of the county's housing agency estimated there were at least seven thousand families in Paradise alone who weren't expected to find housing anywhere in the county "anytime soon."[23] Hundreds of evacuees were camping out in the Walmart parking lot in Chico, without any place else to go.[24]

Whether PG&E ever pays damages to all those affected, FEMA has indicated that the rebuilding process will take years. One product that could help increase local businesses and contractors during the reconstruction process is hempcrete, a cost-efficient construction product that is biodegradable, recyclable, and fire resistant.

In September 2018, the *Cleveland Metro* reported that a hempcrete plant was being constructed to revitalize Glenville, Ohio. According to 2010 census data, nearly 20 percent of the population of Glenville is unemployed, and 31 percent live in poverty. The hempcrete plant is expected to open in June 2019, with entry-level positions starting at $17 an hour ($35,000 annually), plus health insurance, child-care, and transportation benefits.[25] If a similar operation opened in Butte County, it would help multiple industries affected by the Camp Fire, plus provide new work opportunities from agriculture to manufacturing to construction.

HEMPCRETE

Hemp has been used as a building material for hundreds of thousands of years. There are books and online resources dedicated to building homes out of hemp-crete (and other renewable resources), but here are the basic principles behind the product:

- Hempcrete, also known as hemplime and hemp concrete, is a bio-composite made from the inner woody core of the hemp plant (hemp hurds).

- The hemp fibers are mixed with a lime-based binder (hydrated lime/hydraulic lime) and water to form a lightweight, insulating material weighing only about an eighth of the weight of concrete.

- According to an April 2018 article on CNBC.com, there are currently only about fifty homes containing hemp in the United States (in such states as North Carolina, Virginia, Texas, and Hawaii), but hempcrete can be found in hundreds of Canadian and European homes and commercial buildings.[26]

- Hemp concrete is a green building material that is carbon negative. According to a 2016 study, a building made of hempcrete compensates for all of the other processes related to constructing it because hemp offsets the other building materials' carbon emissions.[27]

- Other studies from 2006, 2014, and 2017 state that over time, the lime content actually strengthens the hempcrete. The gain in strength is significant enough to increase the overall durability of hempcrete structures.[28]

- With further regard to its durability, archaeologists have discovered that the Great Wall of China, the Pantheon in Rome, and the plaster on top of the Egyptian pyramids all contain hempcrete.[29]

- Hempcrete offers "high thermal and sound insulation."[30] In European countries, it is used as an infill material for insulation and as actual walls.[31]

- Since hemp hurds are extremely porous and lime has a high level of permeability, hempcrete is breathable. It absorbs and emits moisture to regulate internal humidity, and to avoid trapping moisture (moisture leads to mold growth). It can be used for walls, floors, and roof insulation. According to a 2010 study conducted in at the University of Bath in England on a small, one-story house that used hemp for walls and insulation, the temperature inside the house "stays fairly constant" because the walls "act as a sort of passive air-conditioning system."[32]

- Since hempcrete is breathable, it may not have the same strength as compared to other building materials of the same size and weight. However, it does bond well with other materials, so structural integrity overall can be improved.[33]

- Unlike fiberglass insulation or drywall, using hemp-filled walls is nontoxic, mold-resistant, and pest-resistant.[34] Since hempcrete doesn't contain any materials of interest to insects, it is also free from termite infestation and other pest threats.[35]

A 2017 article from Poland's Lublin University of Technology noted that both flax and hemp can be used to construct low-energy buildings, which are buildings consisting of fully recyclable products.[36] The US Green Building Council estimates that buildings account for about 39 percent of CO_2 emissions in the United States, and Green Buildings, such as LEED-certified buildings, use 32 percent less electricity and save 350 metric tons of CO_2 emissions annually.[37]

While concrete itself is fire resistant, and it can be used in green buildings, the material (due to expansion and contraction) forms large cracks that often need to be repaired. Hempcrete typically has much smaller cracks, and has "self-healing" or autogenous healing properties. When water penetrates the cracks, lime comes to the surface to fill them in. Builders who use hempcrete for interior walls claim the wall system is "significantly more fire resistant" than typical insulated drywall. According to Hempsteads, a hempcrete builder based in the United States:[38]

> *The shiv [hemp] within the wall is coated with lime and is finished with about ¾" of lime stucco on the outside, and ½" on the inside. This has led to increased interest in implementing this material in areas prone to wildfires. The type of detailing that ensures high thermal performance, i.e. the attention to joints at doors, windows, and places where the insulation abuts structural elements, are the same details that make the building even more resistant to fire.*

The concept of using hempcrete to build homes has also taken off in Canada. In Victoria, British Columbia, the "Harmless Home" was built using hemp, lime, and water. The home was referred to as "Lego-inspired" because the walls were built with hempcrete blocks that looked extremely similar to Legos. The hempcrete blocks are manufactured by Just Bio Fiber in Calgary, and the builder stated his green product is carbon negative; the blocks' insulating properties make it mold proof and stronger over time.

"Carbon dioxide is constantly being absorbed into the walls and hardening the walls," said builder Mark Faber.[39] "These walls are going to be lasting hundreds of hundreds of years."

Faber also stated that since the blocks are fireproof, residents in Fort McMurray, an area in British Columbia prone to forest fires, are interested in building homes with his hempcrete product.

"We've done testing with this and held a torch up to it for a quite a long time, up to an hour, and it barely has any impact on it," Faber said of his product in 2017.

In tests, the hempcrete blocks have survived being heated to more than 1,500 degrees,[40] and according to Just Bio Fiber's website, the building material has received a fire-resistant rating of over two hours.[41]

What's also key about hempcrete is the ability to source it locally. If Butte County residents manufacture the hemp near the towns affected by the Camp Fire, then the cost of the building materials (and the cost of building a home in California) could decrease significantly.

In 2012, Bob Clayton, a retired mechanical engineer, built the first hempcrete house in the state of Florida.[42] His 1,640-square-foot home (which included granite finishes, maple cabinets, hickory floors, and other custom upgrades) cost approximately $350,000 to build. However, the cost could've easily been reduced if he didn't have to find specialty contractors knowledgeable about how to build with hemp, and if materials didn't have to be imported. Since hemp wasn't grown in Florida or any nearby state at the time, this impacted the cost to obtain the necessary materials.

Regardless, Clayton stated in *Tampa Bay Times* that his hemp home's electric bill peaked at $90 during the summer.[43] Since Florida and California have intense summer heat, that's a pretty incredible endorsement for how a hemp home's insulation and airflow can significantly decrease electric bills. *I'll take a $90 electric bill instead of a $2,000 electric bill any day.*

HOME ENERGY RATING SYSTEM (HERS)

The nationally recognized industry standard for rating a home's energy efficiency is called HERS (Home Emergy Rating System). The lower the HERS score, the more energy efficient the home is, and the less the homeowner pays in energy bills. A standard new home is about 30 percent more energy efficient than a resale home, based on the HERS score. For example, the average HERS index score in 2016 was 62, which is 38 percent more energy efficient than a home built in 2006.[44] Since hempcrete gets stronger with age, and since hempcrete insulation stays energy efficient over the course of its life, a hemp home built in 2006 would have a better HERS energy rating than a traditional home built the same year.

According to *Hemp Business Journal*, Tim Callahan—who builds ultra-energy-efficient homes with local, natural materials—has claimed that a hempcrete home he built in North Carolina received a HERS score of 17. If that's a standard score for most hemp homes, then their energy efficiency is leaps and bounds over any new build.[45]

The first hemp house ever built in the United States was in Asheville, North Carolina in 2010 by the town's former mayor Ross Martin.[46] According to American Lime Technology, a Chicago-based company offering sustainable, low-carbon building materials (including hempcrete and hemp insulation), the Asheville "Push House"—

- Cost $133 per square foot to build—a phenomenal achievement, considering the house is 3,400 square feet
- Took 2.5 weeks to build
- Reduced the cost of homeowners' insurance by 60 percent

American Lime Technology claims the house will last an average of seven hundred years. The amount of hemp needed to build a house like this? On a mere 1.5 acre plot of land, it takes twelve weeks to grow enough industrial hemp to build a 1,500-square-foot home. Therefore, a builder will only need about 2.3 acres to grow to enough hemp in twelve weeks for a 3,400- square-foot home.[47]

So while homes built with hempcrete may cost slightly more than traditional homes (Tim Callahan estimates hempcrete can add 8 to 12 percent onto the overall price of a home[48]), the savings is in the increased air quality (less VOCs), thermal performance (decreases energy bills), and the low maintenance (less upkeep) of the hemp building materials. Today, there are companies in the United States that build exclusively with hemp and use it for multiple purposes including:

- **Hemp Insulation Batts**

Made from woven hemp fibers, this type of insulation is used in external and internal walls, suspended floors, lofts, and roofs. Since it is biodegradable and recyclable, its much safer in comparison to fiberglass.

According to Adam Block, the vice president of Sunstrand Sustainable Materials (a hemp hurd and raw flax natural fiber insulation company based in Louisville, Kentucky), hemp insulation can also be manufactured as fire-resistant if it's sprayed with a mineral composite.[49]

- **Hemp Medium Density Fiberboard (MDF) (aka Plywood, Particleboard, Hempboard)**

Testing has shown that hemp MDF is stronger than wood MDF and hempboard can be manufactured with the same equipment. Since wood MDF is used extensively in the building industry, using hemp instead of wood can significantly help limit deforestation. Plus, hemp MDF is manufactured without urea-formaldehyde (the adhesive used to bind wood fibers together to make composite boards). This formaldehyde adhesive found in wood composites releases volatile organic compounds (or VOCs) throughout its life span. This means there are fewer toxic substances being used in the overall building process with hemp MDF, since wood MDF is used in almost every living space (kitchen cabinets, baseboard, wall trim, doors, etc.). Therefore, switching to hemp MDF can greatly improve indoor air quality.

Plus, hemp boards can be manufactured be resistant to fire and water damage if the hemp hurds are injected with a phenolic resin,[50] or with a mixture of lignin (a plant binder) and resin.[51] According to Hemp Technologies Global (a US manufacturer of hemp building products): "Hemp Board is comparable in performance to other wood-based board products and is formaldehyde-free! Its unique visual texture can be stained and sealed and supports LEED certification."[52]

TOXIC FORMALDEHYDE AND THE EPA

According to the CDC, newly constructed homes contain more formaldehyde than older homes due to the increased amount of plywood, particleboard, and laminate flooring. Since newer homes are better insulated, that means less air is moving out of the home, and this causes formaldehyde to stay inside the home even longer (unless the windows are open on a consistent basis).[53]

Exposure to formaldehyde inside the home can result in irritations to the eyes, nose, throat, and skin. It also increases breathing problems for people with asthma and COPD (chronic obstructive pulmonary disorder). Children and older adults are more likely to develop these symptoms, and unfortunately, the CDC claims that air filters won't help to lower levels of formaldehyde. The concept of overheating the home to "bake" out the formaldehyde doesn't work, either, as this can raise the overall levels of formaldehyde that products are emitting.

According to the U.S. Department of Health and Human Services, everyday household products including varieties of Elmer's Glue, tile grout, caulks (such as Liquid Nails), paints, cigarettes, and even soaps, cosmetics, shampoos, lotions, sunscreens, fingernail polish, body wash, and cleaning products contain levels of formaldehyde or ingredients that release formaldehyde.[54] Stranger still, formaldehyde can be added as a preservative to food, or it can be produced as a result of cooking.

According to the US Consumer Product Safety Commission, formaldehyde is normally present in both indoor and outdoor air at low levels (less than .03 parts per million), and at this low level, it won't cause severe harm to healthy people.[55] However, in 2003, the National Cancer Institute found a link between formaldehyde and leukemia after conducting a study on industrial plant employees who are exposed to formaldehyde at work.[56] The study followed twenty-five thousand workers at ten industrial facilities over forty years to find that the risk of leukemia increased for those who were exposed to high levels of the chemical.

Then in 2010, another study conducted by the University of California Berkeley supported the National Cancer Institute's findings. Ninety-four workers in China were examined—forty-three worked in facilities with high concentrations of formaldehyde; the other fifty-one workers were never exposed to it. The exposed group had fewer red and white blood cells and platelets, plus a higher prevalence of DNA mutations, which led researchers to conclude that "leukemia induction by formaldehyde is biologically plausible."[57] Following this research and other reports, the International Agency for Research on Cancer, a part of the World Health Organization, formally classified formaldehyde as a cause of leukemia in 2010, stating: "There is *sufficient evidence* in humans for the carcinogenicity of formaldehyde. Formaldehyde causes cancer of the nasopharynx and leukemia."[58]

In the United States, formaldehyde emissions are set by the EPA. Emissions from composite wood products like MDF are regulated by the 2010 Formaldehyde Standards for Composite Wood Products Act, which is Title VI of the Toxic Substances Control Act (TSCA). This act addresses industry standard requirements, but it took the EPA eight years and a court order to roll out the final rules.

In February 2018, the Sierra Club brought a lawsuit against Scott Pruitt to force the EPA to act on creating formaldehyde standards for composite wood, as the agency was tasked to do under the 2010 Formaldehyde Standards for Composite Wood Products Act.[59] The judge decided that the EPA had stalled long enough on enforcing the restrictions,[60] and the agency finally released standards on June 1, 2018, for composite wood panel manufacturers, fabricators of finished goods, importers, distributors, and retailers of composite wood panels and other finished goods containing wood composites.

The final rules were edited in November 2018 and included industry-wide incentives for products if they're manufactured with ultra-low-emitting formaldehyde resins. All companies have until March 22, 2019, to transition to the new standards, which include:[61]

- All finished goods containing composite wood products (wooden toys, flooring, picture frames, plywood, MDF, particleboard, kitchen cabinets, furniture, etc.) must be labeled with a stamp, tag, or sticker stating it complies with TSCA Title VI or the California Air Resources Board.

- The information on the label must include: Name of fabricator or substitute with downstream fabricator, importer, distributor, or retailer; date the finished goods were manufactured (month/year); and statement that the goods are TSCA Title VI compliant.

- A third party must provide certification for laboratory testing and oversight of formaldehyde emissions from manufactured and/or imported composite wood products to ensure they're within EPA standards.

- The emissions standard varies depending on the product, but all rules apply to all composite wood products, imported or domestic.

With Andrew Wheeler now at the helm, the EPA has yet to deliver on a report detailing how formaldehyde can cause leukemia. The health report, called the Integrated Risk Information System (IRIS), was compiled during the Obama administration. Former EPA head Scott Pruitt told Congress in early January 2018 that the report had been completed in fall 2017, but was still being reviewed, and couldn't be released to the public yet.

However, Congress took note of something significantly troubling in the timing of the report's completion and Scott Pruitt's politicization of the EPA. In November 2017, Pruitt removed six academic scientists from the EPA's Science Advisory Board and replaced them with industry advocates.[62] Kimberly Wise White, chair of the American Chemistry Council's Formaldehyde Panel, was one of them. The panel is an industry group representing a number of chemical firms and formaldehyde enthusiasts, including Exxon Mobil, Georgia-Pacific Chemicals, and Hexion, a Koch Industries subsidiary. A press release from the American Chemistry Council states: "Formaldehyde is a basic building block chemical, essential to an estimated $145 billion of the US gross national product."[63]

Product	Emission Standard
Hardwood Plywood – Veneer Core	0.05 ppm of formaldehyde
Hardwood Plywood – Composite Core	0.05 ppm of formaldehyde
Medium-Density Fiberboard	0.11 ppm of formaldehyde
Thin Medium-Density Fiberboard	0.13 ppm of formaldehyde
Particleboard	0.09 ppm of formaldehyde

EPA Formaldehyde Standards under the Composite Wood Products Act. *Source: https://www.epa.gov/formaldehyde/frequent-questions-consumers-about-formaldehyde-standards-composite-wood-products-act*

We're also familiar with the names of the other major chemical companies that produce products with formaldehyde, such as Dow Chemical and even Monsanto. Through recent FOIA requests from Dewayne Johnson's lawsuit against Monsanto, there are several internal emails among executives from 2013 to 2015 that discuss a technical expert who rejected glyphosate's product registration due to the fact that the pesticide contained formaldehyde,[64] an ingredient that wasn't listed and one that the World Health Organization has linked to a cause of leukemia.

According to *Politico*, in 2017, the American Chemistry Council spent more than $7 million to lobby the EPA and Congress "on issues including IRIS, formaldehyde and the policy to limit EPA's use of human health research."[65] (Hexion, one of the largest manufacturers of formaldehyde, also spent tens of thousands of dollars on lobbying efforts in 2018.)

All that lobbying paid off as Kimberly White became a "scientific adviser" for the EPA right around the time when Pruitt claims the formaldehyde report was completed.

Nearly a million jobs "depend on the use of formaldehyde," stated a particular January 26, 2018 letter written by Kimberly White to top EPA officials.[66]

Also under Scott Pruitt, EPA officials took a meeting with the American Chemistry Council's Formaldehyde Panel in January 2018. Two days later, White wrote a letter to the EPA in reference to what was discussed: "As stated in our meeting, a premature release of a draft assessment . . . will cause irreparable harm to the companies represented by the Panel and to the many companies and jobs that depend on the broad use of the chemical."[67]

Given the relationships the EPA has had with other chemical companies, and how corporations like Dow and Monsanto have been able to pass off their own studies as independent scientific research for decades, the most obvious reason for EPA officials to meet formaldehyde manufacturers and supporters like Kimberly White would've been to allow them to edit the EPA's IRIS formaldehyde report.

While the EPA already lists formaldehyde as a "probable carcinogen," the IRIS report is meant to clarify the agency's official standpoint. The EPA's previous assessment in 2010 was criticized by the National Academy of Sciences for failing to draw "clear links" between formaldehyde and leukemia.[68] When the new health study is finally released, it will serve as the foundation for all of the EPA's formaldehyde-related decision making, so if the American Chemistry Council's Formaldehyde Panel has already compromised the science, formaldehyde might continue to be included in nearly every commercial product. As its use expands, it'll be considered an integral part of nearly all-American industries, just as lead was once perceived to be.

On December 10, 2018, Senator Edward J. Markey, a member of the Environment and Public Works Committee, called on the EPA to release the IRIS report, but the agency failed to do so.

"The EPA has succumbed to pressure from industry for far too long, endangering the public's health," stated Senator Markey. "I urge the EPA to ensure that there are no more efforts to delay or block the publication of this assessment."[69]

It will most likely take another court order to get the agency to deliver the goods. In the meantime, there's already a movement to incorporate hemp products and building materials—all of which do not contain formaldehyde—and there are people in positions of influence who have been raising awareness of what can be accomplished by choosing to build with hemp.

The Prince of Wales started developing his own sustainable village in Poundbury, Dorset, in 1993, and the entire town is scheduled for competition in 2022:

- Some of the energy the town uses comes from an anaerobic digester (microorganisms break down biodegradable material and turn it into biogas for heat and electricity).
- The street layout is designed to reduce car travel by increasing green spaces with more walking and cycling opportunities.

- As of April 2017, the British newspaper *The Telegraph* reported that over three thousand people live there currently, with two thousand working in the town's 180 local businesses.[70]

"As our planet becomes overwhelmingly urban, and resources become scarcer, it will no longer be enough just to add gadgets on here and put bolt-ons there," stated Britain's Prince Charles. "We need to rethink the way we plan our homes, shops, schools and their relationship to one another. Such eco-engineering can [be learned] from Nature, from traditional communities and from the best of contemporary technology."[71]

In March 2011, Prince Charles built an eco-home prototype known as the "Prince's House" that used hemp and sheep's wool to insulate the roof and floor;[72] the actual house—built in Watford, Hertfordshire and called "The Prince's Natural House"—was completed in 2012. The building can be constructed for just under £280,000, and it's made with hemp/lime-based bricks and other sustainable materials.[73]

"The Natural House uses natural materials, including clay blocks and lime-based plasters, which reduce the risk of poor air quality," stated Prince Charles when he first unveiled the project. "A breathable wall system avoids the risk of damp and mold accumulation, a key factor in the development of asthma and respiratory problems. Thus, the design is not only low-carbon, it is also health-promoting."[74]

All of the Prince's eco-projects are being built through the Prince's Foundation for Building Community, which focuses on natural ways to decrease carbon emissions and environmental impact without using modern "eco-bling" technology such as solar panels that aren't minimalistic in their environmental impact after considering the manufacturing process (such as mining for lithium). His foundation's spokesperson, Constantine Innemee, explained that the foundation prefers not to use smart technology "add-ons" because if the systems break or go unused, the "theory" that they can reduce a home's energy use "doesn't hold up."[75] Plus some of these products are replaced (thrown away/not recycled) when updated versions become available, so the cycle of waste continues.

HEMP PLASTICS

In November 2018, Prince Charles celebrated his seventieth birthday by raising awareness about the human disaster known as plastic.

"One of my duties has been to find solutions to vast challenges we face over accelerating climate change. . . . However, it seems to take forever to alert people to the scale of the challenge," said Charles in a *Vanity Fair* interview.[76] "Over 40 years ago, I remember making a speech about the problems of plastic and other waste but at that stage nobody was really interested and I was considered old-fashioned, out of touch and 'anti- science' for warning of such things."

In May 2017, Prince Charles and the Ellen MacArthur Foundation launched a $2 million New Plastics Economy Innovation Prize to invite designers and scientists to reinvent the types of plastic packaging that are almost never recycled, ending up in oceans, landfills, or incinerators. This is a significant project, considering nearly 91 percent of the world's plastic is not recycled. Among the winners were companies that found ways to turn plants into plastic-like materials, since cellulose-based raw material (wood, rice straw, textile waste, hemp, etc.) can be turned into food packaging that looks and acts like plastic but doesn't contain toxic chemicals.[77]

In the United States, flexible bioplastic packaging made from hemp is getting a huge boost, now that the 2018 Farm Bill is in effect. Kevin Tubbs, founder of the Denver-based Hemp Plastic Company, takes the by-products (the waste material) from CBD hemp processing and nutritional hemp processing and turns it into bioplastic.

"We expect market dominance in bioplastic hemp," Tubbs told the *Connecticut Post* in November 2018. "Outside of cash crops like hemp hearts and CBD oil, there's a lot of waste in hemp processing. So I started experimenting. Take the waste products and you can do one of two things with them: You can either make hempcrete—a building material—or hemp plastic."[78]

The Hemp Plastic Company is focusing on all aspects of plastic manufacturing, including injection molds, thermal forming, sheet film, blow molding—basically Kevin Tubbs will manufacture anything plastic—from straws to bottles to lamps—using bioplastic hemp. His website states:[79]

> *Our hemp plastics can be used to make parts, toys, packaging, containers and most other plastic products. We can also meet FDA standards for food, pharmaceutical and agricultural products and we can create materials to be compostable, biodegradable, UV proof, fireproof and scratch resistant.*

The reason why hemp is preferred for bioplastic is because 65 to 70 percent of hemp's biomass contains the high cellulose content necessary for bioplastic manufacturing.[80] Flax is also favored because it has 65 to 75 percent cellulose, and kenaf (which has a similar biomass to hemp) has been proven to be an effective substitute for petroleum-based plastics used in electronic products[81] as well as a natural additive to plastics and other materials to make them stronger. Meanwhile, wood (which is currently used for bioplastic) only contains 40 percent cellulose. Although cotton contains the most cellulose—up to 90 percent—it also requires 50 percent more water to grow compared to hemp, and four times more water to process.

So what can be made with fantastic hemp plastic? Here's a handful of specific examples:

- Janice's Kitchen is a Canadian-based company that produces eco-straw alternatives and alternatives to plastic wrap, including a vegan and 100 percent biodegradable option using fabric made from hemp, impregnated with soya wax, and coconut oil.[82]

- Hemp Eyewear, a Kickstarter campaign, developed the first hemp-based frame for sunglasses and prescription lenses. The frames are made from hemp organic plant fiber, and are stronger and lighter than carbon fiber.[83]

- Automobile manufacturers are using hemp composites and hemp plastics for interior door paneling, dashboards, body molding, and hemp-based textiles for interior upholstery. According to *Hemp Business Journal*, BMW, Mercedes, Jaguar, and Volkswagen (plus the Renew Sports Car) are using hemp to decrease the weight of the overall vehicle, to strengthen the components that see the most wear and tear, and to create more fuel-efficient, environmentally friendly cars.[84]

- HempBioPlastic from Kanesis (an Italian company) turns hemp waste into biodegradable plastic that is then used for 3-D printers and injection molding. They also make a Hemp Filament (3-D printing "ink") for 3-D printing that renders higher detail than PLA—another biodegradable material (sourced from sugar canes and cornstarch) used as a 3-D printing filament.[85] There are a few other hemp 3-D printer filament companies including 3D4MAKER—their biodegradable plastic filaments can be recycled or turned into compost as well as 3-D fuel.[86]

- 3-D printers have made everything about building with hemp more exciting. Australian biotech company Mirreco is developing carbon-neutral hemp panels that can be 3-D printed onto floors, walls, and roofs.[87] The company is also focusing on developing actual homes made from hemp with 3-D printers, and creating nanotechnologies from hemp's biomass.

Apis Cor is an impressive 3-D printing company capable of printing whole buildings on site. In 2017, the company demonstrated the ability in a YouTube video: a four-hundred-square-foot-home was built in twenty-four hours (at an approximate cost of $10,000) and included a bedroom, bathroom, and hallway, with all the fixtures including lighting and plumbing.[88] In August 2018, Apis Cor won a top prize[89] at the second of three phases for NASA's 3-D Printed Habitat Competition; the contest's final phase (phase 3) is currently underway.[90]

NASA's 3-D Printed Habitat Challenge seeks to find ways for "America's best and brightest to help us figure out how to build a house on Mars, using advanced 3-D printing technology in the most efficient and sustainable way possible." Those who compete must manufacture a habitat that can be used on another planet, using only 3-D printing technologies to create it. The goal is to create a 3-D printing robot that can be launched to Mars ahead of a manned mission so that the necessary facilities are built, up, and running by the time the astronauts arrive.

NASA has been experimenting with 3-D printers in space since 2014. In November 2018, the agency launched the Refabricator—a 3-D printer/recycling machine—to the International Space Station. The Refabricator can turn waste plastic and previously 3-D printed parts into high-quality 3-D printer filament so that new tools and materials can be created for the astronauts. The Refabricator can recycle the plastic into various objects multiple times. Currently, NASA plans to test the quality of the products back on Earth.[91] If proven successful, the Refabricator could be useful for waste management on missions to the Moon and Mars, and hopefully even before manned missions, the technology can be used on Earth to reuse our mounds of plastic (and microplastic).

Since hemp was federally illegal up until December 2018, none of the 3-D projects under NASA's consideration currently use hemp to print sustainable housing or turn waste plastic into useable items. But hemp has already played a minor role in space travel. In 2012, Canadian astronauts were sent up to the International Space Station with "Holy Crap," a breakfast cereal made from hemp,[92] which gave the astronauts their needed morning nutrition in just two tablespoons.[93]

NASA started another competition for 2019 called the CO_2 Conversion Challenge, which asks scientists and inventors to come up with ways to turn CO_2 into molecules that can be used to produce other molecules on Mars.[94] Since hemp can be used in 3-D printers to create sustainable housing *and* it turns CO_2 into oxygen, perhaps it could be farmed inside of an enclosed habitat (like an eco-dome) so that it fills the enclosure with enough breathable air for the astronauts, creates food for the astronauts, and the biomass can be used for housing, energy/electricity, and many other uses. While hemp has yet to be grown in outer space, it takes very little water to grow, and considering its nutritional benefits, the crop has the potential to be used as one of the most versatile means of survival on the Red Planet.

NOTES

1. John Bacon and Mike Chapman, "All about smoke and sweat and grit: Firefighters battle on in California," *USA Today,* Aug 1, 2018, https://www.usatoday.com/story/news/nation/2018/07/31/carr-fire-monster-9th-most-destructive-california-history/869762002/.
2. Trevor Fay, "Lakehead community meeting addresses Hirz Fire," *KRCR News Channel ABC 7,* Aug 15, 2018, https://krcrtv.com/news/shasta-county/lakehead-community-meeting-addresses-hirz-fire.
3. Geographic Area Coordination Centers, "National Year-to-Date report on fires and acres burned," Dec 21, 2018, https://gacc.nifc.gov/sacc/predictive/intelligence/NationalYTDbyStateandAgency.pdf
4. Emily Shugerman, "Carr Fire: California wildfires will only get worse in the future because of climate change, experts say," *Independent,* July 31, 2018, https://www.independent.co.uk/news/world/americas/carr-fire-latest-california-wildfire-worse-update-global-warming-a8471956.html.
5. Mathias Gafni and Thomas Peele, "Why did fire investigators remove PG&E transmission tower part in Camp Fire probe?" Local News, *Enterprise-Record,* Dec 6, 2018, https://www.chicoer.com/2018/12/05/why-did-fire-investigators-remove-pge-transmission-tower-part-in-camp-fire-probe/.
6. Jaxon Van Derbeken, "Hook on PG&E Tower Eyed as Cause of Deadly Camp Fire," *NBC Bay Area*, Dec 5, 2018, https://www.nbcbayarea.com/news/local/Hook-on-PGE-Tower-Eyed-as-Cause-of-Deadly-Camp-Fire-502035081.html.
7. Dale Kasler, "California orders safety upgrades at PG&E amid Camp Fire scrutiny," Business & Real Estate, *Sacramento Bee,* Nov 29, 2018, https://www.sacbee.com/news/business/article222375680.html.
8. Madeleine Ayer and Scott Glover, "California's largest utility provider's role in wildfires is under scrutiny," CNN investigates, *CNN,* Dec 19, 2018, https://www.cnn.com/2018/12/19/us/camp-fire-pge-invs/index.html.
9. J.D. Morris, "Federal judge asks California attorney general if PG&E committe state crime," California wildfires, *San Francisco Chronicle,* Dec 6, 2018, https://www.sfchronicle.com/california-wildfires/article/Federal-judge-asks-California-AG-if-PG-E-13445959.php.
10. Ibid.
11. Dale Kasler, "Judge in San Bruno criminal case demands answers from PG&E on cause of Camp Fire," Fires, *Sacramento Bee,* Nov 27, 2018, https://www.sacbee.com/news/state/california/fires/article222260460.html.
12. Jeremy Childs, "New lawsuit claims Southern California Edison responsible for Woolsey Fire," *Ventura County Star,* Nov 21, 2018, https://www.vcstar.com

/story/news/2018/11/21/new-lawsuit-claims-energy-utility-responsible-woolsey-fire-southern-california-edison/2083003002./.

13. Matthew Lerner, "Total losses from Camp, Woolsey fires could reach $19 billion: Corelogic," Risk Management, *Business Insurance,* Nov 27, 2018, https://www.businessinsurance.com/article/20181127/NEWS06/912325309/Total-losses-from-Camp-Woolsey-fires-in-California-could-reach-$19-billion-Corel.
14. Eileen Frere, "Woolsey Fire: New lawsuit filed against Southern California Edison on behalf of wildfire's victims," Woolsey Fire, *Eyewitness News ABC 7,* Nov 24, 2018, https://abc7.com/lawsuit-filed-against-socal-edison-on-behalf-of-woolsey-fire-victims/4754820/.
15. Ted Sickinger, "PGE replacing 70,000 electricity meters because of fire risk," Oregon Business News, *The Oregonian,* July 24, 2014, https://www.oregonlive.com/business/index.ssf/2014/07/pge_replacing_some_electricity.html.
16. "Lawsuits claim faulty PG&E smart meters started house fires," *ABC 30 Action News,* Nov 17, 2017, https://abc30.com/lawsuits-claim-faulty-pg-e-smart-meters-started-house-fires/2657513./
17. Ibid.
18. Jefet Serrato, "Robert Kennedy Jr. comes to Chico to help Camp Fire survivors," *Action News Now,* Dec 16, 2018, https://www.actionnewsnow.com/content/video/502891201.html.
19. State of California, "Employment Development Department Unemployment Rate," January 18, 2019, https://www.labormarketinfo.edd.ca.gov/file/lfmonth/chic$pds.pdf
20. Chico Home Prices & Values, *Zillow,* accessed Jan 28, 2019, https://www.zillow.com/chico-ca/home-values/.
21. "Average Salary in Chico, California," *Payscale.com,* accessed Jan 28, 2019, https://www.payscale.com/research/US/Location=Chico-CA/Salary.
22. Ben Christopher and Matt Levine, "Here's why housing costs are so high," Local stories, *NewsReview.com,* Sept 14, 2017, https://www.newsreview.com/chico/heres-why-housing-costs-are/content?oid=25008617.
23. "Up to 2,00 FEMA trailers planned for Camp Fire victims," *Capital Public Radio,* Nov 28, 2018, http://www.capradio.org/articles/2018/11/28/up-to-2000-fema-trailers-planned-for-camp-fire-victims/.
24. Mike Luery, "FEMA promises federal help for Butte County wildfire victims," *KCRA 3*, Nov 14, 2018, https://www.kcra.com/article/fema-promises-federal-help-for-butte-county-wildfire-victims/25107572.
25. Taeler De Haes, "Hempcrete plant expected to bring hundreds of jobs, revitalized Glenville," *ABC 5 News Cleveland*, Sept 14, 2018, https://www

.news5cleveland.com/news/local-news/cleveland-metro/hempcrete-plant-expected-to-bring-hundreds-of-jobs-revitalize-glenville.

26. Bob Woods, "Building your dream home could send you to the hemp dealer," Business of Design, *CNBC.com,* April 20, 2018, https://www.cnbc.com/2018/04/20/building-your-dream-home-could-send-you-to-the-hemp-dealer.html.
27. Ibid.
28. Jami, Tarun & Karade, Sukhdeo & Singh, L. P., "Hemp Concrete—A Traditional and Novel Green Building Material," *International Conference on Advances in Construction Materials and Structures,* March 2018, https://www.researchgate.net/publication/324647603_Hemp_Concrete_-_A_Traditional_and_Novel_Green_Building_Material.
29. Ibid.
30. K. Mikulica and R. Hela, "Hempcrete—Cement Composite with Natural Fibres", *Advanced Materials Research*, Vol. 1124, Sept 2015, pp. 130-134, https://www.scientific.net/AMR.1124.130.
31. Jami, Tarun & Karade, Sukhdeo & Singh, L. P., "Hemp Concrete—A Traditional and Novel Green Building Material," *International Conference on Advances in Construction Materials and Structures,* March 2018, https://www.researchgate.net/publication/324647603_Hemp_Concrete_-_A_Traditional_and_Novel_Green_Building_Material.
32. University of Bath, "Low carbon hemp house put to the test," *ScienceDaily,* September 16, 2010, https://www.sciencedaily.com/releases/2010/09/100915205229.htm.
33. Gregor, Lubo, " Performance of Hempcrete Walls Subjected to a Standard Time-temperature Fire Curve," *Thesis for Master of Engineering*, Aug 2014, https://www.researchgate.net/publication/277533751_Performance_of_Hempcrete_Walls_Subjected_to_a_Standard_Time-temperature_Fire_Curve
34. Wendy Koch, "Hemp homes are cutting edge of green buildings," *USA Today,* Sept 12, 2010, http://content.usatoday.com/communities/greenhouse/post/2010/09/hemp-houses-built-asheville/1#.XArK6JNKi8V.
35. Samuel Agbanlog, "Advantages and Disadvantages of Hempcrete," *Academia,* accessed Jan 28, 2019, https://www.academia.edu/13586258/Advantages_and_Disadvantage_of_Hempcrete.
36. Brzyski, Przemysław et al., "Composite Materials Based on Hemp and Flax for Low-Energy Buildings," *Materials (Basel, Switzerland),* Vol. 10(5), p. 510, May 7, 2017, https://www.ncbi.nlm.nih.gov/pmc/articles/PMC5459053/.

37. US Green Building Council, "Buildings and Climate Change," accessed Jan 28, 2019, https://www.eesi.org/files/climate.pdf.
38. "Will a hemp house catch on fire easier?" *Hempsteads,* 2013, http://hempsteads.info/performance/2013/hemp-fire-resistance.
39. Neetu Garcha, "This 'Lego-inspired' Sooke home will be green—and fire proof—when it's finished: builders," Environment, *Global News,* Oct 4, 2017, https://globalnews.ca/news/3784330/sooke-home-lego-green-fireproof/.
40. Lee Mathews, "'Harmless home' built from Lego-inspired bricks made with hemp," News, *Geek.com,* Oct 29, 2018, https://www.geek.com/news/harmless-home-built-from-lego-inspired-bricks-made-with-hemp-1758380/.
41. "Product Specifications," *Just Biofiber Structural Solutions,* accessed Jan 28, 2019, http://justbiofiber.ca/?page_id=137.
42. "A hempcrete house for Tarpon Springs, FL," *Hempcrete house,* 2012, http://hempcretehouse.coffeecup.com/.
43. Tony Marrero, "Hemp helps building a house in Tarpon Springs, likely first in Florida," News, *Tampa Bay Times,* August 17, 2014, https://www.tampabay.com/news/humaninterest/hemp-helps-build-a-house-in-tarpon-springs-likely-first-in-florida/2193319.
44. Resnet, "Record number of homes HERS rated in 2017 over 227,000 homes HERS rated," *Resnet News,* Jan 9, 2018, http://www.resnet.us/blog/record-number-of-homes-hers-rated-in-2017-over-227000-homes-hers-rated/.
45. Annie Rouse, "Building with Hemp," *Hemp Business Journal,* accessed Jan 28, 2019, https://www.hempbizjournal.com/building-with-hemp-a-carbon-sequestering-energy-efficient-solution/.
46. Robin Jarvis, "This house in North Carolina made entirely of hemp is beyond amazing," *Only in your state,* Jan 14, 2018, https://www.onlyinyourstate.com/north-carolina/hemp-house-north-carolina/.
47. "Martin-Korp Residence, Asheville, NC," *American Lime Technology,* accessed Jan 28, 2019, http://www.americanlimetechnology.com/martin-korp-residence/.
48. Tim Callahan, "Why Hemp Instead…" *Alembic Laboratory Distills: Hemp as a Building Material,* Fall 2015, http://hempsteads.info/wp-content/uploads/2016/07/HEMP-FAQ_160705.pdf.
49. "Let's build on hemp," *Sunstand Sustainable Materials*, Aug 29, 2018, https://www.sunstrands.com/2018/lets-build-on-hemp/.
50. Monocle Man, "Hemp fiberboard poised to replace plywood," *Hemp Connoisseur Magazine,* April 11, 2014, https://www.hcmagazine.com/hemp-fiberboard-poised-to-replace-plywood./.

51. Corey Allen, "Researcher sees future for flax and hemp as particleboard alternative," *Phys.org,* April 11, 2017, https://phys.org/news/2017-04-future-flax-hemp-particleboard-alternative.html.
52. "Hemp Board," *Hemp Technologies Global,* accessed Jan 28, 2019, https://hemptechglobal.com/page15/styled-20/page35.html.
53. "Formaldehyde in Your Home: What you need to Know," *ATSDR,* last modified Feb 10, 2016, https://www.atsdr.cdc.gov/formaldehyde/home/index.html.
54. NIH, "Formaldehyde," *US Department of Health & Human Services*, last modified June 2018, https://hpd.nlm.nih.gov/cgi-bin/household/search?tbl=TblChemicals&queryx=50-00-0.
55. American Cancer Society, "Formaldehyde," *Cancer.org,* last modified May 23, 2014, https://www.cancer.org/cancer/cancer-causes/formaldehyde.html.
56. Hauptmann M, Lubin JH, Stewart PA, et al., "Mortality from lymphohematopoietic malignancies among workers in formaldehyde industries," *J Natl Cancer Inst.,* Nov 5, 2003, Vol 95(21), ppp. 1615-23, https://www.ncbi.nlm.nih.gov/pubmed/14600094.
57. Zhang L., Tang X., Rothman N., et al., "Occupational exposure to formaldehyde, hematotoxicity, and leukemia-specific chromosone changes in cultured myeloid progenitor cells," *Cancer Epidemiol Biomakers Prev.,* Vol 1, pp 80-8, Jan 2010, https://www.ncbi.nlm.nih.gov/pubmed/20056626.
58. IARC Monographs, "Formaldehyde," Jan 2018, https://monographs.iarc.fr/wp-content/uploads/2018/06/mono100F-29.pdf.
59. *Sierra Club and a Community Voice Louisiana v. Scott Pruitt*, No c 17-06293 JSW, Feb 16, 2018, https://earthjustice.org/sites/default/files/files/Formaldehyde%20decision.pdf .
60. Press release, "Judge requires EPA to enforce formaldehyde restrictions in wood products," *Earth Justice,* Feb 16, 2018, https://earthjustice.org/news/press/2018/judge-requires-epa-to-enforce-formaldehyde-restrictions-in-wood-products.
61. EPA, "Formaldehyde," accessed Jan 28, 2019, https://www.epa.gov/formaldehyde/frequent-questions-consumers-about-formaldehyde-standards-composite-wood-products-act.
62. Lisa Friedman, "Pruitt bars some scientists from advising EPA," *New York Times,* Oct 31, 2017, https://www.nytimes.com/2017/10/31/climate/pruitt-epa-science-advisory-boards.html.
63. "ACC Forms New Formaldehyde Panel," *American Chemistry.com,* Aug 24, 2010, https://www.americanchemistry.com/Media/PressReleasesTranscripts/ACC-news-releases/ACC-Forms-New-Formaldehyde-Panel.html.

64. "Monsanto Papers: Secret Documents Page Seven," *Baum Hedlund Aristei Goldman PC,* 2018, https://www.baumhedlundlaw.com/toxic-tort-law/monsanto-roundup-lawsuit/monsanto-secret-documents-page-seven/.
65. Annie Snider, "Sources: EPA blocks warning on cancer-causing chemical," Energy & Environment, *Politico,* July 6, 2018, https://www.politico.com/story/2018/07/06/epa-formaldehyde-warnings-blocked-696628.
66. Ibid.
67. Eric Levitz, "The EPA is hiding proof that a widely used chemical causes leukemia: report," *New York Intelligencer,* July 6, 2018, http://nymag.com/intelligencer/2018/07/the-epa-is-hiding-proof-that-formaldehyde-causes-leukemia.html.
68. Valerie Volcovici, "Pressured by industry, US EPA slows formaldehyde study release: documents," Environment, *Reuters,* May 24, 2018, https://www.reuters.com/article/us-usa-epa-formaldehyde/pressured-by-industry-u-s-epa-slows-formaldehyde-study-release-documents-idUSKCN1IP3EX.
69. Ed Markey press release, "Senator Markey calls on EPA to keep promise and release findings of key study on health impacts of toxic formaldehyde," Dec 10, 2018, https://www.markey.senate.gov/news/press-releases/senator-markey-calls-on-epa-to-keep-promise-and-release-findings-of-key-study-on-health-impacts-of-toxic-formaldehyde_ .
70. Graham Norwood, "Poundbury: a look at Prince Charles' sustainable village in Dorset, on its 30th birthday," Property, *The Telegraph,* April 26, 2017, https://www.telegraph.co.uk/property/buy/poundbury-look-prince-charles-sustainable-village-dorset-30th/.
71. "House that Charles built: step inside the Prince of Wales's hemp-insulated eco-home where even the furniture is recycled," *Daily Mail,* March 12, 2011, https://www.dailymail.co.uk/news/article-1365464/House-Charles-built-Take-tour-Prince-Waless-hemp-insulated-eco-home.html.
72. Ibid.
73. Pilita Clark, "Could this green home be a new model for mass build?" House & Home, *Financial Times,* June 27, 2014, https://www.ft.com/content/463d2c54-f70b-11e3-8ed6-00144feabdc0.
74. "Prince Charles unveils Natural House," *Wales Online,* March 12, 2011, https://www.walesonline.co.uk/news/wales-news/prince-charles-unveils-natural-house-1845643.
75. Pilita Clark, "Could this green home be a new model for mass build?" House & Home, *Financial Times,* June 27, 2014, https://www.ft.com/content/463d2c54-f70b-11e3-8ed6-00144feabdc0.

76. Sam Greenhill, "Prince Charles says he finds it hard not to say 'I told you so' after being dubbed 'old-fashioned' for warning about plastic pollution 40 years ago," *Daily Mail,* Oct 31, 2018, https://www.dailymail.co.uk/news/article-6340089/Prince-Charles-highlighting-plastic-pollution-40-years.html.
77. "VTT Technical Research Centre of Finland," *New Plastics Economy,* accessed Jan 28, 2019, https://newplasticseconomy.org/innovation-prize/winners/vtt-technical-research-centre-of-finland.
78. Christie Lunsford, "The next big thing in 'Green' packaging is hemp bioplastic," *CTPost,* Nov 29, 2018, https://www.ctpost.com/news/article/The-Next-Big-Thing-in-Green-Packaging-Is-Hemp-13430710.php.
79. "Customer Finished Products," *The Hemp Plastic Company,* accessed Jan 28, 2019, https://hempplastic.com/customers/.
80. Zommere G., Vilumsone A., Kalnina D., et al., "Comparative Analysis of Fiber Structure and Cellulose Contents in Flax and Hemp Fibres," *Material Science, Textile and Clothing Technology,* Aug 2013, https://mstct-journals.rtu.lv/article/view/mstct.2013.016.
81. Serizawa S., Lnoue K., Iji M., et al., "Kenaf fiber-reinforced biomass-plastic used for electronic products," *2005 4th International Symposium on Environmentally Conscious Design and Inverse Manufacturing,* Dec 2005, https://ieeexplore.ieee.org/document/1619182.
82. "Janice's Kitchen," 2018, http://www.janiceskitchen.co.za/.
83. "Hemp eyewear," Kickstarter, 2019, https://www.kickstarter.com/projects/hemp-eyewear/innovative-eyewear-for-a-sustainable-future.
84. "Hot consumer products made from hemp," *Hemp Business Journal,* accessed Jan 28, 2019, https://www.hempbizjournal.com/hot-consumer-products-made-from-hemp/
85. "Kanesis," *Hormasa BioPlastics,* 2017, http://horimasabp.com/english/kanesis/.
86. "Hemp 3D printer filament: a sustainable alternative," *3D Materials Business News,* Sept 20, 2018, https://www.3dnatives.com/en/hemp-3d-printer-filament-sustainable-alternative-200920184/.
87. "Join the Revolution," *Mirreco,* 2018, http://mirreco.com/.
88. Charlotte Luxford, "Can 3D printing help solve the global housing crisis?" *Medium,* May 15, 2017, https://medium.com/the-omnivore/can-3d-printing-help-solve-the-global-housing-crisis-42d91bfb08f3.
89. "Latest updates from NASA on 3D-printed habitat competition," *NASA,* Aug 22, 2018, https://www.nasa.gov/directorates/spacetech/centennial_challenges/3DPHab/latest-updates-from-nasa-on-3d-printed-habitat-competition.
90. "NASA's 3D-printed habitat challenge—Phase 3," *Bradley University,* accessed Jan 28, 2019, https://www.bradley.edu/challenge/.

91. "Combination 3D printer will recycle plastic in space," 3-D Printing, *NASA,* Nov 19, 2018, https://www.nasa.gov/mission_pages/centers/marshall/combination-3d-printer-will-recycle-plastic-in-space.html.
92. Chris Young, "The world is beautiful, Canadian astronaut Chris Hadfield on living in space and the view from the window," *National Post,* Nov 24, 2012, https://nationalpost.com/news/canada/the-world-is-beautiful-canadian-astronaut-chris-hadfield-on-living-in-space-and-the-view-from-the-window
93. "Hemp in space," *Kannaway,* accessed Jan 28, 2019, https://kannaway.com/hemp-in-space/.
94. Mike Wehner, "NASA will pay you up to $750,000 to come up with a way to turn CO2 into other molecules on Mars," *Yahoo,* Sept 3, 2018, https://finance.yahoo.com/news/nasa-pay-750-000-come-way-turn-co2-183428634.html

12

OUR FUTURE WITH HEMP

As the 2019 New Year approached, Senator Elizabeth Warren was the first to announce she was opening a Presidential Exploratory Committee for a 2020 presidential run. Reports soon emerged that she supported the idea of a Green New Deal (which at the time was just an idea).

The Green New Deal is a proposal that sets goals within a ten-year window to decrease greenhouse gas emissions from agriculture and manufacturing and expand renewable power sources, including building a smart grid; it also details other progressive ideals such as implementing universal health care and a living wage and increasing the domestic labor market by opening more opportunities for entrepreneurs.[1] While the legislation is broad (broad meaning it may take Congress ten years to actually pass everything in it unless the president is behind it), the fact that American hemp is now legal means that the plant can be a solution to many of the environmental and economic concerns in the Green New Deal, including climate change, pollution, and job opportunities.

Eighty years ago, *Popular Mechanics* published the *New Billion Dollar Crop* article that asserted hemp "will not compete with other American products. Instead, it will displace imports of raw material and provide thousands of jobs for American workers." While American hemp was never given a fair shake back then, the potential for the industry today is overwhelmingly optimistic.

Since hemp can be manufactured into practically anything, the industry has always been perceived as one to directly compete with existing industries and dominate existing markets: hemp products are carbon neutral; hemp fabric is stronger than cotton; hemp paper has less of an impact on the environment compared to wood-pulp methods; hemp biofuel is cleaner for the environment; hemp's phytoremediation potential can remove toxins from the soil and water; its biomass can be used to produce energy; hemp can be turned into concrete and other building materials. While all of these facts are true, none of them describe ways to improve industry—they're describing ways to replace existing industries—which is why hemp has always been stereotyped as a threat. However, in today's economy, hemp is being embraced by some of our oldest institutions for a couple of reasons.

First of all, there isn't nearly enough hemp being grown to take over all existing industries. The amount of processing plants necessary for something like that to occur doesn't even exist.

Secondly, today's industries blend materials and corporations partner with other corporations to make their products stronger. The textile world isn't boxed into strict categories—we have *blends* of cotton, spandex, and polyester—and hemp is already a part of that mix. Add it in as a domestic product, and we can further reduce our carbon footprint. Paper mills aren't going out of business either, but they can incorporate domestic hemp into their products for the same reasons. Hemp plastic is already incorporated into car door panels—the ingredient makes the plastics lighter, which in turn makes the car lighter, which in turn makes it require less fuel, which in turn makes it more attractive to consumers.

No doubt about it, hemp will add more jobs and more products to the economy, but *disrupting* or *displacing* jobs? That's a stereotype that isn't necessarily based in reality, since industries are finding ways to incorporate it into existing products, which in turn makes those existing products less harmful to the environment.

During the last week of December 2018, CNN listed the most important stories of the year. Coincidentally, many of them were problems already addressed in this book. The top weather story was the California wildfires of 2018 (the Camp Fire in particular), and one of the top health stories of 2018 was the FDA's approval of Epidiolex. For the first time in history, the federal government acknowledged cannabidiol was effective in treating seizure disorders.

CNN's Dr. Sanjay Gupta was quick to comment on how effective CBD can be in treating seizures: "Not only can it work, sometimes it's the only thing that does work."

While many have theorized Big Pharma will do whatever it takes to keep CBD and THC illegal, the industry has already integrated them into products. So again, cannabis partners, pairs, synchronizes, works well with others. If hemp is found to be a better alternative, certainly it will dominate a particular market, but again, existing corporations are already diversifying to include hemp. Big Tobacco sees the writing on the wall and is investing in marijuana and hemp businesses; so is the alcohol industry and the beverage industry in general. Cannabis isn't necessarily competition when industries are already working it into their existing product line.

Our future with hemp is promising. American hemp is the beginning of something brand new, with the potential to promote innovative, novel ideas and new technology among some of America's oldest industries. But while we all high-five each other and do a little victory dance for hemp, we can't become complacent.

STATES HAVE THE AUTHORITY TO REVISE EXISTING HEMP LAWS

Although this is being championed as a good thing, state legislatures didn't automatically draw up fair and comprehensive hemp laws as soon as the 2018 Farm Bill was passed. States that didn't already have a hemp pilot program certainly didn't feel pressured to push their agricultural departments to act immediately, either. If the insanity of cannabis prohibition has taught us anything, the time to pay attention is now.

In 2016, Floridians voted to legalize medical marijuana by passing Constitutional Amendment 2. Over 71 percent voted in favor of the bill, but there were hiccups in the weeks leading up to the general election. NORML had to sue Broward County election officials who *ooops* "forgot" to include the constitutional amendment on the mail-in ballots. (Suing them was the only way to ensure replacement ballots would be issued.) The news coverage of the lawsuit may have actually helped the cause as voters were now paying attention to whether or not Constitutional Amendment 2 was on their ballots.

Even though over 71 percent of Floridians voted in favor of medical marijuana, Constitutional Amendment 2 has yet to be fully implemented statewide. While it was designed to pass automatically, as all the stipulations and parameters were already outlined, the state decided to change the law so that only non-smokable marijuana was legal. While activists were busy challenging that, local officials were writing new laws to ban medical marijuana from their cities.

Marijuana businesses and activists also had to sue the State Department of Health, which has been dragging its feet in setting up statewide regulatory framework. As of December 2018, only fourteen companies in the entire state of Florida have a license to dispense marijuana, and there are only eighty-one locations where people can purchase it.[2] (That's eighty-one locations to serve over 21 million people.)

Granted, medical marijuana is federally illegal and hemp is federally legal, but it's up to a state's agricultural department to implement hemp laws, and if state regulators aren't necessarily inclined to do so, it may take some pressure from activists to get the ball rolling (or rolling in the right direction). States can outline the licensing process and application process, what tests are to be done on hemp to ensure it is being grown with .3 percent THC or less, and many other aspects of the cultivation process. If the fees and costs are too high, this will limit how many people can cultivate hemp; it could also push out smaller farms who are relying on a new crop to make ends meet. This has happened before with the Marihuana Tax Act, and if we don't stay informed, it could happen again.

THE FDA'S ROLE IN DETERMINING FEDERAL REGULATIONS

As far as American hemp's future and the future of CBD is concerned, federal regulatory power lies in the hands of the FDA. The FDA can't create laws—that's the responsibility of Congress—but the agency creates the regulations for the laws that Congress passes. FDA commissioner Scott Gottlieb put out a statement in December 2018 to notify the hemp industry that:[3]

1. The FDA declared hempseed-derived food ingredients as GRAS (generally recognized as safe). The notice was given to Fresh Hemp Foods, Ltd. on December 20, 2018. The company now has the FDA's full blessing to use its hemp oil, hulled hempseed, and hempseed protein powder in food products, as long as the ingredient is listed on the product. This means that the FDA already has a system in place for approving hemp for food, as long as it doesn't contain CBD.
2. If a CBD product claims to be used to cure, treat, or prevent a disease in humans or animals, the product must go through the FDA drug approval process to ensure the product actually does this.
3. The FDA will continue to take action against any company selling cannabis products—including CBD—that claim to prevent, treat, or cure serious diseases such as cancer if the company doesn't have FDA-approved scientific research to back up the claims.
4. Because THC and CBD are active ingredients in FDA-approved drugs, they can't be used as ingredients in food or in dietary supplements. (This includes hemp-CBD.)
5. Under FDA guidelines, THC and CBD products cannot be transported across state lines, unless they're an FDA-approved drug.
6. Since the FDA has the authority to change its own regulations, the agency is going to review its requirements to see if there are "pathways" to allow CBD or any cannabis-derived compounds to be permitted in food or dietary supplements. Right now, CBD and THC are classified as pharmaceutical ingredients, but that could change if the FDA decides it wants to change the classification.

What's interesting is that Gottlieb refers to cannabis-derived compounds in an all-inclusive way throughout his memo. He classifies CBD as CBD and THC as THC—regardless of the fact that hemp-CBD is legal and marijuana-CBD isn't:[4]

It's unlawful under the FD&C Act to introduce food containing added CBD or THC into interstate commerce, or to market CBD or THC products as, or in, dietary supplements, regardless of whether the substances are hemp-derived. This is because both CBD and THC are active ingredients in FDA-approved drugs and were the subject of substantial clinical investigations before they were marketed as foods or dietary supplements.

So while we don't know if the CBD and THC used in pharmaceutical products come from hemp or marijuana, it doesn't matter. Epidiolex is patented, and the FDA can't have an active ingredient in a patented product out on the streets.

I do agree that CBD products should be tested to ensure the ingredients are safe and accurate, and cracking down on the snake-oil salesmen is good for everyone in the hemp industry, but the statements pertaining to hemp-CBD not being allowed to cross state lines does go against the legislation Congress just passed. Luckily, Gottlieb threw himself a lifeline by stating he won't ban CBD altogether; he's in the process of looking for "pathways" to allow CBD drinks, lotions, and other CBD products to continue to exist under FDA regulations (whatever those might wind up being).

Yes, CBD is an active ingredient in FDA-approved Epidiolex, but the way in which that particular CBD is manufactured isn't necessarily the same way companies are manufacturing hemp-CBD, so there's wiggle room. As Jonathan Miller, the general counsel for the US Hemp Roundtable (an industry trade group) explained, "CBD is now too big to fail."[5] And industry advocates conclude that the FDA can't shut down hemp-CBD entirely, now that it's actually legal.

After marijuana was legalized at the state level, beer brands including Coors Light, Bud Light/Budweiser were seeing a drop in sales. In Denver, overall beer sales fell by 6.4 percent.[6] To fill the gap, alcohol companies started partnering with cannabis companies to create CBD- and/or THC-infused adult beverages. For instance, in July 2018, Heineken launched Hi-Fi Hops—a cannabis-infused carbonated drink with 5 milligrams of THC and 5 milligrams of CBD. It's currently distributed in California dispensaries; the product tastes like beer, but doesn't contain any alcohol.

Since major corporations are already selling CBD products or investing in cannabis-infused products in states that have already legalized it, my hunch is that the FDA will change its standpoint instead of demanding these products cease to exist. In the past, the FDA and EPA have adopted state guidelines and implemented regulations on a product based on whichever state has the strictest terms (so we might wind up with federal regulations that are similar to Oregon's).

In the meantime, maybe the FDA can do us a solid by regulating the amount of hormones and antibiotics in burger meat because we currently don't have effective federal laws to prevent the overuse of antibiotics—and call me crazy, but that seems a bit more of an imperative compared to regulating CBD out of the market, since the National Institute on Drug Abuse has admitted in Senate caucus hearings that CBD has no adverse effects. In 2018, independent watchdog group U.S. PIRG tested the food meat in twenty-five of the largest burger chains in the United States. In their Chain Reaction IV report findings, twenty-two out of the twenty-five fast-food restaurants were found to have high amounts of antibiotics in their beef and poultry products. Out of the twenty-five fast-food companies, only two did not use antibiotics in their meat: Shake Shack and BurgerFi.[7]

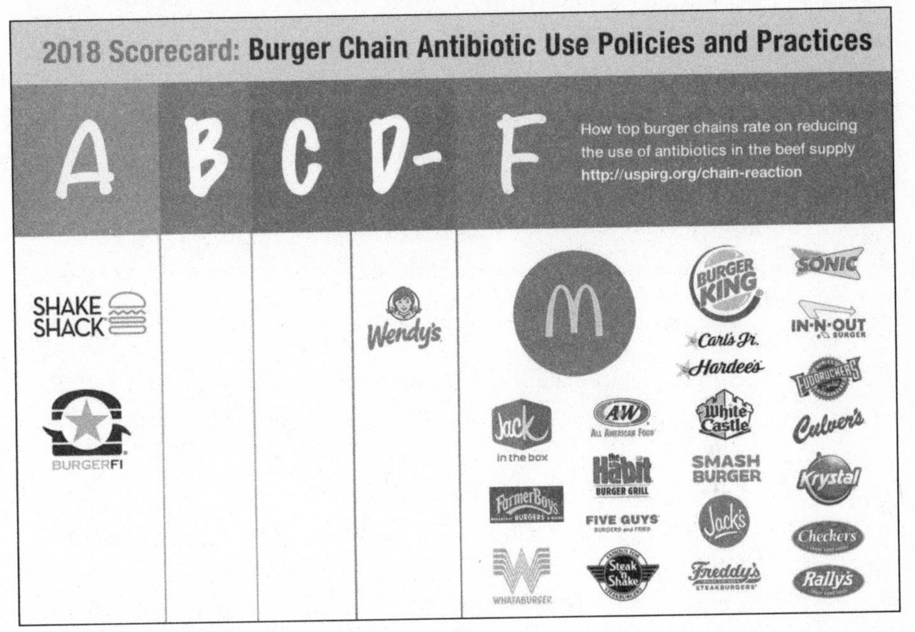

2018 results show that the majority of fast-food chains received a failing grade for antibiotics in their meat sources, 2018 Chain Reaction IV report. *Source: https://www.nrdc.org/resources/chain-reaction-how-top-restaurants-rate-reducing-antibiotics-their-meat-supply and https://www.nrdc.org/sites/default/files/restaurants-antibiotic-use-report-2018.pdf and https://www.nrdc.org/sites/default/files/restaurants-antibiotic-use-report-2018.pdf.*

Even if you don't eat at McDonald's, the company is the largest single purchaser of beef in the United States, and what the corporation allows in its food affects all of us. The beef industry uses six hormones to promote weight gain: estradiol, progesterone, testosterone, zeranol, melengesterol acetate, and trebolone acetate; over 70 percent of the medically important antibiotics sold in the United States go to food animals. This is why the World Health Organization has linked the overuse of antibiotics on poultry and beef farms to antibiotic-resistant "superbugs" in humans.

The more antibiotics used on food products, the more bacteria become immune—plus the more antibiotics we eat, the less effective they are when we need them to get rid of a virus or disease. In 2018, the United States experienced some of the worst foodborne illnesses, all stemming from bacteria, including three *E. coli* outbreaks in lettuce and multiple salmonella outbreaks in products including chicken, frozen shredded coconut, precut melon, ground beef, and ground turkey.[8] The turkey salmonella contamination affected thirty-eight states and led to the recall of more than 164,000 pounds of Jennie-O ground turkey.[9]

Since the Chain Reaction IV report was compiled and released to the public, eighteen of the twenty-five top fast-food chains decided to adopt policies to limit the use of antibiotics in at least one meat category (primarily chicken), so the independent watchdog has given transparency ratings to these restaurants:[10]

- Chick-fil-A, KFC, Jack in the Box, and Papa John's gained points for progress on implementing their commitments. Pizza Hut and Wendy's gained points for making further commitments to reduce antibiotic use in their meat supply chains

- Three chains—Panera Bread, Chipotle, and Chick-fil-A—received an "A" in transparency for their policies to source meat raised without the routine use of antibiotics.

- Chick-fil-A reports that it is on track to have 100 percent of its chicken meet its "No Antibiotics Ever" standard by the end of 2019.

- IHOP and Applebee's each improved from an "F" to a "C" grade for sourcing portions of their chicken and pork supply from producers raising animals without the routine use of antibiotics.

- Starbucks dropped down to a "D" due to a lack of transparency. The company committed to no longer source poultry raised with routine antibiotic use in 2017, but has failed to report on progress toward implementing that goal and did not return this year's survey.

- Seven companies received a failing grade in transparency for not taking any meaningful actions to reduce antibiotic use in any of their meat supply chains: Sonic, Dairy Queen, Olive Garden, Chili's, Arby's, Little Caesars, and Buffalo Wild Wings.

When more companies commit to less antibiotics in their meat sources, the livestock industry is forced to change its practices. While the report proved effective in policing these companies, it hasn't inspired the FDA to do the same. In 2015, the Obama administration released a National Action Plan for Combating Antibiotic Resistant Bacteria. This policy reduced the amount of antibiotics dispensed to humans, but it didn't set any national goals to reduce antibiotic use in the livestock industry—that was left up to the FDA to regulate, and the agency has continued to allow medical antibiotics to be used in animals for other purposes aside from actual medical needs, such as growth promotion.

As stated in chapter 5, hemp definitely has the potential to disrupt American farming, but clearly the entire agricultural industry—from the overuse of carcinogenic glyphosate to antibiotics—could use a little disruption, to say the least.

THE TENTH AMENDMENT

The Tenth Amendment of the US Constitution gave states the right to legalize hemp and marijuana. In the late 1700s, soon after the first US Congress adopted the Constitution, the Tenth Amendment was added to the Bill of Rights. This was done to largely satisfy demands of Anti-Federalists, or those who opposed the formation of the federal government because they feared it would take away the power and legitimacy of states' rights. The Tenth Amendment states:

> *The powers not delegated to the United States by the Constitution, nor prohibited by it to the States, are reserved to the States respectively, or to the people.*

As Alexander Hamilton once said, the supremacy of the federal government is "expressly" confined "to laws made pursuant to the Constitution." If the federal government as a whole (executive, legislative, or judiciary branch) takes any actions outside of the realm of the Constitution, then those actions can be voided by the states.[11]

To put this simply: states have the power to overturn federal law or create laws that contradict it if the laws fall outside the parameters of the Constitution. Since cannabis isn't mentioned in the Constitution, states were able to legalize hemp and marijuana within their borders and effectively nullify the Controlled Substances Act.

One example of state law superseding federal law is California's decision go above and beyond the FDA's regulations of antibiotics and hormones in livestock. As of January 2018, only a licensed veterinarian can administer antimicrobial drugs to livestock in the state of California if it is necessary to treat a disease or infection, if it is necessary to control the spread of disease or infection, or if it's necessary in relation to surgery or a medical procedure.[12] The livestock industry as a whole—from beef to poultry to pork—can no longer use antimicrobial drugs to promote weight gain or for any other non-medical/non-disease-prevention purpose. The Keep Antibiotics Effective Act in Maryland is a similar law that goes beyond federal law by greatly restricting the use of antibiotics in livestock. While the Chain Reaction IV authors note that the policies of both laws haven't fully gone into effect yet, an ordinance policy passed by San Francisco has produced results. Since 2015, grocery chains must notify San Francisco's Department of the Environment which antibiotics "are used in their meat supply chain, for what purpose, how much, and on how many animals" so that consumers can be aware of what they're buying.[13]

Though a state (and even local government) can pass stricter regulations than the FDA and EPA, the Tenth Amendment doesn't grant the power for a state to overturn FDA regulations or pass regulations that contradict the FDA. The only exception is if it can be argued that Congress never gave the FDA the power to regulate a particular industry or if the FDA's regulatory measures are "overreach."

In 1996, the tobacco industry was successful in overturning the FDA's power to regulate cigarettes and smokeless tobacco using the overreach argument.[14] The victory didn't last long because it motivated anti-smoking advocates to pressure Congress to create legislation that would grant the FDA full authority. The Family Smoking Prevention and Tobacco Control Act of 2009 put the agency in charge of every aspect of tobacco products—from how they're manufactured to how they're marketed. In 2016, Congress extended the law to e-cigarettes, hookahs, and all other nicotine products that didn't previously fall under the FDA's direct authority.[15]

While we're still waiting to see what the FDA's final decisions are on regulating hemp-CBD, it's possible that the agency can be sued for overreach if its "pathways" don't provide legal access to all aspects of hemp, as Congress outlined in the 2018 Farm Bill. If a federal judge sides against the FDA, then Congress would either have to do nothing (indicating agreement with the judge's ruling) or pass further legislation to clarify what the FDA can/can't regulate. Let's hope it doesn't come to that.

THE FUTURE OF AMERICAN HEMP

Whatever is decided by state and federal regulators, there are actions we can take on a daily basis to influence American hemp policy. The first step is to take what you know about hemp and share it with others because educating people is the first step in developing more allies.

The problem with federal hemp legislation is that it's attached to the 2018 Farm Bill and the Agricultural Improvement Act of 2018—both bills are set to expire in five years. So by 2023, Congress will be reviewing this legislation and making changes to it. While I don't think Congress is going to dial back hemp legalization, the more progress we make with integrating the industry, the better.

I realize activism can be a time-consuming endeavor, even for the most passionate hemp enthusiast. Most of us work for a living; we all have bills to pay and family obligations. While marching around Washington, DC, at various pro-hemp rallies may seem like the most visible way to support hemp, is it isn't exactly possible for everyone. Research has proven that protests alone aren't effective because additional steps need to be taken after the protest is over for real change to occur. The easiest way to get the word out about hemp's benefits is to start small:

- **Social Media Influencer**

Does everyone in your social media circle know what hemp can do? While in many cases, posting memes on social media doesn't really do anything to make an impact for a movement, this isn't necessarily true for hemp.

When a new hemp technology or product comes out, share it on social media. Post memes or articles about the benefits of hemp. Pick a topic discussed in this book and put your thoughts out there on Facebook or Twitter and feel free to use any of the endnotes to back up your opinions (a majority of the links in the notes are primary sources).

After Canada legalized marijuana, the Toronto PD launched a Twitter meme campaign to encourage Canadians to stop snitching about marijuana use. *Source: Twitter*

After Canada legalized recreational marijuana, the Toronto Police Department took a novel approach on Twitter by posting memes. They wanted to get the word out that reporting someone for smoking marijuana isn't a valid 9-1-1 call because it's now perfectly legal. Instead of just writing that in a tweet (boring), they made jpg versions of actual 9-1-1 calls that were not emergencies (such as asking for directions) as a comparison. The tweets went viral because they were a funny way of showing that law enforcement was behind the new legislation.[16]

When you share a meme about hemp, more points for you if it's funny, just make sure the meme is accurate. Google any quotes attributed to a historical figure or celebrity, and make sure your facts are backed up by scientific research. The last thing you want is a troll telling you something in your hemp post isn't true and for that troll to be right.

- **Talk About Hemp**

If you're able to discuss hemp on social media, try bringing it up in casual conversation. Yes, it's a political issue, but it's also safe because it's legal now. For instance, if there's legislation on the ballot that benefits hemp, don't ask your friends and family if they're going to vote for it. Instead, tell them why you're going to vote for it. Explain the benefits. Then be prepared. They might have questions. They might not be fully informed. And that's okay.

If they aren't fully informed, don't don't *don't* add to the divisiveness by calling them stupid. Don't call them a horsehead. Don't attack the problem. Educate. Besides, if you've already been posting pro-hemp memes all over social media, people might come to you directly and ask your opinion. This exact scenario happened to my husband during the 2018 midterm elections. We live in Missouri and there was a choice of three ballot initiatives to legalize medical marijuana. Our friends, family, and neighbors all asked him which one he thought was the best, and they voted based on his feedback. My husband doesn't post on social media often, but when he does, he usually posts articles relating to the insanity of cannabis prohibition. Prior to the midterms, he posted the three initiatives and why he was going to vote in favor of only one of them—Amendment 2 (which did pass, by the way). The closer we came to November 6, the more people asked for his opinion. Why? Because he had already established a history of sharing relevant, valid, and factual information about cannabis.

- **Annoy Your Legislatures**

If a certain piece of legislation affects the legalization of hemp, use your phone! Call your people! Bonus points if you post about it on social media along with the phone numbers for your state reps, or bring up the issue with friends and family.

You could also tweet at your representatives nonstop, asking when your phone call will be returned, or tell them why it's important to vote pro-hemp. They probably won't tweet back, but the social media guy handling the account might take note and say something to someone about the importance of a particular hemp bill.

If enough people call their representatives about a particular issue, secretaries and interns pass along the message.

The masses are raising their pitchforks and those pitchforks are on fire!

Yes, this may be considered annoying the crap out of someone until they do what you want, or, as I like to call it, using those internet trolling skills to actually accomplish something.

- **Don't Forget to Buy Hemp Products!**

If any form of confrontation is too much to handle, then the easiest way to promote hemp is to purchase hemp products. Make a silent statement with what you choose to buy, what you choose to wear, and where you choose to shop. This might seem a bit silly, but the more people who buy hemp and support the industry, the more justification we have that hemp is a necessity.

By simply purchasing hemp products, you're also doing your part by lessening our carbon footprint. You're helping the environment and hemp businesses at the same time, and big bonus points if you buy American hemp products. No better feeling than supporting a domestically grown and manufactured product.

If you have a choice between ordering a hemp smoothie or whey smoothie, make a statement. You don't have to announce why you're ordering the hemp smoothie, but if you notice a store no longer carries a hemp product, inquire about it. Sometimes stores will make the effort to restock items if enough customers request them.

- **Stay Informed About the Issues**

As fun as it was to write this book, the material is sure to be dated as hemp legislation evolves. Most hemp activist groups have free newsletters and mailing lists that you can sign up for online. That's the best way to keep up to date. Websites such as the National Hemp Association, the Ministry of Hemp, and Vote Hemp are great resources. On advocacy sites such as Vote Hemp, local issues are constantly updated, so you'll know how hemp legalization is progressing in your state, and if there are any direct actions you can take. Websites will even provide you with the email addresses for your state and local reps, and provide a direct link for you to email them.

FINAL THOUGHTS

While he was campaigning for president, Franklin D. Roosevelt admitted that alcohol prohibition had failed. The government had proven itself incapable of enforcing sobriety, and this was only one of the reasons why Roosevelt wanted to repeal the Eighteenth Amendment:

1. During the Great Depression, Americans needed jobs, and by legalizing the alcohol industry, a whole new job market would emerge.

2. The second reason was tax dollars. Americans were willing to spend money to acquire alcohol, and since alcohol was illegal, they were buying it tax-free. That meant the government was losing money at a time it couldn't afford to.

3. While the government may have meant well by outlawing alcohol, FDR didn't think it was the government's job to control the people's temperance. That was up to a person's personal beliefs and convictions.

4. The government actually increased all the negative aspects of alcohol by outlawing it, and spent massive amounts of money in the process by trying to shut down bootleggers, the mob, and other criminals involved in creating, selling, and distributing alcohol illegally.

"We all agree that temperance is one of the cardinal virtues," Roosevelt stated in a 1932 campaign speech.[17] "But the methods adopted . . . with the purpose of achieving a greater temperance by the forcing of Prohibition have been accompanied in most parts of the country by complete and tragic failure."

According to the May 1930 issue of *Popular Science Monthly,* the Prohibition Commissioner estimated that in 1919 (the year before Prohibition became law) the average American spent $17 per year on alcoholic beverages. By 1930—because law enforcement diminished the supply and because the risk involved in obtaining alcohol increased the price—the average American spent $35 per year on alcohol. There was no inflation in the United States during this period, so the price literally doubled as a result of prohibition . . . plus the illegal alcohol industry made an average of $3 billion per year in untaxed income.[18]

The failures of alcohol prohibition mirror many of the failures of the war on drugs. Incidentally, the same president who repealed alcohol prohibition also started cannabis prohibition by signing the Marihuana Tax Act into law. The unfortunate fact that hemp was even included in the drug war and was labeled as a Schedule I narcotic represents a truly sad, long chapter in US history. Hopefully, the freshly minted 2018 Farm Bill is the tipping point that ultimately free not only hemp but marijuana from its Schedule I classification sooner rather than later.

Throughout the ages, hemp has proven to be a fundamental part of many different cultures and civilizations. According to the 1913 USDA yearbook, a written work known as the "Lu Shi," which dates back to the Sung Dynasty (500 AD), asserts that Emperor Shen Nung taught the people of China to cultivate "*ma*" (the Chinese word for hemp) as early as the twenty-eighth century BC.

"Hemp was probably the earliest plant cultivated for the production of a textile fiber," the authors of the USDA yearbook theorize.[19]

Hemp built this great nation, and many other civilizations that came before us; it has deep roots in American history. As Thomas Jefferson once said: "Hemp is of first necessity to the wealth and protection of the country." Now that hemp is finally legal, we have the ability to leave our children a better world than the one we inherited, and I sincerely hope we don't let them down.

NOTES

1. "Draft Text for Proposed Addendum to House Rules for 116th Congress of the United States," accessed Jan 24, 2019, https://docs.google.com/document/d/1jxUzp9SZ6-VB-4wSm8sselVMsqWZrSrYpYC9slHKLzo/preview
2. Jim Saunders, "Florida fights ruling on medical marijuana law," *Sun Sentinel,* Dec 26, 2018, https://www.sun-sentinel.com/news/politics/fl-ne-nsf-medical-marijuana-fight-20181226-story.html.
3. FDA, "Statement from FDA Commissioner Scott Gottlieb, M.D., on signing of the Agriculture Improvement Act and the agency's regulation of products containing cannabis and cannabis-derived compounds," News & Events, *US Department of Health and Human Services,* Dec 20, 2018, https://www.fda.gov/NewsEvents/Newsroom/PressAnnouncements/ucm628988.htm.
4. Ibid.
5. Jenni Avins, "CBD is 'too big to fail,' and US agencies are getting on board," Yes We Cannabidiol, *Quartz,* Dec 27, 2018, https://qz.com/1509527/the-fda-and-usda-are-working-toward-legal-cbd/.
6. Rich Schettino, "These are the top 5 beverage companies investing in cannabis," Finance, *Pot Network,* Aug 14, 2018, https://www.potnetwork.com/news/these-are-top-5-beverage-companies-investing-cannabis.
7. Matthew Wellington, "Chain Reaction IV: Burger Edition," US Pirg Education Fund, accessed Jan 24, 2019, https://uspirg.org/feature/usp/chain-reaction.

8. CDC, "Foodborne Outbreaks," *Food Safety, Multistate Foodborne Outbreak Investigations, Centers for Disease Control and Prevention, last modified Jan 25, 2019,* https://www.cdc.gov/foodsafety/outbreaks/multistate-outbreaks/outbreaks-list.html.
9. Debra Goldschmidt, "More than 164,000 pounds of ground turkey recalled; 52 more people sick in deadly salmonella outbreak," Health, *CNN,* Dec 22, 2018, https://www.cnn.com/2018/12/21/health/turkey-salmonella-outbreak-recall-bn/index.html.
10. US Pirg, "Chain Reaction IV: Burger Edition," Oct 2018, https://uspirg.org/sites/pirg/files/ChainReaction4_Report-10_17_18.pdf.
11. "The Supremacy Clause," *Tenth Amendment Center,* accessed Jan 24, 2019, https://tenthamendmentcenter.com/the-supremacy-clause/.
12. "SB-27 Livestock: use of antimicrobial drugs. (2015-2016)," Text, *California Legislative Information,* Oct 10, 2015, https://leginfo.legislature.ca.gov/faces/billNavClient.xhtml?bill_id=201520160SB27.
13. US Pirg, "Chain Reaction IV: Burger Edition," Oct 2018, https://uspirg.org/sites/pirg/files/ChainReaction4_Report-10_17_18.pdf.
14. New York Times News Service, "Appeals Court Panel Overturns FDA's Regulation of Tobacco," *Chicago Tribune,* Aug 15, 1998, https://www.chicagotribune.com/news/ct-xpm-1998-08-15-9808150098-story.html.
15. FDA, "The Facts on the FDA's New Tobacco Rule," Consumer Updates, *US Department of Health and Human Services*, June 16, 2016, https://www.fda.gov/ForConsumers/ConsumerUpdates/ucm506676.htm.
16. German Lopez, "Toronto police to Canadians: stop snitching on your neighbors about marijuana," Policy and Politics, *Vox,* Oct 17, 2018, https://www.vox.com/policy-and-politics/2018/10/17/17989448/canada-marijuana-legalization-toronto-police-911.
17. Franklin D. Roosevelt, Campaign Address on Prohibition in Sea Girt, New Jersey Online by Gerhard Peters and John T. Woolley, *The American Presidency Project*, accessed Jan 24, 2019, https://www.presidency.ucsb.edu/node/289317.
18. E.E. Free, "The Prohibition Commissioner Tells Where America Gets its Booze: An Interview with Dr. James M. Doran," *Popular Science Monthly,* Vol. 116, No. 5, May 1930, pp 19–21, https://books.google.com/books?id=OigDAAAAMBAJ&pg=PA19.
19. USDA, "Yearbook of the United States Department of Agriculture 1913," Washington Government Printing Office 1914, p 288, http://antiquecannabisbook.com/chap2B/China/USAgri1913.htm.